DESIGN AND CONSTRUCTION OF SANITARY AND STORM SEWERS

PREPARED BY
A JOINT COMMITTEE OF THE
AMERICAN SOCIETY OF CIVIL ENGINEERS
AND THE WATER POLLUTION CONTROL FEDERATION

Copyright 1969 by the AMERICAN SOCIETY OF CIVIL ENGINEERS and the WATER POLLUTION CONTROL FEDERATION

HISTORY OF THE MANUAL

The original manual was the result of seven years' work by a joint committee of ASCE and WPCF (then FSIWA) members and was copyrighted and published in 1960. The members of the full committee on the original manual included:

ASCE	WPCF
Richard R. Kennedy	C. Gordon Gaither
Joseph C. Lawler	H. Sidwell Smith
Ray E. Lawrence	Leland L. Spahr
Raymond R. Ribal	Charles R. Velzy
Bernal H. Swab,* *Chairman*	Samuel I. Zack

In addition, 35 task force members assisted the full committee on the original manual and an editing committee, composed of the following, reviewed and arranged the material for publication:

H. H. Benjes   J. C. Lawler
S. W. Jens   H. S. Smith
R. R. Kennedy   A. L. Tholin
            B. H. Swab,* *Chairman*

The original manual was well received and both ASCE and WPCF valued it as one of their most important publications.

In 1964 after the results of an extensive poll indicated that it would be worthwhile to consider revisions to the original manual, both WPCF and ASCE approved the formation of a Joint Committee on Revision of Manual of Practice of Design and Construction of Sanitary and Storm Sewers. The following were members of the full committee:

Roy Aaron	Richard D. Pomeroy
Paul L. Andrews	Lincoln W. Ryder
John S. Autry	Bernal H. Swab *
David G. Chase	Royal C. Thayer
Clarence E. Cuyler	Charles R. Velzy
Glenn E. Hands	Cay G. Weinel, Jr.
Paul A. Kuhn	Lloyd W. Weller
Cecil M. Pepperman	

Joseph C. Lawler, *Chairman*

A preliminary outline was prepared and subjects assigned. Each member selected additional Task Force members to assist in the preparation of the revisions to individual chapters. These members included:

Aaron Aarons	Harold D. Meisenheimer
Robert D. Bargman	Elmer M. Miller
Cloyd D. Beerup	Richard E. Morris, Jr.
Fred D. Bowlus *	Myron K. Nelson
Nathaniel H. Calder	William A. O'Leary
Richard O. Davis	Charles A. Parthum
Jacob Feld	James W. Porter
Walter E. Garrison	Robert Lee Smith
John C. Guillou	Merlin G. Spangler
George K. Hasegawa	Joshua Sprague
George M. Hite	Robert D. Tolf
Earl R. Howard	Allan H. Wymore
H. Wilfred Izzard	Robert M. Zimmerman
Stifel W. Jens	Henry M. Zukowski
Gordon S. Magnuson	

Two overall Committee meetings were held, after which an Editing Committee was appointed, consisting of Messrs. Paul A. Kuhn, Charles A. Parthum, Cecil M. Pepperman, Richard D. Pomeroy, Cay G. Weinel, Jr., Lloyd W. Weller, and Joseph C. Lawler, *Chairman*. The Editing Committee reviewed and arranged the material for publication. Detailed editing and production of the revised edition were accomplished by WPCF.

Two drafts of the revised manual were prepared and distributed to members of both Societies, other engineers, and manufacturers. Many valuable suggestions were received.

The final draft of the revised edition was approved for publication by the WPCF on October 12, 1967, and by the ASCE on February 19–20, 1968.

---

* Deceased.

TABLE OF CONTENTS

1. **ORGANIZATION AND ADMINISTRATION OF SEWER PROJECTS** .. 1
   - A. Introduction 1
   - B. Definition of Terms and Classification of Sewers 1
   - C. Phases of Project Development 2
     1. Preliminary or Investigative 2
     2. Design 3
     3. Construction 3
     4. Operation 3
   - D. Interrelation of Project Development Phases 3
   - E. Parties Involved in the Design and Construction of Sewer Projects. 4
     1. Owner 4
     2. Engineer 5
     3. Contractor 5
     4. Other Parties 6
   - F. Role of Parties in Each Phase 6
     1. Preliminary 6
     2. Design 7
     3. Construction 7
     4. Operation 7
   - G. Control of Sewer System Use 8

2. **SURVEYS AND INVESTIGATIONS** 9
   - A. Introduction 9
   - B. Types of Information Required 9
     1. Physical 9
     2. Developmental 10
     3. Political 10
     4. Financial 10
   - C. Sources of Information 11
     1. Physical 11
     2. Developmental 11
     3. Political 12
     4. Financial 12
   - D. Surveys for Different Project Phases 12
     1. Preliminary Surveys 12
     2. Design Surveys 13
     3. Construction Surveys 13
   - E. Investigations 13

3. **QUANTITY OF SANITARY SEWAGE** 15
   - A. Introduction 15
   - B. Design Period 15
   - C. Design Flows 16
     1. Population Estimates 17
     2. Land Usage 18
     3. Tributary Areas 19
     4. Per Capita Sewage Flow 19

	5. Flow Estimate Based on Population and Flow Trends	20
	6. Distribution of Population	24
	7. Commercial Contributions	24
	8. Industrial Contributions	25
	9. Institutional Contributions	27
	10. Cooling Water	28
	11. Stormwater and Groundwater Contributions	28
	12. Minimum and Peak Flows of Sanitary Wastewater	32
D. Recapitulation		36
4. QUANTITY OF STORMWATER		41
A. Introduction		41
B. Rational Method		42
	1. Area	43
	2. Rainfall	44
	3. Runoff Coefficient	49
	4. Application of Rational Method	53
	5. Modifications of the Rational Method	54
C. Other Methods of Runoff Determination		55
	1. Hydrograph (Overland-Flow) Method	55
	2. Inlet Method	57
	3. Unit Hydrograph Method	61
D. Overland Flow		61
E. Retention Basins for Reducing Storm Sewer Inflow		61
5. HYDRAULICS OF SEWERS		67
A. Introduction		67
B. Terminology and Symbols		68
C. Hydraulic Principles		69
	1. Type of Flow	69
	2. One-Dimensional Method of Flow Analysis	70
	3. Energy Losses	72
	4. Pressure-Conduit Flow Analysis	72
	5. Open-Channel Profile Analysis	73
D. Flow Friction Formula		76
	1. General	76
	2. Darcy-Weisbach Equation	76
	3. Kutter and Manning Formulas	78
	4. Hazen-Williams Formula	80
E. Coefficients for Friction Formulas and Factors Affecting Them		82
	1. General	82
	2. Conduit Material	83
	3. Size of Conduit and Reynolds Number	85
	4. Shape of Conduit and Hydraulic-Elements Graphs	86
F. Scouring Velocities		88
	1. Self-Cleaning Velocities	88
	2. Velocity for Equal Cleansing	89
G. Non-Uniform Flow Problems		91
	1. Normal and Critical Depths	91
	2. Drawdown and Backwater Curves	92
	3. Hydraulic Jump	98
	4. Culverts	98
	5. Stormwater Inlets	100
	6. Transitions	106

  7. Bends 108
  8. Junctions 109
  9. Vertical Drops and Other Energy Dissipators............. 110
  10. Measuring and Regulating Devices...................... 112

6. **DESIGN OF SEWER SYSTEMS**............................ 119
 A. Introduction 119
 B. Energy Concepts of Sewer Systems........................ 119
 C. Combined vs. Separate Sewers............................ 120
 D. Layout of System.. 121
 E. Curved Sewers 122
 F. Type of Conduit 123
 G. Ventilation 123
 H. Sewer Design in Relation to Sulfide Generation............ 124
 I. Depth of Sewer 127
 J. Minimum and Maximum Velocities and Design Depths of Flow.. 128
  1. Minimum Velocities 128
  2. Maximum Velocities 129
  3. Design Depth of Flow................................ 130
 K. Infiltration 130
  1. Infiltration-Exfiltration Test Allowance................ 130
  2. Infiltration-Exfiltration Testing 131
 L. Design for Various Conditions............................ 131
 M. Storm Channel Design................................... 133
 N. Intercepting Sewers 136
 O. Relief Sewers 137
 P. Organization of Computations 138

7. **APPURTENANCES AND SPECIAL STRUCTURES**.............. 141
 A. Introduction 141
 B. Manholes 141
  1. Objectives 141
  2. Manhole Spacing 141
  3. General Shape and Dimensions....................... 142
  4. Frame and Cover.................................... 142
  5. Steps 144
  6. Channel and Bench.................................. 144
  7. Manholes on Large Sewers........................... 145
  8. Shallow Manholes 146
  9. Connection Between Manhole and Sewer.............. 146
  10. Construction Materials 146
 C. Bends 147
 D. Junctions 148
 E. Drop Manholes 149
 F. Terminal Cleanout Structures............................ 149
 G. Building Sewers 150
 H. Check Valves and Relief Overflows....................... 152
 I. Stormwater Inlets 153
  1. Curb Inlets 153
  2. Gutter Inlets 155
  3. Combination Inlets 155

J. Catch Basins 156
K. Siphons 156
   1. Single- and Multiple-Barrel Siphons 156
   2. Profile 157
   3. Air Jumpers 157
   4. Sulfide Generation 159
   5. Provision for Draining 159
L. Sanitary Sewage Diversion and Stormwater Overflow Devices 160
M. Flap or Backwater Gates 161
N. Sewers Above Ground 162
O. Underwater Sewers and Outfalls 163
   1. Ocean Outfalls 163
   2. Other Outlets 164
P. Measuring Wastewater Flows 165
   1. General Methods 165
   2. Temporary Metering Stations 167
   3. Permanent Metering Stations 168

8. MATERIALS FOR SEWER CONSTRUCTION 171
A. Introduction 171
B. Asbestos-Cement Pipe 172
C. Brick Masonry 173
D. Clay Pipe 173
E. Concrete Pipe 174
F. Iron and Steel 176
   1. Cast Iron Pipe 176
   2. Ductile Iron Pipe 177
   3. Fabricated Steel Pipe 177
G. Organic Materials 179
H. Corrosion Protection 180
   1. General 180
   2. Concrete and Asbestos-Cement Protection 180
   3. Iron and Steel Pipe Protection 181
   4. Other Linings 181
I. Joints 181

9. STRUCTURAL REQUIREMENTS 185
A. Introduction 185
B. Loads on Sewers Due to Gravity Earth Forces 185
   1. General Method 185
   2. Types of Load Conditions 186
   3. Loads for Trench Conditions 187
   4. Loads for Embankment Conditions 193
   5. Loads for Jacked Conduits and Certain Tunnel Conditions ... 200
   6. Loads for Tunnels 204
C. Loads on Sewers Due to Superimposed Loads 205
   1. General Method 205
   2. Concentrated Loads 205
   3. Distributed Loads 207
   4. Conduits Under Railway Tracks 208
   5. Conduits Under Rigid Pavement 209

TABLE OF CONTENTS

 D. Supporting Strength of Rigid Conduits 209
  1. Laboratory Test Strength 209
  2. Pipe Bedding 210
  3. Backfill 210
  4. Field Supporting Strength 210
  5. Load Factor 210
  6. Supporting Strength in Trench Conditions 211
  7. Supporting Strength in Embankments 215
 E. Supporting Strength of Flexible Pipes 217
  1. General Method 217
  2. Corrugated Metal Pipe 219
  3. Plastic Pipes 220
 F. Factor of Safety 221
  1. General 221
  2. Rigid Pipes 221
  3. Flexible Pipes 222
 G. Design Relationships 222
  1. Rigid Pipe 222
  2. Flexible Pipe 223
 H. Charts for Determining Earth Loads on Buried Conduits 223
 I. Recommendations for Attaining Design Load Supporting Strength 225
  1. Factor of Safety 225
  2. Effect of Trench Sheeting 226

10. CONSTRUCTION CONTRACT DOCUMENTS 229

 A. Introduction 229
 B. Plans 230
  1. Purpose 230
  2. Field Data 230
  3. Plan Preparation 230
  4. Contents 231
  5. Record Drawings 237
 C. Specifications and Related Documents 237
  1. Purpose 237
  2. Arrangement 238
  3. Bidding Requirements 239
  4. Contract Forms 241
  5. General Conditions 242
  6. Specifications 242
  7. Supplementary Information—Not a Part of Contract Documents 243
  8. Standard Specifications 243
  9. Sewer Specifications Check List 244

11. CONSTRUCTION METHODS 247

 A. Introduction 247
 B. Construction Surveys 247
  1. General 247
  2. Preliminary Layouts 248
  3. Setting Line and Grade 248
  4. Tunnel Construction 250
 C. Site Preparation 251

- D. Open-Trench Construction 251
  1. Trench Dimensions 251
  2. Excavation 251
  3. Sheeting and Bracing 254
  4. Dewatering 258
  5. Foundations 260
  6. Pipe Sewers 262
  7. Cast-in-Place Concrete Sections.............................. 265
  8. Backfilling 267
  9. Surface Restoration 269
- E. Tunneling 270
  1. General Classification 270
  2. Auger or Boring 270
  3. Jacking 270
  4. Mining Methods 271
- F. Special Construction 280
  1. Railroad Crossings 280
  2. Crossing of Principle Traffic Arteries...................... 281
  3. Stream and River Crossings................................... 281
  4. Outfall Structures 281
- G. Sewer Appurtenances 283
- H. Open Channels 284
- I. Construction Records 285

12. WASTEWATER AND STORMWATER PUMPING STATIONS......... 287
- A. Introduction 287
- B. Station Capacity 288
- C. Wet-Well Design 289
- D. Screening Devices 293
  1. Bar Screens 293
  2. Communitors 295
  3. Water Supply 296
- E. Pumping Equipment 296
  1. Types in General Use.. 296
  2. Centrifugal Pumps 296
  3. Pneumatic Ejectors 300
  4. Pump Selection 300
  5. Pump Drives 310
- F. Piping and Valves 315
- G. Electrical Equipment 315
  1. General 315
  2. Voltage Selection 316
  3. System Selection 317
  4. Transformers 319
  5. Switchgear 319
  6. Unit Substations 320
  7. Motor-Starting Equipment 320
  8. Motor Overload Protection 321
  9. Controls 322
  10. Cable 323
  11. Conduits and Fittings 323
- H. Prefabricated Pumping Stations 324

TABLE OF CONTENTS

    I. Chlorination 324
        1. General 324
        2. Application Points 324
        3. Equipment and Controls 326
        4. Safety 327
   J. Appurtenances 327
        1. Wash Rooms 327
        2. Meters and Gages 327
        3. Water Supply and Water-Seal Equipment................. 327
        4. Hoisting Equipment 328
        5. Fencing 329
        6. Landscaping 329
        7. Decoration and Finish 329
        8. Heating and Ventilating 329
        9. Building Drainage 330
      10. Lighting 330
      11. Safety Features 330
      12. Paints and Protective Coatings.......................... 331

INDEX xiii

FOREWORD

This is the first (1969) revised edition of the Manual of Practice on "Design and Construction of Sanitary and Storm Sewers" which was originally published in 1960 as a joint effort of the American Society of Civil Engineers and the Water Pollution Control Federation to present a compilation of current practice in this field.

In 1964 a joint committee was formed of members of the Water Pollution Control Federation and the American Society of Civil Engineers to consider and draft revisions to the Manual. The efforts of this committee are presented in this first revised edition.

The Manual should be considered as an aid to the practicing engineer as a check list of items to be considered in a sanitary sewerage or storm drainage project, as represented by acceptable current procedure. It is not intended as a substitute for engineering experience and judgment, or as a treatise replacing standard texts and reference material.

In common with other manuals prepared on special phases of engineering, this Manual recognizes that this field of engineering is constantly progressing with new ideas and methods coming into use. It is hoped that users will present any suggestions for improvement to the Water Pollution Control Federation or the Sanitary Engineering Division of the American Society of Civil Engineers for possible inclusion in future revisions to keep this Manual current.

The members of the Committee thank the reviewers of the Manual for their assistance by submitting their suggestions for improvement.

The initial printing was exhausted and the second printing made in the Spring of 1970. Errors and omissions evident from the 1969 edition were corrected.

CHAPTER 1. ORGANIZATION AND ADMINISTRATION OF SEWER PROJECTS

A. INTRODUCTION

Sewer systems are essential to the public health and welfare in all areas of concentrated population and development. Every community produces water-borne wastes of domestic, commercial, and industrial origin and is subject to the runoff of stormwater. Sewers perform the vitally needed functions of collecting these waters and conveying them to points of discharge or disposal.

The design and the construction of sewers occupy a large and important segment in the field of civil engineering. But unlike most other public works, a properly functioning sewer system is virtually invisible. Thus, its importance to the community frequently is overlooked.

Storm- and sanitary-sewer projects pass through various stages of design and construction. Consequently, the best efforts of a number of specialists must be combined to achieve the desired results at least cost.

This calls for an understanding of the objectives of the several phases of the project and of the responsibilities and interests of the various parties involved. It is toward this end that the discussions in this chapter are oriented.

B. DEFINITION OF TERMS AND CLASSIFICATION OF SEWERS

The user of this manual is referred to the "Glossary—Water and Wastewater Engineering" (1). With some modifications, the following basic definitions are as given in that publication:

Sewer.—A pipe or conduit that carries wastewater or drainage water.

Wastewater.—The spent water of a community. From the standpoint of source it may be a combination of the liquid and water-carried wastes from residences, commercial buildings, industrial plants, and institutions, together with any groundwater, surface water, and stormwater that may be present. In recent years the word wastewater has taken precedence over the word sewage.

Sanitary Sewer.—A sewer that carries liquid and water-carried wastes from residences, commercial buildings, industrial plants, and institutions, together with minor quantities of storm, surface, and groundwaters that are not admitted intentionally.

Storm Sewer.—A sewer that carries stormwater and surface water, street wash and other wash waters, or drainage, but excludes domestic wastewater and industrial wastes. Also called storm drain.

Combined Sewer.—A sewer intended to receive both wastewater and storm or surface water.

Storm-Overflow Sewer.—A sewer used to carry the excess of storm flow from a main or intercepting sewer to an independent outlet.

Relief Sewer.—(a) A sewer built to carry the flows in excess of the capacity of an existing sewer; (b) A sewer intended to carry a portion of the flow from a district in which the existing sewers are of insufficient capacity, and thus prevent overtaxing the latter.

Building Sewer.—In plumbing, the extension from the building drain to the public sewer or other place of disposal; also called house connection.

Lateral Sewer.—A sewer that discharges into a branch or other sewer and has no other common sewer tributary to it.

Branch Sewer.—A sewer that receives wastewater from a relatively small area and discharges into a main sewer serving more than one branch-sewer area.

Submain Sewer.—A sewer into which the wastewater from two or more lateral sewers is discharged and which subsequently discharges into a main, a trunk, or other collector.

Main Sewer.—In larger systems, the principal sewer to which branch sewers and submains are tributary; also called trunk sewer. In small systems, a sewer to which one or more branch sewers are tributary. In plumbing, the public sewer to which the house or building sewer is connected.

Intercepting Sewer.—A sewer that receives dry-weather flow from a number of transverse sewers or outlets and frequently additional predetermined quantities of stormwater (if from a combined system), and conducts such waters to a point for treatment or disposal.

Outfall Sewer.—A sewer that receives wastewater from a collecting system or from a treatment plant and carries it to a point of final discharge.

Separate sanitary and storm sewers are desirable and used with few exceptions in new systems. The major advantages of separate systems are the protection of watercourses from pollution and the exclusion of stormwater from the treatment system with a consequent saving in treatment plant construction and operating cost. Combined sewers frequently are encountered in older communities where it may be extremely difficult or costly to provide separate systems.

C. PHASES OF PROJECT DEVELOPMENT

Conception and development of typical sewer projects comprise the following phases:

1. Preliminary or Investigative

The objective of this phase is to establish the broad technical and economic bases for policy decisions and final designs. The importance of this phase cannot be overemphasized. Inadequate preliminary work will be detrimental to all succeeding phases and may endanger the successful completion of the project or cause the owner to undertake planning which would not produce the most economical or efficient result. This phase usually culminates in an engineering report which includes items, such as:

(a) Statement of the problem and review of existing conditions.
(b) Capacities and conditions required to provide service for the design period.
(c) Method of achieving the required service—if more than one method is available, an evaluation of each alternative method.
(d) General layouts of the proposed system with indication of stages of development to meet the ultimate condition when the project warrants stage development.
(e) Establishment of applicable engineering criteria and preliminary sizing and design to permit preparation of construction and operating cost estimates of sufficient accuracy to provide a firm basis for feasibility determination, financial planning, and consideration of alternative methods of solution.
(f) Various available methods of financing and their applicability to the project.

It must be recognized that the preliminary engineering report is not a detailed working design or plan from which a sewer project can be constructed. Indeed, such detail is not necessary to meet the objectives of the preliminary or investigative phase. Nonetheless, proper preliminary engineering is the fundamental step of final planning.

2. Design

The design phase of a sewer project ends with the completion of construction plans and specifications. These documents form the basis for bidding and performance of the work; and they must be clear and concise. Design, therefore, consists of the elaboration of the preliminary plan to include all details necessary to construct the project.

3. Construction

This phase involves the actual building of the project according to the plans and specifications previously prepared. Obviously, construction will be facilitated if the preceding work has been done thoroughly and competently.

4. Operation

Although this manual is devoted to matters of design and construction, the operation of a sewerage system is an important aspect in the development of such projects.

D. INTERRELATION OF PROJECT DEVELOPMENT PHASES

Since all phases of sewer projects are interrelated, the following points are applicable:

1. The capacity, arrangement, and details of a sewer system will not be

satisfactory unless the preliminary or investigative phase is current and properly completed.

2. Adequate preliminary engineering and estimating are essential to sound financial planning, without which subsequent phases of the project may be placed in jeopardy.

3. Inadequate design or improperly prepared plans and specifications can lead to confusion in construction, higher costs, failure of the project to meet intended functions, or actual structural or hydraulic failure of component parts.

4. Proper execution of the construction phase is vitally necessary to produce the quality and features provided by adequate design. Moreover, the value of the design can be lost by incompetent or careless handling of the construction phase.

5. All engineering projects have certain features requiring operation and maintenance. Unless they are anticipated and provided for, the usefulness of the project will be impaired.

E. PARTIES INVOLVED IN THE DESIGN AND CONSTRUCTION OF SEWER PROJECTS

Engineering projects, including sewers, are the result of the combined efforts of the several interested parties. The owner, engineer, and contractor are the most important ones. The project attorney, financial consultant, various regulatory agencies, and other specialists also are involved to varying degrees. The nature and general responsibilities of these individuals or organizations are as follows:

1. Owner

The owner's needs initiate the project and he provides the necessary funds. The owner is a party to all contracts for services and construction and may act directly or through any duly authorized agent. The owner most often refers collectively to the citizens of a governmental unit whose affairs may be handled by various legislative and administrative bodies. But the owner may also be a private group.

When the owner is a governmental unit, its business may be conducted by one of the following, depending on the organization of the unit and the laws controlling its operations:

(a) City councils or similar bodies, carrying out sewer projects as only one of many duties for the given unit.

(b) A special commission or board of a governmental unit dealing with more limited areas of interest than usually are handled by a city council. Such boards or commissions may be concerned with sewer projects alone or with a governmental unit's general utility system. The geographical limits of responsibility of such boards or commissions coincide with those of the parent governmental unit.

(c) A specially constituted district whose geographical limits may or may not coincide with those of other governmental units and whose affairs are administered by a separate and distinct administrative board or commission. Such units commonly are referred to as "districts"—for example, the Metropolitan Sanitary District of Greater Chicago. Often the responsibilities of such districts are limited to main trunk sewers, intercepting sewers, and treatment facilities, leaving local sewers as the responsibility of the individual governmental units within the area served by the larger district.

The fund-raising powers of the first two bodies usually are regulated by the same laws which apply to financing by the parent governmental unit. But fund-raising powers for a specially constituted district may show considerable variation from those of the governmental units which they override.

Temporary private ownership of sewer projects sometimes is encountered in the development of new subdivisions in which the developer constructs the sewer system and later transfers title to the appropriate governmental unit in accordance with local regulation. In some instances, sewerage systems and treatment works are under the permanent ownership of private utility companies.

2. Engineer

The engineer has the responsibility of supplying the owner with basic information needed to make project-implementing policy decisions, detailed plans and specifications necessary to bid and construct the project, consultation, and general and resident inspection during construction, and supervisory services necessary for the owner to establish satisfactory operation and maintenance procedures. These responsibilities are all of a professional character and must be discharged in accordance with ethical standards by qualified engineering personnel.

The engineer may be a single individual who performs all services on small projects. More often it is an organization which supplies the combined services of many people. Thus, the engineer may be defined as the person or organization employed or retained by the owner to provide design and consulting services and to inspect the results of the work performed by the contractor.

Engineering for sewer projects may be performed either by engineering departments which are a part of a governmental unit or by private engineering firms retained by the owner for specific projects. In many instances sewer projects are a joint effort by both types of organizations.

3. Contractor

The contractor performs the actual construction work under the terms of the contract documents prepared by the engineer. The construction agreement is between the contractor and the owner, not the engineer.

The functions of the contractor may be carried out by an owner's employees especially organized for construction purposes, but such practice for sewer projects of any magnitude is not common.

4. Other Parties

Other parties who may enter into sewer project development, particularly in the United States, are as follows:

(a) Legal Counsel. All public works projects are subject to local and state laws; and competent legal advice is required to assure compliance with these laws and the avoidance of setbacks because of legal defects in the project. Special legal counsel also may be required in connection with the financing of the project, particularly where a bond issue is involved.

(b) Financial Consultant. Advisory services with respect to project financing often are required and may be provided as a separate and specialized service. Such services occasionally are provided as part of a general financing agreement with a financing agency.

(c) Regulatory Agencies. The most frequently encountered regulatory body is the state health department or in some states a specially designated water pollution control agency which usually adopts minimum standards pertaining to sanitary features of design and whose approval of proposed design, plans, and specifications for sanitary sewer projects is required. Other regulatory bodies having jurisdiction may include agencies, such as:

1. Municipality or sewer district.
2. Local, regional, or state planning commissions.
3. Federal bureaus concerned with water pollution control.
4. The Corps of Engineers, United States Army, or state agencies having functional control of navigable waters.

F. ROLE OF PARTIES IN EACH PHASE

The role of the owner, engineer, and contractor with respect to each other in the different phases of the project is discussed subsequently.

1. Preliminary

The owner and the engineer are the principal parties involved in the preliminary phase of sewer projects, although the engineer may look to the contracting industry for special advice and consultation on construction methods or conditions peculiar to a given project which affect cost and design and on which a contractor may have specialized knowledge. It must be recognized that all policy decisions relating to the project, arranging for financing, etc., rest solely in the hands of the owner, even though he must depend on the engineer and legal and financial advisors for advice and guidance in making decisions.

2. Design

The design phase, up to the time of soliciting and receiving construction bids, involves both the owner and engineer. Designs prepared by the engineer usually are subject to the approval of the owner. The engineer may recognize preferences of the owner and be guided by these preferences when they are consistent with good engineering practice. The engineer must recognize and conform to the legal and procedural requirements governing each project.

3. Construction

The construction phase involves the owner, engineer, and contractor in a unique relationship. Contracts for engineering are between the owner and the engineer; whereas contracts for construction are between the owner and the contractor. The engineer is recognized as the agent of the owner but not so well recognized is the responsibility of the engineer toward the contractor. The engineer may provide services, such as staking and layout, approval of materials, inspection of work, and processing of payment estimates, all of which are of vital interest to the contractor. The engineer also must maintain rigid impartiality between the contractor and owner and must protect the contractor's interest when circumstances arise which, in his opinion, require a decision in favor of the contractor rather than the owner. This position of the engineer as an umpire in relations between the owner and the contractor places great responsibility on him for the maintenance of high ethical standards. Competent and thorough work in preliminary and design stages obviously will minimize problems to be encountered in the construction phase.

In his relationship with the contractor, the engineer must exercise his authority on behalf of the owner to ensure the completion of the work substantially in accordance with the requirements of the contract documents. At the same time, he must avoid direct supervision of the contractor's construction operations. If he does control or direct the acts of the contractor, he may become involved as a third party in any legal action brought against the contractor.

4. Operation

The engineer has a final obligation to the owner in providing full information on the intended functioning of all parts of a project. Nevertheless, it is the owner's staff which must assume responsibility for operation at the time the project or any part of it is completed and accepted by him. The engineer, and to some extent the contractor, may be expected to give advice and help in the initial stages of operation, at least to the extent of removing deficiencies which may appear, but the owner must provide a competent staff for operating and maintaining the completed project. In some cases, the engineer, by special agreement with the owner, may provide advisory services in connection with operation and maintenance procedures for a period of time after initial operation.

G. CONTROL OF SEWER SYSTEM USE

Of all public utilities, sewer systems probably are the most abused through misuse. This situation results from a misconception that a sewer can be used to carry away any unwanted substance or object that can be put into it. The absence of adequate regulations setting forth proper uses and limitations of the system and the lack of enforcement of existing regulations by those responsible for operation of the system tend to foster such a misconception. Abuse of the sewer system can result in extensive damage and compound the problems of wastewater treatment. Without proper maintenance and control, a sewer system may become a hazard to the public safety and increase operating costs unnecessarily.

The following are common consequences of sewer system misuse:

(a) Explosion and fire hazards resulting from discharge of explosive or flammable substances into the system.
(b) Sewer clogging by roots and accumulations of grease, grit, and miscellaneous debris.
(c) Physical damage to sewer systems resulting from discharge of corrosive or abrasive wastes.
(d) Surface- and groundwater overload resulting from improper connections to sanitary sewers.
(e) Watercourse pollution resulting from discharge of sewage to storm sewers.
(f) Interference with sewage treatment resulting from extreme wet-weather flows or from wastes not amenable to normal treatment processes.

In the organization of sewer projects, as well as the management of the completed systems, provision must be made for the controlled use of the sewers and enforcement of appropriate regulations. A comprehensive report on sewer ordinances is contained in WPCF Manual of Practice No. 3 (2). This publication will be helpful in checking the adequacy of existing regulations or in preparing new ones.

References

1. "Glossary—Water and Wastewater Engineering." Amer. Soc. Civil Engr., New York, N.Y. (1969).
2. "Regulation of Sewer Use." Manual of Practice No. 3, Water Pollution Control Federation, Washington, D.C. (1968).

CHAPTER 2. SURVEYS AND INVESTIGATIONS

A. INTRODUCTION

Surveys and investigations produce the basic data needed for the conception and development of an engineering project. Their fundamental importance requires that they be carried out competently and thoroughly if an effective project is to result.

The term "survey" as used herein refers to the process of collecting and compiling information necessary to develop any given phase of a project. In one sense it may include observations relating to general conditions affecting a project, such as historical, political, physical, and fiscal matters. On the other hand, a survey may comprise the precise instrument measurements which are necessary for the engineering design.

The term "investigation" often is used interchangeably with "survey." Its use in this manual, however, usually refers to the assimilation and analysis of the data produced by surveys to arrive at policy and engineering decisions.

Surveys and investigations for the preliminary phase of a project are broad in nature, with emphasis on covering all factors relating to a project and determining the relative importance of each. Surveys and investigations for design and construction phases are more precise and detailed, usually being limited by the scope of the project.

Methods of conducting a survey vary widely, depending on the phase of development under consideration and the objectives. Intelligent surveys require broad knowledge of the particular field and an understanding of the problems to be solved in the phase of the project for which the survey is being conducted. Knowledge of the various aspects of sewer design as set forth in other chapters of this manual will lead to recognition of the specific information to be obtained in a survey for any given project. The objectives of the surveys for the several project phases and the type of information required for each phase are discussed in this chapter.

B. TYPES OF INFORMATION REQUIRED

Several different kinds of information, applicable in varying degrees to the different project phases, may be collected during the course of surveys for a typical project. These include:

1. Physical

(a) Topography, surface conditions, details of paving to be disturbed, underground structures, subsoil conditions, water table details.

(b) Details of existing system to which a proposed sewer may connect.
(c) Locations of streets, alleys, or unusual obstructions; required rights-of-way; and all similar data necessary to define physical features of a proposed sewer project.
(d) Pertinent information relative to possible future extension of the proposed project by annexation or service agreements with adjacent communities or areas.

2. Developmental

(a) Population trends and density in area to be served.
(b) Type of development, i.e., residential, commercial, or industrial.
(c) Quantity and strength of wastes from industrial contributors and location of outlets.
(d) Water use data and flow gagings, where appropriate, to establish flow rates from existing similar areas, and meteorological and hydrological data as related to storm sewers.
(e) Historical and experience data relating to existing facilities which may affect proposed sewers.
(f) Comprehensive or regional plans of other agencies, especially planning and zoning groups.
(g) Location of future roads, airports, parks, industrial areas, etc., which may affect the routing and location of sewers.
(h) Capacity and condition of existing sewer system and stormwater drainage courses.
(i) Other pertinent data necessary to establish the required design criteria and capacity for the given project.

3. Political

(a) Present political boundaries and probability of annexation of adjacent areas.
(b) Possible service agreements with adjacent communities; feasibility of multi-municipal or regional system.
(c) Existence and effectiveness of industrial waste ordinances regarding pretreatment or limitations on the concentrations of damaging substances.
(d) Requirements for new waste ordinances to achieve desired results.
(e) Effectiveness and adequacy of present political subdivision to undertake the project; desirability of a new organization to sponsor same.

4. Financial

(a) Information relative to existing policies, obligations, or commitments bearing on financing of proposed sewers or drainage improvements.
(b) Amounts and retirement schedule of outstanding bonds and unobligated bonding capacity available for proposed project.

(c) Availability of federal or state assistance through grants or loans.
(d) Taxable valuation, existing tax levies, and any limits affecting proposed project.
(e) Schedule of existing sewer service rates and revenues yielded by them, or schedule of water rates and revenues.
(f) Property plats as required for sewer assessments and special methods of assigning assessments.
(g) Local construction and operating conditions affecting cost.
(h) All similar data necessary to establish a feasible financing program for the proposed project.

C. SOURCES OF INFORMATION

Possible sources of the different types of information sought by surveys for sewer projects include:

1. Physical

(a) Existing maps and system plans, including United States Geological Survey topographic maps, city plats and topographic maps, state highway plans and maps, tax maps, local utility records and plans.
(b) Aerial photographs.
(c) Instrument surveys, including approximate surveys by such devices as hand level and aneroid barometer which may be useful for preliminary work.
(d) Photographs of complex surface detail to supplement instrument surveys and photographs to show detail of existing systems.
(e) Borings and test pits, either by hand or by machine, for determining subsurface soil and water conditions, also sounding-rod probing for underground structure location and an indication of soil conditions.

2. Developmental

(a) Census reports.
(b) Planning and zoning reports and maps.
(c) General area examination to note type, degree, and density of development.
(d) Canvass of significant industry to determine type and amount of waste.
(e) Flow gagings and sampling in existing sewers to establish flow characteristics from similar areas.
(f) Records of water pumpage and water use.
(g) Rainfall and stream-flow records in or near the area (for storm-sewer projects).
(h) Design basis and operatonal characteristics of existing sewers from system records.

(i) Criteria of regulatory agencies having jurisdiction over the project.
(j) Engineering reports or studies of related projects in the area.

3. Political

(a) Municipal and state laws.
(b) Conferences with owner as well as other officials.
(c) Comprehensive plans established by planning agencies.
(d) Local and area meeting reports and minutes.

4. Financial

(a) Pertinent records of the owner's fiscal officer.
(b) Auditor's or treasurer's records relating to tax levies.
(c) Operating statements and reports of the sewer, water, and other utility departments.
(d) Ordinances or laws governing outstanding bonds and procedures for financing and contracting the proposed project.
(e) Assessment plats and schedules of prior projects to show methods in use in the locality.
(f) Tax maps showing subdivisions and ownership of property to be affected by special assessments.

D. SURVEYS FOR DIFFERENT PROJECT PHASES

The objectives of the different phases of project development must be understood if the survey to be conducted is to be meaningful. Typical phases of project development are discussed in Chapter I. Discussion of the objectives and nature of surveys for each phase follows:

1. Preliminary Surveys

This phase of development is concerned with the broad aspects of the project, including such things as required capacity, basic arrangement and size, probable cost, and methods of financing. Accordingly, information is required in sufficient detail to show general physical features affecting layout and general design. The need as well as the feasibility of the project development must be considered at this time. When a connection to an existing sewer system is proposed, the preliminary survey must contain flow data for use in estblishing the design capacity and sewer layout.

Extreme precision and detail are neither necessary nor desirable in this phase, but all data obtained must be reliable. The type and extent of the information needed for any given project usually will become apparent as the work progresses and may vary widely depending on the size and complexity of the project. Since preliminary surveys are used to develop the information on which the findings of the engineering report are based, they must be thoroughly and competently prosecuted. Sufficient allowance must be made for items affecting the total cost of the project, such as a rising trend in construction costs and unusual quantities of difficult excavation. Otherwise, costs will be underestimated.

Occasionally, photogrammetric methods may be advantageous in obtaining a portion of the data needed in making a preliminary survey.

2. Design Surveys

Surveys for this phase form the basis for engineering design as well as the preparation of plans and specifications. Design surveys are concerned primarily with obtaining physical data rather than development or financial information. In contrast with the preliminary surveys, they must contain all the detail and precision the designer needs to correlate his design and the resulting construction plans with actual field conditions. Design surveys involve the use of surveying instruments in establishing the accurate location of pertinent topographic features. Photogrammetric methods also may be used in obtaining this information.

Presumably, the preliminary phase will have established the general extent of the project so that the area to be covered by the design survey can be defined. However, further definition of location within the general area may be required during the course of the design survey. The design survey also may extend to some degree beyond the proposed construction limits in order that future expansion will be facilitated.

Obviously, accurate surveys are required to produce accurate designs and plans. Vertical control usually is established by setting bench marks throughout the project, the elevations of which have been checked by level circuits to within 0.01 ft (0.003 m). Although property and street lines often are used for horizontal control, in large projects control traverses may be considered desirable. A special problem frequently encountered in design surveys is the nature and extent of existing underground structures which must be cleared or displaced by the new sewer. Such information, insofar as practicable, must be obtained during the design survey to establish rights-of-way, minimize utility relocation costs, obtain better bids, and prevent changing or rerouting of lines once the project is started. Where the accurate location of important structures cannot be ascertained by other means and conflict is possible, excavation to determine location, elevation, and detail at the point of crossing may be warranted.

3. Construction Surveys

Surveys for this phase are concerned almost exclusively with physical aspects. They are required to establish control for line and grade, to check conformity of construction, and to run settlement levels on existing structures immediately adjacent to the construction where necessary.

E. INVESTIGATIONS

The foregoing dealt with information that must be gathered before the design and construction of a sewer project can be undertaken. Now attention is directed to the investigative procedures and studies that are an essential element of every engineering project and closely related to the general surveys. Moreover, investigations are based largely on the information produced by the surveys.

Investigations may take many forms, but always are directed toward determining the most feasible and practical methods of achieving a desired result. On small sewer projects, they may involve no more than an on-the-spot decision to use conventional minimum standards for a simple gravity extension to an existing system. Larger projects, on the other hand, may have several alternatives, all of which must be considered. Projects involving relief of existing systems, for example, usually require extensive studies before the design capacity and the method of correction can be ascertained.

The following questions are typical of those to be resolved by investigation:

(a) What is extent of area to be served and what is the pattern of present and future land use? Has an area zoning plan been adopted? How does the area relate to a regional sewer plan?

(b) What general arrangement of the system will best fill the need? What easements are required for this arrangement?

(c) What part of combined flow shall be intercepted for treatment from an existing combined system? How can such combined flow be reduced?

(d) What are the estimated present and future flows?

(e) What storm frequency or pattern is to be used for storm sewer design?

(f) Shall sanitary sewers all discharge to one point for treatment or shall treatment be provided at more than one point?

(g) How will requirements of other agencies (state and county highway departments, railroads) dictate specific locations for crossings, rights-of-way, installation, and details of materials or construction?

(h) Are deep gravity sewers, gravity sewers with circuitous routing, or wastewater pumping stations to be used?

(i) What material or materials are to be used for sewer construction?

(j) What will the project cost to construct and operate? How can these costs best be financed? What taxes, assessments, and service charges will be required? What are the sources of funds?

(k) Does the project necessitate the establishment of a new authority, for example, a sanitary district?

Consideration of these and similar matters usually is carried out in the preliminary phase of the project. The summary of the investigations together with the conclusions and recommendations resulting therefrom constitutes the preliminary engineering report. Sound answers to the foregoing types of questions are of paramount importance to the project. Finally, the findings must be presented in a manner that will acquaint the residents and officials of the community with the details of the project and assist them in deciding on the merits of the recommendations.

CHAPTER 3. QUANTITY OF SANITARY SEWAGE

A. INTRODUCTION

Separate sanitary sewers are provided primarily to carry the spent water supply of a community, including its industrial wastes, to a point of treatment or ultimate disposal. But in the interest of keeping this volume to a minimum, all waters free from harmful or objectionable impurities should be excluded, insofar as practicable. Connection of roof, yard, and foundation drains to the sanitary sewers should be strictly prohibited. Building construction and grading practices which permit surface water to enter basements or crawl spaces and subsequently gain access to the sanitary sewer through illegal drains similarly should be controlled.

The sewer capacity to be provided must be determined from careful analysis of the present and probable future quantities of domestic and industrial wastewaters, groundwater infiltration, and extraneous flows entering through foundation or basement drains and similar sources.

In some areas, peak discharges in the sanitary system may result from extraneous flows and be greatly in excess of peak water usage.

Every political subdivision having a sewer system and anticipating future growth either in terms of population or commercial and industrial expansion should have a long-range plan for the installation of sewers within its service area. The plan must be flexible, current, and coordinated with the master plan of the community and possibly of adjacent areas. For any given area the plan should contain, with appropriate modifications, most of the engineering data discussed in this chapter.

A plan of this type will permit the orderly and timely expansion of facilities on a sound financial basis, without resort to costly "crash" programs.

B. DESIGN PERIOD

The length of time throughout which the capacity of a sewer will be adequate is referred to as the design period. It must be established prior to the design of the sewer. Once determined, consideration then must be given to the quantity of wastewater to be handled. Because the flow is largely a function of population served, population density, and water consumption, lateral and submain sewers should be designed for peak flows corresponding to the population at saturation density as set forth in the community's master plan or otherwise predicted. In systems relatively free of unwanted surface or groundwater, anticipated maximum water-use rates and population density may be used as a guide in determining maximum sewage flows. Nonetheless, a critical review is due any apparent relationship between water use and wastewater discharged. Indeed, where extraneous flows are a major factor, the number of structures or extent of area served or to be served may be more significant than the population or maximum water use.

Trunk sewers, interceptors, and outfalls commonly are designed for the peak flow expected at least 25 to 50 yr in the future. Based on answers to a questionnaire, 16 of 38 states were found to require 50 yr, 2 states required at least 25 yr, and 20 had no specific requirements. In many instances, these design periods are not sufficiently long to extend to the date of population saturation. For large sewers, past and future trends in population, water use, and existing wastewater flows must be studied before a design period can be selected.

In the case of trunk sewers serving relatively undeveloped lands adjacent to metropolitan areas, it may not be feasible to construct initial facilities for more than a limited period. Nevertheless, easements and rights-of-way for future trunks should be secured during the original construction program or as far in advance of development of the area as possible.

C. DESIGN FLOWS

Design flows for sanitary sewer systems can be separated into two categories: those in which the sewage flow is the major factor and only minor allowances need be made for extraneous flows; and those in which extraneous flows dominate. In the latter case, the sewage contribution may be a minor factor in the maximum rate of flow. Clearly, the connection of roof, yard, areaway, and foundation drains to sanitary sewers must be prohibited if this second category is to be avoided. Yet it must be recognized that established custom, as well as topographical and political considerations, may make all or part of this approach impractical.

The designer, however, must determine which of these situations, or what combination of them, exists or will exist. For example, extraneous flows in a relatively new sanitary sewer system in one midwestern suburban community (1) were found to be as high as 0.02 cfs/acre (0.08 cu m/min/ha) or in excess of 1,300 gpd/cap (4.94 cu m/day/cap). Average dry-weather flows on the other hand were less than 70 gpd/cap (0.266 cu m/day/cap). A study carried out in another community by other investigators uncovered peak flows of 0.042 cfs/acre (0.18 cu m/min/ha) and 0.033 cfs/acre (0.14 cu m/min/ha), respectively, in two suburban subdivisions (2). Under conditions this severe it may not be practicable to convey wet-weather maximum flows to central treatment plants; other means for handling them may be required (3).

Construction costs of sewers designed to accommodate extreme rates of extraneous flow naturally are greater than for those carrying only sanitary wastes; but unless this differential is brought to the attention of the governing body, there is little hope that corrective action will be taken or even considered.

In the design of sanitary sewers and treatment works where wastewater contributions govern, the following flows are important: daily minimum and maximum, daily mean, and the peak. This last flow may be defined as the mean rate during the maximum 15 min for any 12-month period. It should be recognized that flows normally will vary greatly during the design period.

The mean daily flow of sewage and groundwater is derived from analysis of a full year's operating data, whenever possible. This information also may aid in estimating minimum, maximum, and peak flows, where they cannot be measured, as well as the size and operating cost of sewage treatment works and pumping stations. The daily minimum and maximum discharges for the initial and final years of the design period are of value in determining treatment plant capacities. Peak flow estimated for the end of the design period determines the hydraulic capacity of sewers and some treatment plant conduits as well as ultimate pump capacities. Minimum flows, both for initial and final conditions, are related directly to the design of sewers to insure proper cleansing and the location and extent of protective measures required to protect sewers against sulfide corrosion.

A sanitary sewer has two main functions: to carry the peak discharge for which it is designed and to transport suspended solids so that deposits in the sewer are precluded. It is essential, therefore, that the sewer have adequate capacity for the peak flow and that it function at minimum flows, both initially and finally, without nuisance.

1. Population Estimates

In estimating the sanitary sewage fraction of the average daily flow, it is customary to multiply the future tributary population by the probable per capita sewage contribution. Obviously, the accuracy of the population estimate is extremely important in this computation. No effort should be spared in assembling and studying as much population information as possible. It usually is rewarding to make inquiries about United States census reports, chamber of commerce studies, voter registration lists, school census statistics, and installation records from local telephone, electricity, gas, water, and sewer utilities. Sometimes, city or area planning studies may be available in which population estimates have been worked out with great care and in considerable detail.

In addition to their immediate value in estimating flows, estimates of future population may influence the choice of the design period. For example, a rapidly growing population may make the use of a long design period uneconomic.

Future population trends depend on many factors, some known and some unknown. Known factors include location with respect to transportation facilities for workers, raw materials, and manufactured products; possible expansion of present industries; availability of sites for residential, commercial, or industrial development; civic interest in community growth; availability of other utility services at reasonable rates;

and real estate values. Nevertheless, population estimates based on all known factors can be upset by extraordinary events such as the discovery of some new natural resource in the vicinity or the decision by a large manufacturer to locate in the community.

Methods used in the past for population predictions have included:

(a) Arithmetical increase per year or per decade;
(b) Uniform percentage rate of growth based on recent census periods;
(c) Decreasing percentage rate of increase;
(d) Graphical comparison with the growth of other similar but larger cities;
(e) Graphical extension of the curve of past growth; and
(f) Verhulst's theory (logistic trend method) (4).

Details of these methods are discussed in textbooks on sewer design. An example of a suitable method of population prediction is illustrated in Figure 1. Probable future growth is indicated by two limiting lines, as shown, because of the difficulty of predicting future population with any great degree of accuracy.

Demographers, professionals in the field of population forecasts, tend to use somewhat more sophisticated methods of population projections such as ratio and correlation, component analysis, and employment forecasts (5).

2. Land Usage

City or urban area planning studies, where available, are a valuable guide in determining future land use. Present land uses as given by zoning regulations and the history of migrations of population from one district to another also are obviously useful guides in studying population

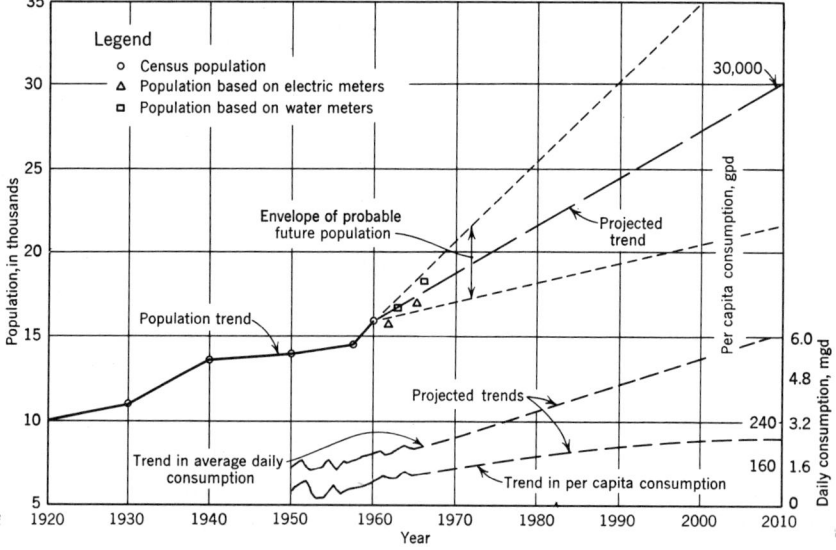

FIGURE 1.—Population and water consumption trends. (Gpd/cap $\times$ 3.8 = l/day/cap; mgd $\times$ 3,785 = cu m/day.)

changes and distribution within a community. The present trend toward decentralization of business and industry, and the development of suburban shopping centers, has and will continue to have a complicating influence on population predictions.

In some cities, areas have been classified as renewal districts and will be rebuilt as housing projects. This type of information usually will be contained in planning studies; and while such changes in land use may not materially affect the population in that area, they will affect the quantity of sanitary sewage to be generated because better facilities in the newer apartments will tend to increase water consumption.

3. Tributary Areas

Tributary areas for sewer design may be limited by natural topography, political boundaries, or economic factors. The sewer design for small drainage areas normally should provide sufficient capacity to serve the entire area. In larger drainage areas there is usually an economic advantage in providing adequate capacity for a certain period of time and then constructing relief sewers when and where needed as the growth pattern becomes better established.

Political boundaries sometimes will determine the extent of a project if legal restrictions prevent the financing of construction beyond the limits of the municipality or district. Nonetheless, sewer capacities should be based on total tributary area if possible since political boundaries and legal restrictions may change.

The economics of project financing also may place limits on the design; that is, the design may be restricted to the political boundaries or even a smaller area if the forecasted population growth rate is too low.

4. Per Capita Sewage Flow

Dry-weather flow quantities usually are slightly less than the per capita water consumption because water is lost through leakage, lawn sprinkling, swimming pools, etc. In arid regions the mean sewage flow may be as little as 40 percent of the average water consumption. On the other hand, industrial wastes or sewage from users of individual water supplies may result in average sewage quantities greater than measured per capita water usage. Consideration also should be given to changes in water-use habits and new household appliances, such as garbage grinders, dishwashers, and automatic clothes washers, in arriving at probable future per capita sewage contributions. All records suggest an increased use of water if for no other reason than its increased availability, and possibly because of the increased availability of sewers.

The average daily per capita domestic wastewater flows used for design purposes in a number of geographic locations are found in Table I. Most of these flows include only nominal allowance for extraneous flows, but there are exceptions. Figures 2 and 3 are design curves used in Austin and Dallas, Tex. Both show that as the service area increases, the per acre contribution decreases.

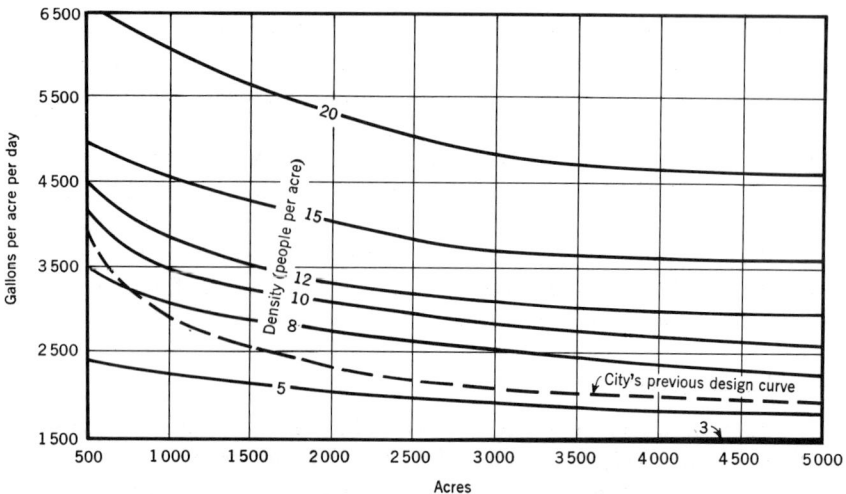

FIGURE 2.—Design flows used in Austin, Tex. (Gpd/acre × 0.00935 = cu m/day/ha; acre × 0.405 = ha.)

5. Flow Estimates Based on Population and Flow Trends

Figure 1 shows a typical plot of past census populations of a small city and estimates of the population made from the number of electric meters and water meters for each year subsequent to the last regular census year. The records of the average daily and per capita water consumption which may be used in lieu of measured sewage flows also are shown for the previous 15-yr period.

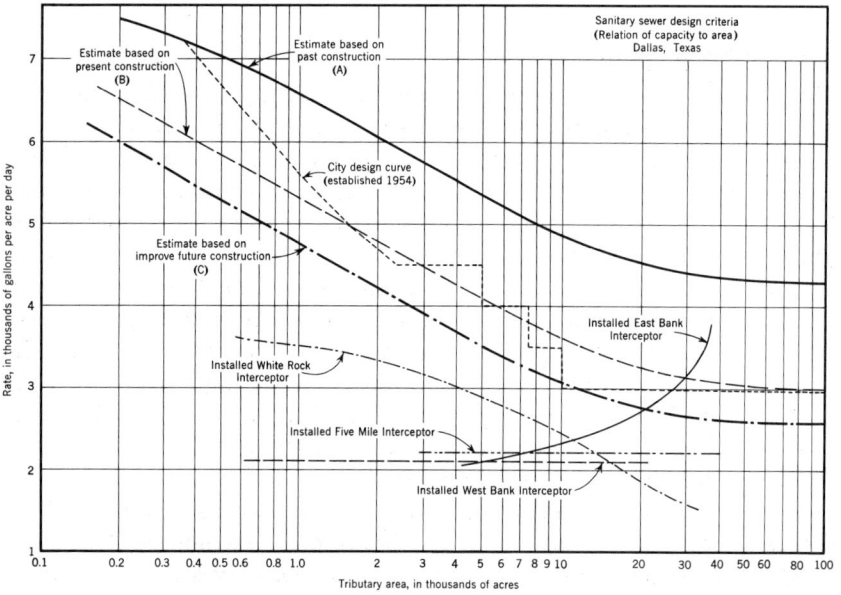

FIGURE 3.—Design flows used in Dallas, Tex. (Gpd/acre × 0.00935 = cu m/day/ha; acre × 0.405 = ha.)

TABLE I.—Some Typical Design Flows

City	Year and Source of Data	Average Rate of Water Consumption (gpd/cap)	Population Served (1,000's)	Average Sewage Flow (gpd/cap)	Sewer Design Basis (gpd/cap)	Remarks
Baltimore, Md.	...	160	1,300	100	135 × factor	Factor 4 to 2
Berkeley, Calif.	...	76	113	60	92	
Boston, Mass.	...	145	801	140	150	Flowing half full
Cleveland, Ohio	1946 (6)	...	...	100	...	
Cranston, R. I.	1943 (6)	...	...	119	167	
Des Moines, Iowa	1949 (6)	...	...	100	200	
Grand Rapids, Mich.	...	178	200	190	200	
Greenville County, S. C.	1959	110	200	150	300	Service area includes City of Greenville. Sewers 24 in. and less designed to flow ½ full at 300 gpd/cap; sewers larger than 24 in. designed to have 1-ft freeboard
Hagerstown, Md.	...	100	38	100	250	
Jefferson County, Ala.	...	102	500	100	300	
Johnson County, Kan.	1958					
Mission Township Main Sewer Dist.		70	70	60	1,350	Most houses have basements with exterior foundation drains
Indian Creek Main Sewer Dist.		70	30	60	675	Most houses have basements with interior foundation drains

TABLE I.—Continued

City	Year and Source of Data	Average Rate of Water Consumption (gpd/cap)	Population Served (1,000's)	Average Sewage Flow (gpd/cap)	Sewer Design Basis (gpd/cap)	Remarks
Kansas City, Mo.	1958	...	500	60	675 1,350	For trunks and interceptors. For laterals and sub-mains. Many houses have basements and exterior foundation drains
Lancaster County, Neb.	1962	167	148	92	400	Serves City of Lincoln.
Las Vegas, Nev.	...	410	45	209	250	
Lincoln, Neb. (Lateral Dists).	1964			60		For lateral sewers max flow by formula: Peak flow = $5 \times$ avg flow $\div$ (Pop in 1,000's)$^{0.2}$
Little Rock, Ark.	...	50	100	50	100	
Los Angeles, Calif.	1965	185	2,710	85	*	* 85 gpd residential multiplied by peak factor. See Fig. 6
Los Angeles County Sanitation Dist.	1964	200	3,500	70**	...	** Domestic flow only, ranges from 50 to 90 gpd/cap depending on cost of water, type of residence, etc. Domestic plus industrial averages 90 gpd
Greater Peoria, Ill.	1960	90	150	75	800 8,500	Based on 12 persons per acre for lateral and trunk sewers, respectively
Madison, Wis.	1937 (6)	...	...	...	300	Maximum hourly rate
Milwaukee, Wis.	1945 (6)	...	...	125	...	All in 12 hr 250 gpd/cap rate
Memphis, Tenn.	...	125	450	100	100	
Orlando, Fla.	...	150	75	70	190	

QUANTITY OF SANITARY SEWAGE

Location	Year (Ref)					Remarks
Painesville, Ohio	1947 (6)	...	...	125	600	Includes infiltration and roof water
Rapid City, S. D.	...	122	40	121	125	New York State Board of Health Standard
Rochester, N. Y.	1946 (6)	...	...	...	250	Sewer design 150 gpd/cap plus 600 gpd/acre infiltration. Sewers 24 in. and less designed to flow ½ full; sewers larger than 24 in. designed to have 1-ft freeboard
Santa Monica, Calif.	...	137	75	92	92	
Shreveport, La.	1961	125	165	...	...	
St. Joseph, Mo.	1960	...	85	125	450	Main sewers
					350	Interceptors
Springfield, Mass.	1949 (5)	...	...	...	200	150 gpd/cap was used on a special project
Toledo, Ohio	1946 (5)	...	...	...	160	
Washington, D. C. Suburban Sanitary Dist.	1946 (5)	...	...	100	2 to 3.3 × avg	
Wyoming, Mich.	1960	150	50	82+	400	+Calculated actual domestic sewage flow, not including infiltration or industrial flow

Note: Gal × 3.785 = l; gpd/acre × 0.00935 = cu m/day/ha; ft × 0.3 = m; in. × 2.54 = cm.

The population and water consumption are predicted up to the year 2010. The zigzag nature of past population growth indicates the inappropriateness of using any mathematical curve to forecast the future population. It is probable that the future population will, in the engineer's judgment, be included within the envelope shown by means of dotted lines on Figure 1. The upper line represents the maximum past rate of growth and the lower line, the minimum. In the illustration, a projected trend has been adopted yielding a predicted population of 30,000 at the end of the design period. Should the actual population growth follow the upper boundary of the envelope, the capacity of the works should be reached by about 1990; and, if the population growth should follow the lower boundary, the facilities should be adequate much beyond the year 2010.

The past trend in per capita water consumption in the example shows a slow increase; and the future trend in average daily consumption has been estimated by multiplying the projected trend in per capita consumption by the projected trend in population.

6. Distribution of Population

The design of a sewer depends on the prediction of the future population densities throughout the tributary area. Population density for the entire community may be estimated by dividing the future population by the total area of the community to be served by sewers, with proper deductions for parks, playgrounds, uninhabitable swamps, lakes, ponds, and rivers. The future design population density for a portion of the community normally should be greater than the existing average for the community as a whole up to the limit of the saturation density. Of course, it cannot be assumed that all small areas will reach saturation densities of population; yet, there is no way of predicting which areas will. Accordingly, it is desirable that all sewers serving a small area be designed on the basis of saturation density (7).

The saturation population will range from 2 or 3/acre (5 or 7.5/ha) for very large lots, to about 30/acre (75/ha) for small lots zoned for single-family residences. It may be as high as 50/acre (125/ha) for areas zoned for two-family residences on relatively small lots, and may exceed 1,000/acre (2,500 ha) for multistory apartment buildings. This extreme range indicates that the best estimate of the ultimate saturation population for any area may vary greatly.

7. Commercial Contributions

Wastes from commercial areas usually are estimated in terms of gpd/acre, based on a study of contributions from existing developed areas and comparable data from other cities. Design for commercial areas can well be more generous because of the high costs and inconvenience of later replacement or provision of relief sewers and the usually higher damage potential from surcharge. In smaller communities with no significant commercial development, allowances for commercial area sewage often can be assumed to be included within the per capita flows developed for domestic sewage.

Commercial areas in larger cities are often larger and developed more completely. They attract workers and customers from relatively long distances, will have more fixtures with greater usage, and thus generate a greater sewage flow per acre. The number of stories in commercial structures also will affect the wastewater quantities.

Table II lists actual allowances for contributions from commercial areas in various cities. Note that the variation ranges from 4,500 to more than 160,000 gpd/acre (39 to 1,500 cu m/day/ha).

Table III gives ranges of average sanitary sewage quantities from commercial areas, approved by some public agencies.

Coin-operated laundries and car washes may contribute substantial flows particularly where units are large or numerous.

Actual flow measurements (8) at apartment projects in Houston, Tex., in the early 1960's account for the results shown in Table IV. Additional data for apartments in Chicago are contained in Table XI.

In large cities, the transient population must be considered, as is shown in the following example. In the southern half of the Borough of Manhattan in New York City, the population is composed of:

(a) Residents—those who live in the area, regardless of where they work.
(b) Workers—those who work in the area, regardless of where they reside.
(c) Transients—those who visit the area during the day for business, recreation, or other reasons.

In a study area reaching from West 14th Street to the Battery, between the Hudson River and Broadway, the several sanitary sewage quantities for the categories shown above are tabulated in Table V.

8. Industrial Contributions

Industrial wastewater quantities may vary from little more than the normal domestic rate to tens of thousands gpd/acre (Table II). The type of industry to be served, the size of the plant, the type of supervision, and the method of on-site treatment of the waste, considering in all cases the present and future circumstances, are important factors in estimating wastewater quantities. Peak discharge to the sewer sometimes can be reduced by detention tanks or basins.

The "Industrial Wastes Guide" of the Ohio River Pollution Control Report (10) contains data on quantities of wastes that can be expected from many industries. Average domestic sanitary sewage contributed from industry may vary from 8 to 25 gpd/cap/shift (30 to 95 l/cap/shift).

TABLE II.—Sewer Capacity Allowances for Commercial and Industrial Areas

City	Year and Source of Data	Commercial Allowance (gpd/acre)*	Industrial Allowance (gpd/acre)
Baltimore, Md.	1949 (6)	6,750–13,500; 135 gpd/cap, resident pop.	7,500 min
Berkeley, Calif.	—	—	50,000
Buffalo, N. Y.	— (9)	60,000	—
Cincinnati, Ohio	— (9)	40,000	—
Columbus, Ohio	1946 (6)	40,000; excess added to residential amt.	—
Cranston, R. I.	1943 (6)	25,000	—
Dallas, Tex.	1960 (8)	30,000 added to domestic rate for downtown; 60,000 for tunnel relief sewers	—
Detroit, Mich.	— (8)	50,000	—
Grand Rapids, Mich.	—	Offices, 40–50 gpd/cap; hotels, 400–500 gpd/room; hospitals, 200 gpd/bed; schools, 200–300 gpd/room	250,000
Hagerstown, Md.	—	Hotels, 180–250 gpd/room; hospitals, 150 gpd/bed; schools, 120–150 gpd/room	—
Houston, Tex.	1960 (8)	Peak flows: Offices, 0.36 gpd/sq ft; retail, 0–20 gpd/sq ft; hotels, 0.93 gpd/sq ft	—
Las Vegas, Nev.	—	Resort hotels, 310–525 gpd/room; schools, 15 gpd/cap	—
Los Angeles, Calif.†	1965	Commercial, 11,700: Each; hospitals, 0.75 mgd; schools, 0.12 mgd; universities, 0.73 mgd	15,500
Los Angeles County Sanitation Dist.	1964	10,000 avg; 25,000 peak	—
Kansas City, Mo.	1958	5,000	10,000
Lincoln, Neb.	1962	7,000	—
Memphis, Tenn.	—	2,000	2,000
Milwaukee, Wis.	1945 (6)	60,500	—
St. Joseph, Mo.	1962	6,000	—
St. Louis, Mo.	1960 (8)	90,000 avg; 165,000 peak	—
Santa Monica, Calif.	—	Commercial, 9,700; hotels, 7,750	13,600
Shreveport, La.	—	3,000	—
Toledo, Ohio	1946 (6)	15,000 avg; 30,000 peak	—
Toronto, Ont.	1960 (8)	63,500, downtown sewers	—

* Except as otherwise noted.
† Values used for future planning; individual studies made for specific projects.
Note: Gpd/acre $\times$ 0.00935 = cu m/day/ha; gpd/sq ft $\times$ 0.0408 = cu m/day/sq m; gal $\times$ 3.785 = l.

QUANTITY OF SANITARY SEWAGE

TABLE III.—Average Commercial Flows

Type of Establishment	Avg Flow (gpd/cap)
Stores, offices, and small businesses	12 to 25
Hotels	50 to 150
Motels	50 to 125
Drive-in theaters (3 persons per car)	8 to 10
Schools (no showers), 8-hr period	8 to 35
Schools (with showers), 8-hr period	17 to 25
Tourist and trailer camps	80 to 120
Recreational and summer camps	20 to 25

Note: Gal × 3.785 = l.

TABLE IV.—Measured Flow from Apartment Projects, Houston, Tex.

Stories per Structure	Dwelling Units per Acre	Flow (gpd/acre)		
		Average	Peak Day	Peak Hour
2	16.6	3,430	3,690	14,900
2	20.3	4,340	5,880	15,000
15	80.3	8,720	10,200	20,000

Note: Acre × 0.405 = ha; gpd/acre × 0.00935 = cu m/day/ha.

TABLE V.—Estimated Average Wastewater Flows in a Section of New York City

Contributor	Population	Water Consumption (gpd/cap)	Wastewater Flow (mgd)
Residents	67,000	200	8.55
Workers	409,000	30	8.56
Transients	392,000	15	3.80

Note: Gal × 3.785 = l; mgd × 3,785 = cu m/day.

9. Institutional Contributions

Institutional sewage contributions generally are more determinable, since the various factors affecting them are known. The ultimate population, area, number of buildings, individual habits, etc., are fixed largely almost as soon as the location is selected. Records generally are available and the sewer use pattern may be well established, especially for governmental institutions. Because similar institutions often are reasonably well related insofar as flows are concerned, a study of existing facilities may permit estimates to be made for new or unserved ones. However, care must be taken to adjust for variations in building and lot sizes when flows are expressed in areal units. Institutions often are located in residential areas, thereby concentrating population and thus adding to the sewage flow.

Measurements made at St. John's Hospital, St. Louis, Mo., in the spring of 1960 are shown in Table VI.

The hospital had 525 beds and a staff of 750 (3 shifts). Rates of flow were recorded by instruments attached to each water meter supplying the hospital. It was estimated that approximately 93 percent of the water was returned to the sewers. Electric power and steam used by the hospital were furnished by others.

TABLE VI.—Water Use at St. John's Hospital, St. Louis, Mo.

Item	Water Use Rate (gpd/bed)	Ratio of Designated Use to Avg
Avg daily	300	1
Max day	375	1.25
Avg 5-hr max	530	1.70
5-hr max	675	2.25
2-hr max	800	2.70

10. Cooling Water

All single-pass unpolluted cooling water used either for air-conditioning or industrial processes should be kept out of sanitary sewers.

Once-through cooling water for air-conditioning equipment is used at the rate of 1.5 to 2 gpm (5.7 to 7.6 l/min) per ton of refrigeration. A 5-ton unit would use approximately 14,000 gpd (53 cu m/day). It would not take many such units to overcharge small sanitary sewers seriously. The application of demand water charges per ton of non-conserving air-conditioning equipment and developments to eliminate or minimize water use suggest that the discharge of cooling water may become a less serious problem in the future.

One ton of refrigeration is the heat absorption capacity required to melt one ton (2,000 lb or 908 kg) of ice in 24 hr. This is equal to a heat exchange rate of 12,000 Btu/hr (3,024 kg-cal/hr).

11. Stormwater and Groundwater Contributions

Sanitary sewer design quantities must include an allowance for non-waste components which inevitably become a part of the total flow. Proper design and construction will reduce the quantity of unwanted water which will enter the sewer through cracked pipes, defective joints, faulty manholes, and submerged manhole covers. But they will not eliminate infiltration from improper house connections, illegal connections, or other defects on private property. These aspects normally are controlled through enforceable regulations or ordinances. Experience demonstrates that reasonable control of the construction and connection of house sewers, as well as the prohibition of roof and areaway drains and similar connections can be accomplished. But in some areas it may not be feasible to prohibit foundation drains; and in certain terrains it may not be practical to control building and yard grading sufficiently to prevent wholesale entrance of water into basements during severe storms.

It is important that the appropriate agency formulate regulations which will assure that extraneous flows are kept within economically justifiable limits. The usual procedure is to balance the costs of control against the benefits obtained. Costs, for example, may include extra expenses for yard grading and building construction, or alternate means of disposal of foundation drainage. Financial benefits include the differential value of reduced sewer sizes and the present worth of all future excess pumping station and treatment plant capacities and operating costs. There is a point of balance for each sewer system beyond which the cost of further reduction of extraneous flow will not be offset by equal savings. The designing engineer should locate this point as nearly as possible and design accordingly.

A more detailed discussion of extraneous flows and their control follows:

(*a*) **Leakage into Manholes; Roof and Areaway Drainage; Surface Runoff into Basements and Crawl Spaces.** The runoff from impervious areas represented by roofs and pavement, ordinarily very large in proportion to sanitary flows, should be kept out of sanitary sewers by enforced regulation. Tests (11) made on manhole covers submerged in only 1 in. (2.54 cm) of water indicate that the leakage rate per manhole may be from 20 to 75 gpm (76 to 265 l/min), depending on the number and size of holes in the cover. Although such leakage may contribute quantities of stormwater several times in excess of the average sanitary flow, it can be minimized by using solid covers with half-depth pick holes. A few illicit roof drain connections also can overcharge smaller sewers. Rainfall of 1 in./hr on 1,200 sq ft of roof area, for example, would contribute water at about the rate of 12 gpm (rainfall of 1 cm/hr on 100 sq m equals 16.7 l/min). Direct entry of surface runoff into basements or drained crawl spaces through window wells, areaways, basement garages, or directly through foundation walls can result in flows of extreme magnitude. Regulations should be adopted and enforced to prevent or at least severely limit conditions of this sort. Since compliance with the regulations may increase the cost of yard grading and building construction, determined and continuing resistance should be anticipated. The designer, therefore, must evaluate the situation and make allowances for such amounts of manhole leakage, roof water, and surface runoff as in his judgment will be unavoidable under the probable enforcement conditions for the specific area under design.

(*b*) **Foundation Drainage.** Foundation drainage should be barred from sanitary sewer systems by adequate regulations, and, like roof and yard drainage, should be diverted to a storm sewer system. Again, complete enforcement of regulations will seldom occur and allowances must be made for illegal connections. Expected quantities of flow from foundation drain connections may vary from insignificant to prohibitive amounts; they must be evaluated for each system. In the Kansas City, Mo., area an average allowance of 1.25 gpm (4.75 l/min) per house is made for foundation drainage.

(*c*) **Infiltration.** Sanitary sewers must be designed to carry unavoidable amounts of groundwater infiltration or seepage in addition to the peak sanitary flows and unexcludable quantities of stormwater.

Groundwater gains entrance to sewers through pipe joints, broken pipe, cracks or openings in manholes, and similar faults. Defective Y-branches are known to have contributed appreciable percentages of total infiltration.

Prior to the introduction of compression-type joints, the bulk of infiltration, except in sewers containing excessive amounts of broken pipe, entered at faulty joints. Many sanitary sewers have been built with either cement-mortar, or hot-poured or cold-installed bituminous joints. None of these jointing materials is entirely satisfactory because of the initial difficulty in making a tight joint and its deterioration with time. Fortunately modern jointing practice and the use of compression-type joints make it possible to reduce leakage from this source drastically. Most leakage into new systems now can be traced to defects in foundations or pipe strengths, or to faulty construction. A detailed discussion of joints and jointing materials is found in Chapter 8.

Poorly laid house connections may be extremely important sources of excessive infiltration since these lines often have a total length greater than the collecting sewers. House connections have been found to contribute as much as 90 percent of the total infiltration into a system. Because inspection and workmanship sometimes are found wanting when it comes to house connections on private property, some cities require pressure tests to be conducted. Moreover, there is a need for suitable public control of these connections in every community, including specifications and an insistence on proper construction practices.

Existing sewerage systems frequently are very leaky. Infiltration rates as high as 60,000 gpd/mile (140 cu m/day/km) of sewer have been recorded for systems below groundwater, with rates up to and exceeding 1 mgd/mile (2,450 cu m/day/km) for short stretches.

Infiltration and exfiltration tests and allowances for new installations are discussed in Chapter 6.

As with all other sources of unwanted water, infiltration must be kept to a minimum if the cost of pumping and treating sewage is to be minimized (12).

Excessive amounts of infiltration also can result in increased pipe sizes or the supplementing of existing sewers.

In the design of extensions to existing systems, past practices and trends in infiltration allowances should be considered. A study (13) reported in 1955 shows that by far the majority of stipulated allowances fell within the ranges shown in Table VII.

In Table VIII are additional data from a study concluded in 1965 (14).

TABLE VII.—Infiltration Specification Allowances

Pipe Diam (In.)	Infiltration Permitted	
	(gpd/mile)	(gpd/in. diam/mile)
8	3,500 to 5,000	450 to 625
12	4,500 to 6,000	375 to 500
24	10,000 to 12,000	420 to 500

Note: In. $\times 2.54 =$ cm; gpd/in. diam/mile $\times 0.000925 =$ cu m/day/cm diam/km.

TABLE VIII.—Variation of Infiltration Allowances among Cities

Number of Cities Reporting	Allowance (gpd/in. diam/mile)
4	1,500
4	1,000
1	800
2	700
1	600
63	500
11	450 to 300
16	250 to 150
21	100
5	50

Note: Gpd/in. diam/mile $\times$ 0.000925 = cu m/day/cm diam/km.

Comparing the data of Tables VII and VIII, it appears that specified infiltration allowances have not been reduced significantly in the 10-yr interval between the reports. With non-compression type joints it is possible to meet the average specification allowance of 500 gpd/in. diam/mile (0.465 cu m/day/cm diam/km) in workmanship, but this low infiltration rate is not likely to be maintained where the system is in groundwater. The reasons are discussed in the section on joints in Chapter 8.

The selection of a capacity allowance to provide for infiltration should be based on the physical characteristics of the tributary area, the type of pipe and joint to be used, and the type and condition of the joints and pipes in the existing contributory sewers. For small to medium-sized sewers (24 in. and smaller; 61 cm) it is common to allow 30,000 gpd/mile (71 cu m/day/km) for the total length of main sewers, laterals, and house connections, without regard to sewer size. Others make an allowance of from 10,000 to 40,000 gpd/mile (24 to 95 cu m/day/km), depending on sewer size and job conditions. This design infiltration allowance is added to the peak rate of flow of wastewater and other components to determine the actual design peak rate of flow for the sewer.

Seepage allowances are for average conditions where a portion of the length of the sewers is above the groundwater table and a portion below. If a substantial portion is to be permanently below groundwater, a larger allowance for infiltration should be made or special watertight joints specified.

A survey of municipal infiltration allowances (14) is summarized in Table IX.

Design allowances for infiltration normally are greater than infiltration-exfiltration test allowances. The infiltration-exfiltration tests are performed when the sewer is new. The design allowance is based normally on the anticipated condition of the sewer when it is nearing the end of its useful life.

TABLE IX.—Infiltration Design Allowances for Several Cities

City	Allowance (gpd/acre)
Seattle, Wash.	1,100
Bay City, Tex.	1,000
Lorain, Ohio	3,000
Marion, Ohio	750
Ottumwa, Iowa	600
West Springfield, Mass.	2,000
Alma, Mich.	140

Note: Gpd/acre $\times$ 0.00935 = cu m/day/ha.

12. Minimum and Peak Flows of Sanitary Wastewater

The flow of sanitary wastewater (not including the groundwater infiltration or unavoidable surface-water contributions) will vary continuously throughout any one day, with extreme low flows usually occurring between 2 and 6 AM and peak flows occurring during the daylight hours. The groundwater-surface water component, on the other hand, remains practically constant throughout any one day except during and immediately following periods of rainfall.

The ratio of the peak flow of the sanitary component to the average for the day will range from less than 130 percent for some large sewers to more than 200 percent for smaller laterals. Moreover, the ratio of the maximum daily flow at the end of the design period to the minimum daily flow at the beginning of the period may range from less than two to more than five, depending largely on the rate of growth of the area served by the sewer. Hence, the range of flows for which a sewer must be designed, that is, peak flow to extreme minimum, will vary from less than 3 to 1 for large sewers serving stable populations to more than 20 to 1 for small sewers serving growing populations where domestic wastewater is the major component of the total flow. The ratios may be much greater where extraneous flows are the governing factor.

The records of existing sewerage or water systems are rarely complete enough to permit estimates of the sanitary sewage component of the minimum flow or the peak flow. On a broader basis, Figures 4, 5, and 6 are examples of the variations in peak and minimum rates of flow for situations in which dry-weather wastewater flows are expected to govern. Figure 4 shows the ratios of peak and minimum flows to average daily sewage flow recommended for use in design by various authorities.

A recent comprehensive treatment of variations in sanitary sewage by Geyer (15) sets forth a detailed method for estimating flows and flow variations in systems where water use and nominal extraneous flow are the governing factors.

Figure 5, based on dry-weather maximums, is the modification of a chart originally prepared for the design of sewers for a group of 18 cities and towns in the Merrimack River Valley, Mass. The ratios given are

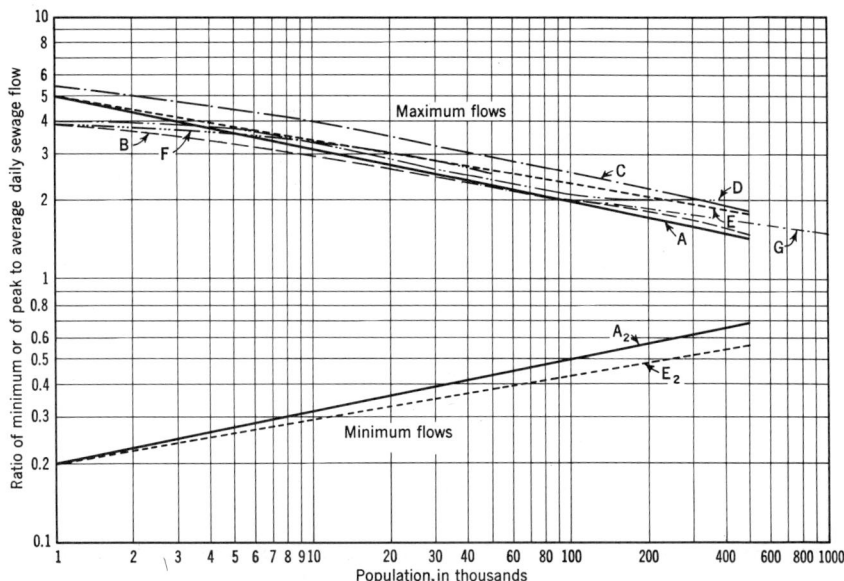

* Curve A source: Babbitt, H. E., "Sewerage and Sewage Treatment." 7th Ed., John Wiley & Sons, Inc., New York (1953).
Curve A_2 source: Babbitt, H. E., and Baumann, E. R., "Sewerage and Sewage Treatment." 8th Ed., John Wiley & Sons, Inc., New York (1958).
Curve B source: Harman, W. G., "Forecasting Sewage at Toledo under Dry-Weather Conditions." *Eng. News-Rec.* **80**, 1233 (1918).
Curve C source: Youngstown, Ohio, report.
Curve D source: Maryland State Department of Health curve prepared in 1914. In "Handbook of Applied Hydraulics." 2nd Ed., McGraw-Hill Book Co., New York (1952).
Curve E source: Gifft, H. M., "Estimating Variations in Domsetic Sewage Flows." *Waterworks and Sewerage*, **92**, 175 (1945).
Curve F source: "Manual of Military Construction." Corps of Engineers, United States Army, Washington, D.C.
Curve G source: Fair. G. M., and Geyer, J. C., "Water Supply and Waste-Water Disposal." 1st Ed., John Wiley & Sons, Inc., New York (1954).
Curves A_2, B, and G were constructed as follows:

Curve A_2, $\dfrac{5}{P^{0.167}}$

Curve B, $\dfrac{14}{4 + \sqrt{P}}$

Curve G, $\dfrac{18 + \sqrt{P}}{4 + \sqrt{P}}$

in which P equals population in thousands.

FIGURE 4.—Ratio of extreme flows to average daily flow compiled from various sources.*

approximately correct for a number of other municipalities in the same general area.

Figure 6 was developed by the Bureau of Engineering, City of Los Angeles, and has been in use since 1962.

Many state regulatory agencies (14 out of 38 state boards of health) have set 400 gpd/cap (1.5 cu m/day/cap) for laterals and 250 gpd/cap (0.95 cu m/day/cap) for trunk sanitary sewers as the minimum acceptable design flow rates where no actual measurements or other pertinent data are available. These minimum values assume the presence of a normal quantity of infiltration, but make no allowance for flows from foundation drains, roofs, yard drains, or of unpolluted cooling water. Additional design quantities should be added where conditions favoring excessive

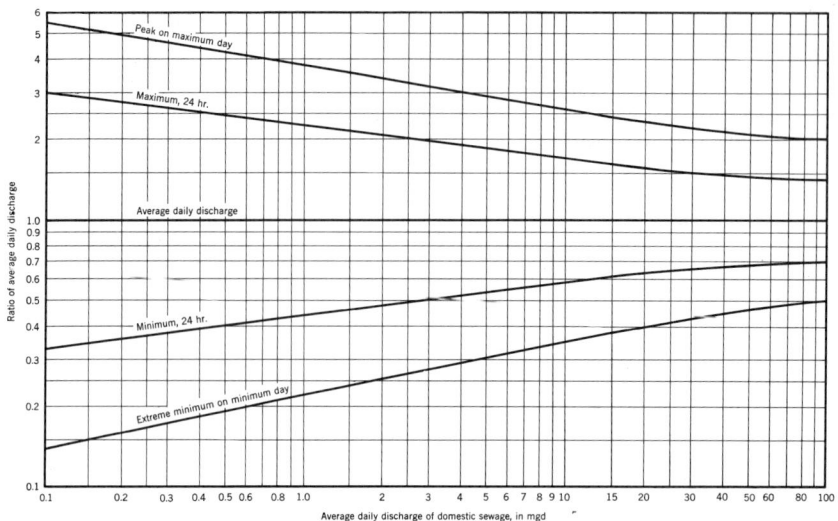

FIGURE 5.—Ratio of extreme flows to average daily flow in New England. (Mgd × 3,785 = cu m/day.)

infiltration or other extraneous flows are present. Also, provision must be made for industrial wastes which are to be transported by the sewers.

(*a*) **Fixture-Unit Method of Design.** Estimate of peak sewage flows for facilities such as hospitals, hotels, schools, apartment buildings, and office buildings may be made by the "fixture-unit" method (16). For these facilities flows approaching the peak rates often occur during the daylight hours. If the velocities in sewers designed for these flows are adequate for self-cleansing, deposits during the night hours will be resuspended and no nuisance should result. The United States of America Standards Institute National Plumbing Code, USASI A40.8–1955, defines "fixture-unit flow rate" as "the total discharge flow in gallons per minute of a single fixture divided by 7.5 which provides the flow rate of that particular plumbing fixture as a unit of flow. Fixtures are rated as multiples of this unit of flow." It further defines "fixture-unit" as a "quantity in terms of which the load-producing effects on the plumbing system of different kinds of plumbing fixtures are expressed on some arbitrarily chosen scale." From the first of these it can be seen that a fixture-unit is approximately 1 cfm (0.028 cu m/min).

Table X shows the fixture-unit value for various plumbing fixtures and groups of fixtures. Based on these, the discharge rate for the average single-family house or apartment is about 12 fixture units or three per person for a family of four.

Figure 7 shows the probable peak rates of discharge from systems consisting of various numbers of fixture-units, as taken from probability studies by Hunter (16). The probable peak rate of discharge from a system serving 1,000 persons at three fixture-units per person is, for example, about 440 gpm (28 l/sec). This compares to results obtained from Figure 5 for a 1,000-person system at an average daily discharge of

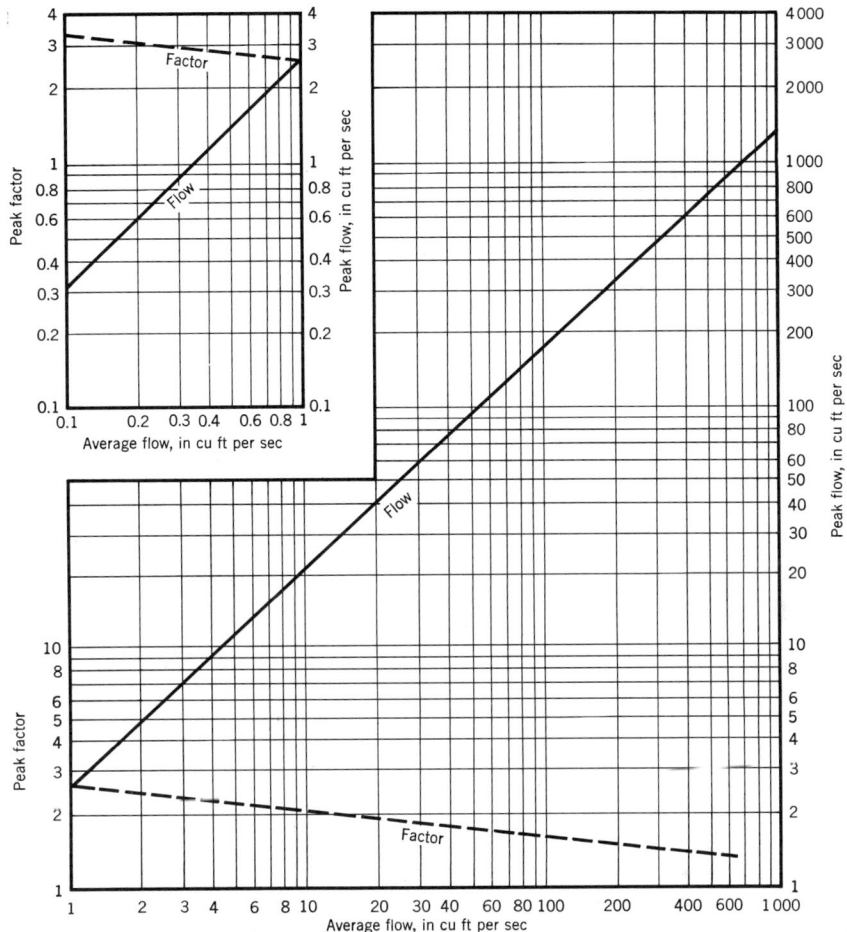

FIGURE 6.—Ratio of peak flow to average daily flow in Los Angeles.
(Cfs × 1.7 = cu m/min.)

100 gal/cap (380 l/cap). The peak rate of discharge estimated from Figure 5 is approximately 0.56 mgd (2,120 cu m/day), or 390 gpm (25 l/sec).

Although the preceding example exhibited close agreement between a peak flow estimated by the fixture-unit method and that computed on a per capita flow basis, it must be remembered that the fixture-unit method is based on a probability projection. This projection contains a number of assumptions as to the average number of fixture-units and average water use per capita. It also is based on a distribution of water use or water use habits representing an average or normal population. Large variations in conditions such as the number of fixture-units per capita and the average use per capita will affect the flow per fixture-unit and cause actual flows to vary from the flows that would be estimated from Figure 7. In the same manner, an extremely homogeneous population or a highly regulated population can cause a marked variation between the flows experienced and those predicted.

TABLE X.—Fixture-Units per Fixture or Group *

Fixture Type	Fixture-Unit Value as Load Factors
1 bathroom group consisting of tank-operated water closet, lavatory, and bathtub or shower stall	6
Bathtub † (with or without overhead shower)	2
Bidet	3
Combination sink-and-tray	3
Combination sink-and-tray with food-disposal unit	4
Dental unit or cuspidor	1
Dental lavatory	1
Drinking fountain	½
Dishwasher, domestic	2
Floor drains	1
Kitchen sink, domestic	2
Kitchen sink, domestic, with food waste grinder	3
Lavatory	1
Lavatory	2
Lavatory, barber, beauty parlor	2
Lavatory, surgeon's	2
Laundry tray (1 or 2 compartments)	2
Shower stall, domestic	2
Showers (group) per head	3
Sinks	
Surgeon's	3
Flushing rim (with valve)	8
Service (Trap standard)	3
Service (P trap)	2
Pot, scullery, etc.	4
Urinal, pedestal, syphon jet, blowout	8
Urinal, wall lip	4
Urinal stall, washout	4
Urinal trough (each 2-ft section)	2
Wash sink (circular or multiple) each set of faucets	2
Water closet, tank-operated	4
Water closet, valve-operated	8

* From United States of America Standards Institute National Plumbing Code, USASI A40.8–1955.

† A shower head over a bathtub does not increase the fixture value.

Note: For a continuous or semi-continuous flow into a drainage system, such as from a pump, pump ejector, air-conditioning equipment, or similar device, two fixture units shall be allowed for each gpm of flow.

Maximum water usage in apartment projects in Chicago (17) was considerably below that predicted by the fixture-unit method. The results of this study which covered apartment projects for the elderly, for low income, large family groups, and middle-income families are shown in Table XI.

D. RECAPITULATION

The quantity of wastewater which must be transported is based on full consideration of the following:

(a) The design period during which the predicted maximum flow will not be exceeded, usually 25 to 50 yr in the future.

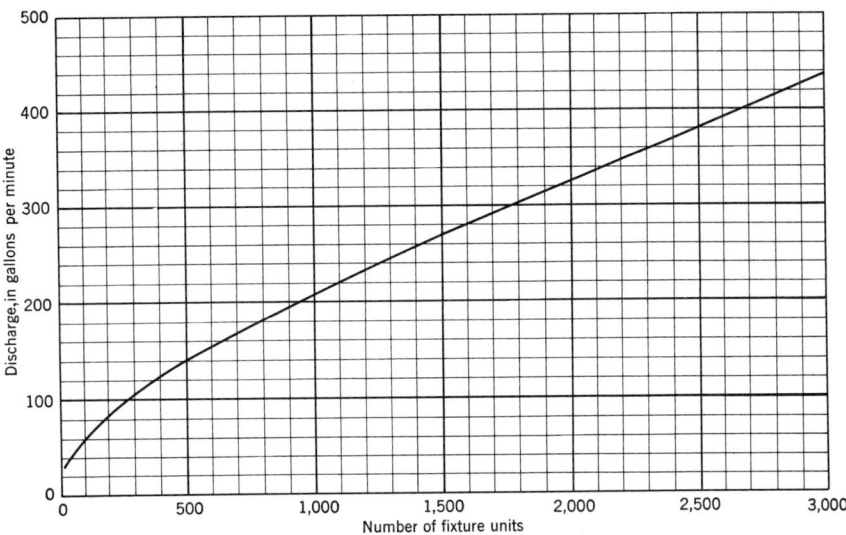

FIGURE 7.—Relation of peak discharge to fixture-units. (Gpm × 0.0631 = l/sec.)

(b) Domestic sewage contributions based on future population and future per capita water consumption, or if a more satisfactory parameter than water consumption is available, that parameter should be used. Careful analysis should be made of population distributions and the relationship of maximum and minimum to average per capita sewage flows. The fixture-unit method of estimating peak rates should be employed for small populations, giving due care to the forecasting of the probable number of fixture units and water use per capita. When large areas are to be considered, the peak rate of flow per capita or per acre sometimes is decreased as area and population increase.

(c) In some instances, maximum flow rates may be determined almost entirely by extraneous flows, the source of which may be foundation, basement, roof, or areaway drains, storm runoff entering through manhole covers, or infiltration. Foundation, roof, and areaway drain connections to sanitary sewers should be prohibited. Proper construction and yard grading practices should be mandatory. Nevertheless, there may be times when strict prohibition may not be feasible or even practicable. In any event, some storm and surface water will get into separate sanitary sewers and a judgment allowance, therefore, must be made.

(d) Commercial area contributions are sometimes assumed to be adequately provided for in the peak allowance for per capita sewage flows in small communities. A per acre allowance for comparable commercial areas based on records is the most reasonable approach for larger communities.

(e) Industial waste flows should include the estimated employee contribution, estimated or gaged allowances per acre for industry as a whole, and estimated or actual flow rates from plants with process wastes which may be permitted to enter the sanitary sewer.

TABLE XI.—Maximum Water Demands in Selected Chicago Apartments *

No. Apts	Population		Fixture Units			Maximum Demand (gpm)							
						Predicted by Hunter Curve (16)			Revised † or Observed Demands			
	Total	Per Apt	Total	Per Apt	Per Cap	Total	Per Apt	Per Cap	Total	Per Apt	Per Cap	Per Fixture Unit

a. Apartments for the Elderly

116	191	1.65	1,400	12.1	7.3	240	2.07	1.26	125	1.08	0.66	0.089
181	299	1.65	2,000	11.1	6.7	320	1.77	1.07	200	1.10	0.67	0.100
252	416	1.65	2,850	11.3	6.8	410	1.63	0.99	280	1.11	0.66	0.098
129	213	1.65	1,200	9.3	5.6	260	2.02	1.22	160	1.24	0.75	0.133
198	327	1.65	2,175	11.0	6.6	350	1.77	1.07	220	1.11	0.67	0.101
151	249	1.65	1,500	9.0	6.0	290	1.92	1.16	175	1.16	0.71	0.117
200	330	1.65	2,200	10.0	6.7	350	1.75	1.06	185	0.93	0.56	0.084
482	795	1.65	7,200	14.9	9.1	625	1.30	0.79	275	0.57	0.35	0.038
157	259	1.65	1,550	9.9	6.0	300	1.91	1.16	190	1.21	0.73	0.123
151	249	1.65	1,500	9.9	6.0	290	1.92	1.16	175	1.16	0.71	0.117

b. Apartments for Low-Income Families

480	3,312	6.9	5,280	11.0	1.6	640	1.32	0.194	282	0.59	0.09	0.054
140	728	5.2	1,540	11.0	2.1	275	1.96	0.378	151	1.08	0.21	0.098
474	2,940	6.2	5,214	11.0	1.8	638	1.35	0.217	311	0.66	0.11	0.060

c. Apartments for Middle-Income Families

| 318 | 670 | 2.1 | 3,600 | 11.3 | 5.4 | 470 | 1.48 | 0.70 | 307 | 0.97 | 0.46 | 0.085 |

* Chicago Housing Authority.
† Data predicted from test data.
Note: Gpm × 0.0631 = l/sec.

(*f*) Institutional wastes are usually domestic in nature although some industrial wastes may be generated by manufacturing at prisons, rehabilitation centers, etc. Peak and minimum design flow rates from persons in the institution are discussed elsewhere (4).

(*g*) Air-conditioning and industrial cooling waters, if permitted to enter sewers, may amount to 1.5 to 2.0 gpm (5.7 to 7.6 l/min) per ton of nonwater-conserving cooling units. Unpolluted cooling water should be kept out of separate sanitary sewers.

(*h*) Infiltration may occur through defective pipe, pipe joints, and structures. The probable amount should be evaluated carefully. Design allowances should be larger (under some circumstances very much larger) than those stipulated in construction specifications for which acceptance tests are made very soon after construction. Under-evaluation of infiltration is one reason why some sewers have become overloaded.

(*i*) The relative emphasis given to each of the foregoing factors varies among engineers. Some have set up single values of peak design flow rates for the various classifications of tributary area, thereby integrating all contributory items. It is recommended, however, that maximum and minimum peak flows used for design purposes be developed step by step, giving appropriate consideration to each factor which may influence design.

References

1. Weller, L. W., and Nelson, M. K., "A Study of Stormwater Infiltration into Sanitary Sewers." *Jour. Water Poll. Control Fed.*, **35,** 762 (1963).
2. Tolf, R. D., "Foundation Footing Drains in Separate Sanitary Sewers." Presented at Annual Meeting, Central States Water Poll. Control Assn. (1961).
3. Weller, L. W., and Nelson, M. K., "Diversion and Treatment of Extraneous Flows in Sanitary Sewers." *Jour. Water Poll. Control Fed.*, **37,** 343 (1965).
4. McLean, J. E., "More Accurate Population Estimate by Means of Logistic Curves." *Civil Eng.*, **22,** 2, 35 (1952).
5. McJunkin, F. E., "Population Forecasting by Sanitary Engineers." *Jour. San. Eng. Div., Proc. Amer. Soc. Civil Engr.*, **90,** SA4, 31 (1964).
6. Stanley, W. E., and Kaufman, W. J., "Sewer Capacity Design Practice." *Jour. Boston Soc. Civil Engr.*, 317 (1953).
7. Greeley, S. A., and Stanley, W. E., "Capacity Factors for Unequal Population Development." In "Handbook of Applied Hydraulics." 2nd Ed., p. 1017, McGraw-Hill Book Co., Inc., New York (1952).
8. Johnson, R. E. L., "Development of Sanitary Sewer Design Criteria for the City of Houston, Texas." *Jour. Water Poll. Control Fed.*, **37,** 1597 (1965).
9. Babbit, H. E., "Sewerage and Sewage Treatment." 6th Ed. John Wiley & Sons, Inc., New York.
10. "Industrial Wastes Guide." House Document No. 266, 78th Congress, Ohio River Poll. Control Rept., Part III, Suppl. D., Washington, D. C. (1943).
11. Rawn, A M, "What Cost Leaking Manhole?" *Waterworks and Sewerage,* **84,** 12, 459 (1937).
12. Horne, R. W., "Control of Infiltration and Storm Flow in the Operation of Sewerage Systems." *Sew. Works Jour.*, **17,** 209 (1945).
13. Velzy, C. R., and Sprague, J. M., "Infiltration Specifications and Tests." *Sewage and Industrial Wastes,* **27,** 245 (1955).

14. Anon., "Municipal Requirements for Sewer Infiltration." *Pub. Works,* **96,** 6, 158 (1965).
15. Geyer, J. C., and Lentz, J. L., "An Evaluation of the Problems of Sanitary Sewer System Design." Rept., Federal Housing Admin., Washington, D.C.
16. Hunter, R. B., "Methods of Estimating Loads on Plumbing Systems" Rept. BMS 65, National Bureau of Standards, Washington, D. C. (1940.)
17. Braxton, J. S., "Water Pressure Boosting Systems, Evaluation of Water Usage and Noises." *Cons. Engr.,* **XXIV,** V, 112 (1965).

CHAPTER 4. QUANTITY OF STORMWATER

A. INTRODUCTION

Stormwater runoff is that portion of the precipitation which flows over the ground surface during and for a short time after a storm.

Simply stated, stormwater drainage structures conduct runoff from places where it is not wanted to the nearest acceptable discharge point, all in sufficient time to avoid unacceptable amounts of damage and inconvenience.

An economic balance is necessary between the cost of structures and the direct and indirect costs of possible damage to property and inconvenience to the public over a long period of years. Only rarely is lack of urban drainage capacity the cause of loss of life, but it may cause considerable property damage.

In estimating stormwater quantities, it is helpful to keep in mind that flow into stormwater conduits is mainly by gravity from collecting surfaces of various characteristics such as rough or smooth, pervious or impervious, through swales and gutters, into inlets, and thence to conduits or channels of progressively larger capacities.

Consideration should be given to the effect of increased urbanization and to changing public opinion about the need for better drainage. Basin-wide planning is essential, particularly in urban areas, to prevent piece-meal drainage construction which at a later date may be found incompatible.

The rate of stormwater runoff to be used in the design of storm sewers is difficult to evaluate. Precipitation, which causes runoff, is highly variable. Storm runoff is almost wholly that part of the rainfall which is not lost by infiltration into the soil or left in surface depressions and on plant surfaces to evaporate. Surface and subsurface conditions affecting these losses are subject to complex variables, both natural and artificial.

Numerous empirical runoff formulas once were used widely, but largely have been abandoned in spite of their simplicity. They give satisfactory results only for the locality and specific conditions for which they were developed, and do not permit the designer to exercise his judgment with respect to the component variables.

There are two basic approaches to the problem, first, that of computing runoff as related to rainfall through a proportionality factor, and second, that of estimating residue from rainfall after abstractions for infiltration and retention losses and for temporary detention in transit. The first has been applied for many years in the rational method introduced in 1889, and the second is basic to the development of accurate and economically applicable methods, described later.

In both, a design frequency is assigned. Design frequency, as used herein, is the frequency with which a given event is equaled or exceeded on the average, once in a period of years. Probability of occurrence, which is the reciprocal of frequency, is preferred by some engineers. Thus, a 5-yr frequency event would be expected to be equaled or exceeded 20 times in 100 yr; or may be said to have a 0.20 probability or a 20-percent chance that the event will be equaled or exceeded in any particular year.

For the rational method the frequency of the stormwater runoff is assumed to be equal to that of the frequency of the average rainfall intensity selected. In the other methods discussed in this chapter, runoff is determined on the basis of a design storm of assumed rainfall intensity pattern. Although this pattern may not be assigned a frequency of occurrence, the total precipitation within the selected time period may be assigned a frequency of occurrence, or design frequency.

Experience has shown the rational method to yield satisfactory results for relatively small areas if properly applied. More than 90 percent of the engineering offices throughout the United States that replied to a questionnaire in 1956 on storm-sewer design practice indicated use of this method with satisfactory results for urban drainage areas.

At the first Engineering Foundation Research Conference on Urban Hydrology Research (1965) it was pointed out that the rational method still was used widely although generally criticized. Because of its widespread use, the rational method must be considered current practice, and the first part of this chapter is devoted to discussion of it.

The need for methods which will conform more closely to actual events occurring during storms long has been recognized. The science of hydrology has made progress in recent years with the result that newer methods, with the data necessary for their practical application, are beginning to appear.

B. RATIONAL METHOD

In the rational method, known as the Lloyd-Davies method in the United Kingdom, runoff is related to rainfall intensity by the formula,

$$Q = CiA \dots\dots\dots\dots\dots\dots\dots\dots\dots\dots 1$$

in which Q is the peak runoff rate in cfs, C is a runoff coefficient depending on characteristics of the drainage area, i is the average rainfall intensity in in./hr, and A is the drainage area in acres (1.008 cfs = 1 in. depth of rainfall applied at a uniform rate in 1 hr to an area of 1 acre; cfs $\times$ 1.7 = cu m/min).

The rational method is based on the following assumptions:

1. The peak rate of runoff at any point is a direct function of the average rainfall intensity during the time of concentration to that point.

2. The frequency of the peak discharge is the same as the frequency of the average rainfall intensity.

3. The time of concentration is the time required for the runoff to

become established and flow from the most remote part of the drainage area to the point under design.

The latter assumption applies to the part most remote in time, not necessarily in distance. In the rational method, average intensities have no time sequence relation to the actual rainfall pattern during the storm. The intensity-duration curve used in this method is not a time sequence curve of precipitation. Failure to recognize this point has led to erroneous usage of the rational method in attempts to evaluate the effects of antecedent precipitation and infiltration.

The determination of values for the coefficient, C, is difficult because this factor must represent many variables including infiltration, ground slope, ground cover, surface and depression storage, antecedent precipitation and soil moisture, shape of drainage area, overland flow velocity, etc.

Although the basic principles of the rational method are applicable to large drainage areas, reported practice generally limits its use to urban areas of less than 5 sq miles (13 sq km) (1). Development of data for application of hydrograph methods usually is warranted on larger areas.

For the larger areas, storage and subsurface drainage flow cause an attenuation of the runoff hydrograph so that rates of flow tend to be overestimated by the rational formula method unless these are taken into account.

In spite of the limitations of the rational method, long-time experience has resulted in practical definition of its variables; and its generalized representation of runoff requires the designer to use his judgment in evaluating the component factors. Thus, the rational method, if understood and correctly used, represents a definite improvement over the earlier empirical formulas and is a useful engineering tool.

1. Area

The area tributary to any point under consideration in a storm-sewer system must be determined. It may be measured accurately and is the only element of the rational method subject to precise determination.

Boundaries of the drainage area may be established by field surveys or from suitable maps or aerial photographs.

The complete drainage area is subdivided into component parts, each tributary to a point of inlet. This requires a preliminary layout of the system and tentative location of inlet points. Rearrangement of the system layout or of inlet location often is indicated as the design proceeds. This requires reorganization of component parts of the main drainage area to conform to the system layout and inlet scheme finally adopted.

Drainage area information should include also the following:

(a) Land use—present and predicted future—as it affects degree of protection to be provided and percentage of imperviousness.

(b) Character of soil and cover as they may affect the runoff coefficient.

(c) General magnitude of ground slopes which, with previous items and shape of drainage area, will affect time of concentration.

2. Rainfall

(a) **Rainfall Intensity Factors.**—Determination of rainfall intensity, i, for storm-sewer design involves consideration of the following factors:
1. Average frequency of occurrence.
2. Intensity-duration characteristics of rainfall for selected average frequency of occurrence.
3. Time of concentration.

(b) **Rainfall Frequency.**—The average frequency of rainfall occurrence used for design determines the degree of protection afforded by a given storm sewer system. This protection should be consistent with the amount of damage prevented. But in practice, cost-benefit studies usually are not conducted for the ordinary urban storm drainage project. Judgment supported by records of performance in other similar areas is usually the basis of selecting the design frequency.

The range of frequencies used in engineering offices is as follows:
1. For storm sewers in residential areas, 2 to 15 yr, with 5 yr most commonly reported.
2. For storm sewers in commercial and high-value districts, 10 to 50 yr, depending on economic justification.
3. For flood protection works, 50 yr or more.

Other factors which may affect choice of design frequency include:
1. Use of less frequent, more intense rainfall for design of those parts of the system not economically susceptible to future relief.
2. Use of less frequent, more intense rainfall for design of combined sewers than for separate storm sewers because of basement flooding and consequent greater damage which may occur with overloaded combined sewers.
3. Use of less frequent, more intense rainfall for design of special structures such as expressway drainage pumping systems where runoff exceeding capacity would seriously disrupt an important facility. Design frequencies of 50 yr or more may be justified in such cases, particularly in small drainage areas, even though the project may be located in a district justifying only 5-yr frequency for normal drainage.
4. Adoption of less intense, greater frequency rainfall than normal but commensurate with available funds so that some degree of protection can be provided.

It should be apparent that the cost of storm sewers is not directly proportional to design frequency. Rousculp (2) cites studies of effects of various factors on sewer cost and shows that sewer systems designed for 10-yr frequency storms may cost only about 6 to 11 percent more than systems designed for 5-yr frequency storms depending on sewer slope, the lesser increase applying to steeper sewers.

(c) Intensity—Duration—Frequency Relationship.—Basic data for rainfall intensity-duration-frequency are derived from gage measurement of rainfall. Length of record, gage accuracy and placement, and density significantly affect reliability of rainfall data. Processing of data for runoff determination involves statistical procedures yielding time-intensity curves for different frequencies. There are several methods available for the preparation of point rainfall data for intensity-duration-frequency analysis. These are discussed in many of the references, including those on data processing (3) (4) (5).

In the analysis of point rainfall intensity-duration data, there are two approaches, the annual-duration and partial-duration series. In the annual-duration series, for each duration selected, the heaviest rainfall that occurred in each year is listed, with no tabulation of lesser intensities during that same calendar year, even though some of them might be greater than rainfall intensities that occurred in other years of record. In the other treatment of the data, partial-duration series, all rainfall intensities above a practical minimum for each duration for the entire period of record are included.

The first of these methods gives the probability of occurrence that the maximum rainfall in any one year for a specified duration will equal or exceed a given intensity. The second method gives the frequency or the number of occurrences in a given period of time that a rainfall of given intensity and duration will be equaled or exceeded.

Table XII presents empirical factors to convert annual-duration series of rainfall frequency data to partial-duration series based on information in the U.S. Weather Bureau's Technical Paper No. 40 (6). Partial-duration series data are recommended for use in storm-sewer design to preserve the meaning of the stated frequency of occurrence.

Sherman (7) has injected the concept of "extended duration" which has been used widely and is recommended for developing rainfall intensity-frequency data. In this method, if the total precipitation in a storm is sufficient to show significant average rates for periods longer than the actual duration of the rainfall, such storms are included in the data compilations as multiple events as though they had continued for the longer times. For example, a rainfall amount for an actual 50-min storm would be listed not only as a 50-min storm, with its corresponding in-

TABLE XII.—Empirical Factors for Converting Annual-Duration Series to Partial-Duration Series

Return Period (yr)	Multiplier
2	1.14
5	1.04
10	1.01

tensity, but also as a 60-min or 90-min, or longer duration storm with a corresponding lesser intensity.

In practice, the collection and analysis of basic rainfall data by the designer are limited to extensive projects. Rainfall data compiled and processed by the U.S. Weather Bureau, Department of Agriculture, and similar governmental agencies are used widely where size of project or lack of local records does not justify complete local statistical analysis. Charts prepared by Yarnell (8) often were used in the past for more frequent rainfalls normally encountered in storm-sewer design and in the region east of the Rocky Mountains where orographic influence on rainfall may not be significant. More recent data are contained in technical papers of the Weather Bureau (6) (9) (10).

Technical Paper No. 40 (6), which should be used in preference to the earlier issues, presents charts of rainfall depth for the contiguous United States for durations of 30 min to 24 hr and return periods from 1 to 100 yr. Rainfall intensity-duration curves for the selected locality and frequency may be prepared from these and similar charts. Figure 8 is a curve of this type.

Because of sharp local variations in rainfall in mountain and valley regions west of the 105th meridian, additional reference may be made to Technical Papers No. 28 (11) and 38 (10) of the Weather Bureau or to local data and experience.

Rainfall intensity-duration relationships for a given frequency and less than a 2-hr duration may be expressed mathematically (7) by an equation of the form:

$$i = \frac{A}{t+b} \text{ or } i = \frac{A}{(t+b)^n}, \text{ or } i = \frac{A}{t^n+b} \dots \dots \dots \dots 2$$

in which i is the average intensity, t is the duration, A and b are abstract constants, and n is an exponent. Values for A, b, and n may be obtained from logarithmic transformation of the equation, using observed rainfall or Weather Bureau data for the locality. Derivation of rainfall formulas for New York City is described by Bleich (12) who found the exponential form to give best correlation. Several reported formulas in use in contiguous localities produce widely varying results. If rainfall formulas are to be used properly, they should be derived from and be consistent with rainfall experience in the given locality. Chow (13) also has developed general formulas and design charts for finding rainfall intensity frequency (14).

The relationship of rainfall intensity-duration, storm pattern, time of concentration, and average intensity is depicted in Figure 9. Only the average intensity for a given duration may be determined from the rainfall intensity-duration curve, Figure 9a. The storm pattern must be known to determine at what time within the total storm period the given average intensity may have occurred, Figure 9b. This figure illustrates the general case in which the beginning of the period (time) of concentration would not fall at the start of precipitation.

(d) Time of Concentration.—An estimate of the time of concentra-

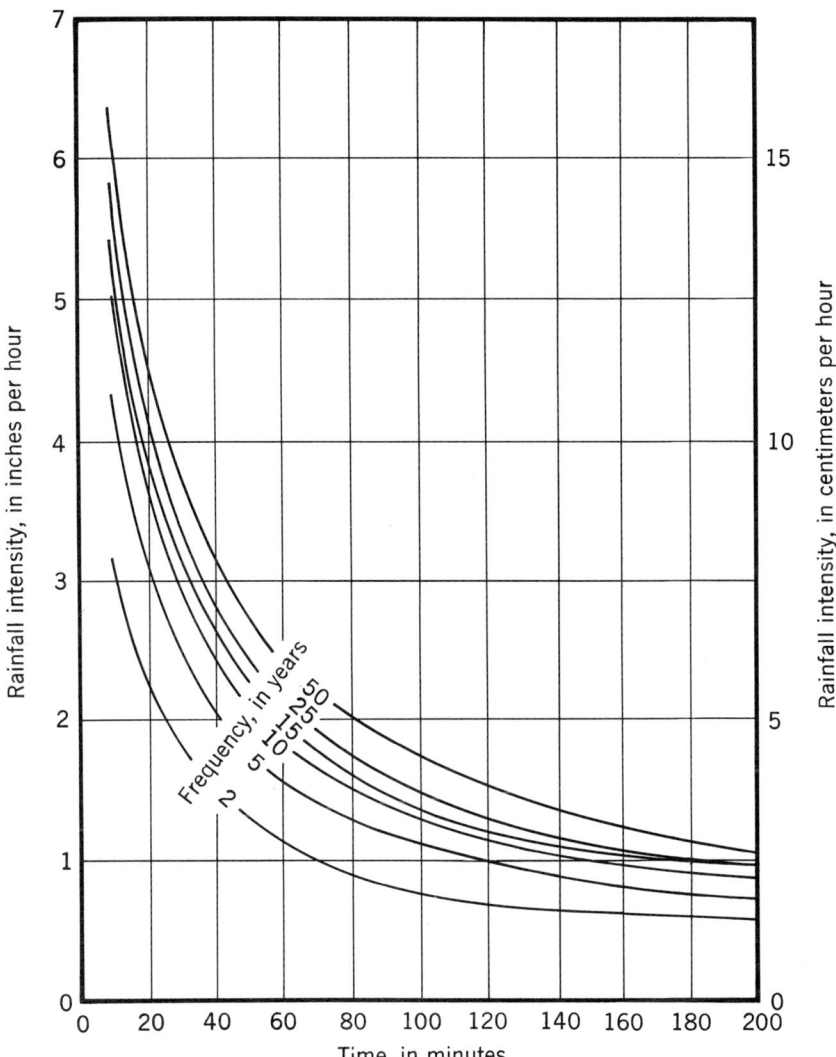

FIGURE 8.—Intensity-duration rainfall curves, Boston, Mass.

tion to the point under consideration is made so that the average rainfall rate may be determined. For urban storm sewers, the time of concentration consists of the inlet time plus the time of flow in the sewer from the most remote inlet to the point under consideration.

Time of flow in the sewer may be estimated closely from the hydraulic properties of the conduit. Inlet time is the overland flow time for runoff to reach established surface drainage channels such as street gutters and ditches, and travel through them to the point of inlet.

Inlet time will vary with surface slope, nature of surface cover, and length of path of surface flow, as well as with the variables influenced by antecedent rainfall intensity and duration such as infiltration capacity

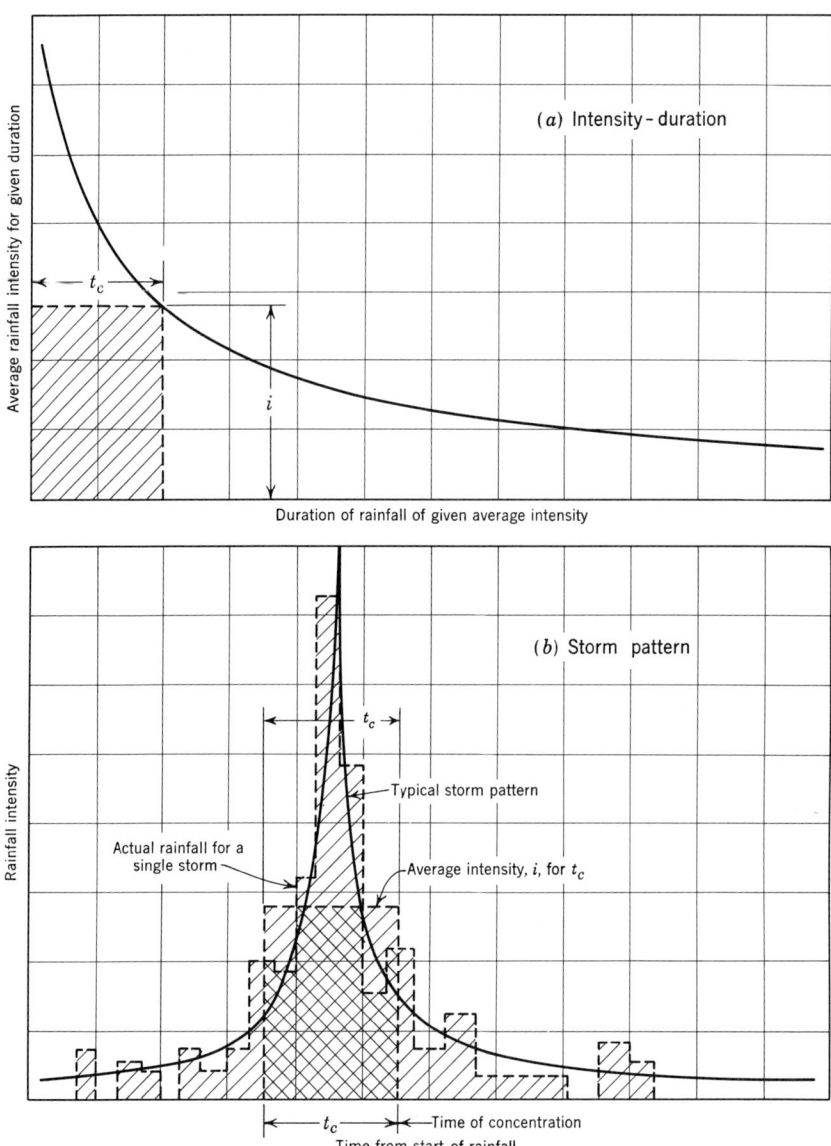

FIGURE 9.—**Rainfall-intensity duration curve and storm pattern;** $i =$ average intensity of rainfall, $t_c =$ time of concentration.

and depression storage. In general, the higher the rainfall intensity, the shorter the inlet time.

Reported inlet times used for design vary from 5 to 30 min, with 5 to 15 min most commonly used. In densely developed areas, where impervious surfaces shed their water directly to storm sewers through closely spaced inlets, an inlet time of 5 min often is reported. In well-developed districts with relatively flat slopes, an inlet time of 10 to 15 min is common. In flat residential districts with widely spaced street inlets, inlet times of 20 to 30 min are customary.

(e) **Other Factors Affecting Design Rainfall Intensity.**—Diminution of point rainfall values for larger areas may result in lower average intensities. Early consideration of the relationship between areal distribution and average depth of rainfall from a storm was given by Marston (15).

Figure 10, taken from Technical Paper No. 40 (6) and based on data gathered from 20 dense networks of gages in various regions of the United States east of the 90th meridian, gives factors for converting the average point rainfall to the average over the area of interest.

Indications from studies by Huff and Neill in 1957 (16), based on a network of gages in Illinois, are that the reduction in average rainfall depth for increase in area should be somewhat smaller than given by Figure 10. Reductions based on their data would result in more conservative estimates for design.

In irregularly shaped drainage areas a part of the area having a shorter time of concentration, and thereby subject to a higher intensity rainfall, may cause a greater runoff rate at a design point than that contributed by the entire area with its longer concentration time and correspondingly lower intensity of rainfall.

3. Runoff Coefficient

(a) **General Considerations.**—The runoff coefficient, C, is the variable of the rational method least susceptible to precise determination. Its use in the formula implies a fixed ratio for any given drainage area, whereas, in reality, the coefficient accounts for abstractions or losses between rainfall and runoff which may vary for a given drainage area as influenced by differing climatological and seasonal conditions.

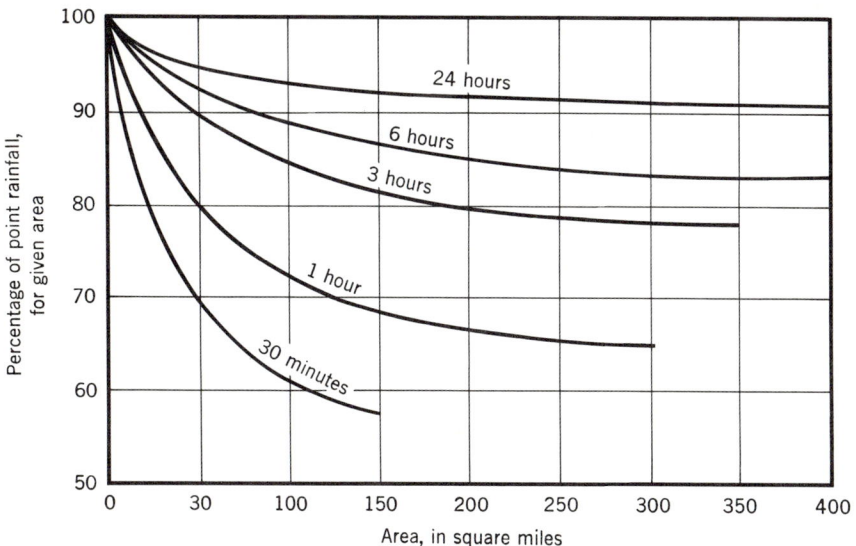

FIGURE 10.—Area-depth curves.

These losses, together with their order of magnitude as observed by various investigators, include:

1. **Interception by Vegetation.** This is not usually significant in urban drainage but may range from 0.01 to 0.5 in. (0.03 to 0.13 cm) in forest areas, depending on type of cover.
2. **Infiltration into Permeable Soils.** The ability of a soil to absorb water and percolate it to deeper groundwater is affected by certain events before and during a given storm, such as compaction of the surface, in-washing of finer sediments, and swelling of clays or colloidial soils. The Hydrology Handbook (3) gives the following range of values of infiltration capacity of various types of bare soils after 1 hr of continuous rainfall:

Soil Group	Infiltration (in./hr)
High (sandy, open structured)	0.50 to 1.00
Intermediate (loam)	0.10 to 0.50
Low (clay, dense structured)	0.01 to 0.10

Note: In. × 2.54 = cm.

The handbook also notes the profound influence of ground cover showing that bare-soil infiltration capacity can be increased from 3 to 7.5 times with good permanent forest or grass cover, ranging down to little or no increase with poor row crops. Antecedent precipitation also affects soil infiltration capacity, but few quantitative data are available to evaluate this factor.

3. **Retention in Surface Depressions.** The first excess rainfall fills depressions essentially present in all surfaces. Retention in forest litter may be as much as 0.3 in. (0.08 cm); in good pasture, 0.2 in. (0.05 cm); and in smooth cultivated land, 0.05 to 0.10 in. (0.13 to 0.3 cm). In urban areas of moderate grade, recent gagings show retention to be about 0.05 in. (0.13 cm) for impervious surfaces. Retention has been assumed to be 0.10 in. (0.3 cm) for surfaces such as lawns and normal urban pervious surfaces.
4. **Evaporation and Transpiration.** These are of little significance for the short rainfall duration encountered in urban storm drainage design.

Reported practice indicates widespread use of overall coefficients found by experience to produce acceptable results, with limited effort to make a quantitative evaluation of component factors. Generally satisfactory experience might seem to warrant this approach, but the designer is urged to acquaint himself with the phenomena represented by the runoff coefficient in order to exercise his independent judgment. Runoff coefficients are discussed in detail in many of the references cited for this chapter.

(b) Average Coefficients.—The use of average coefficients for various surface types, which are assumed not to vary through the duration of the storm, is common. The range of coefficients, classified with respect to the general character of the tributary area reported in use, is:

Description of Area	Runoff Coefficients
Business	
Downtown	0.70 to 0.95
Neighborhood	0.50 to 0.70
Residential	
Single-family	0.30 to 0.50
Multi-units, detached	0.40 to 0.60
Multi-units, attached	0.60 to 0.75
Residential (suburban)	0.25 to 0.40
Apartment	0.50 to 0.70
Industrial	
Light	0.50 to 0.80
Heavy	0.60 to 0.90
Parks, cemeteries	0.10 to 0.25
Playgrounds	0.20 to 0.35
Railroad yard	0.20 to 0.35
Unimproved	0.10 to 0.30

It often is desirable to develop a composite runoff coefficient based on the percentage of different types of surface in the drainage area. This procedure often is applied to typical "sample" blocks as a guide to selection of reasonable values of the coefficient for an entire area. Coefficients with respect to surface type currently in use are:

Character of Surface	Runoff Coefficients
Pavement	
Asphaltic and Concrete	0.70 to 0.95
Brick	0.70 to 0.85
Roofs	0.75 to 0.95
Lawns, sandy soil	
Flat, 2 percent	0.05 to 0.10
Average, 2 to 7 percent	0.10 to 0.15
Steep, 7 percent	0.15 to 0.20
Lawns, heavy soil	
Flat, 2 percent	0.13 to 0.17
Average, 2 to 7 percent	0.18 to 0.22
Steep, 7 percent	0.25 to 0.35

The coefficients in these two tabulations are applicable for storms of 5- to 10-yr frequencies. Less frequent, higher intensity storms will require the use of higher coefficients because infiltration and other losses have a proportionally smaller effect on runoff. The coefficients are based on the assumption that the design storm does not occur when the ground surface is frozen.

(*c*) **Coefficients Varying with Time.**—Figure 11 shows the variation of the runoff coefficient with respect to length of time of prior wetting, according to Horner (17). Several design offices report use of these or similar data, sometimes in combination with the zone principle to reflect effect of antecedent precipitation on the runoff coefficient. These curves are based on depression storage and infiltration beginning at the start of precipitation. Since the average intensity used in the rational method is not fixed in a time sequence, the applicable runoff coefficients cannot be determined directly from Figure 11.

Such data have been used incorrectly by assuming the beginning of rainfall and the start of the concentration period to be coincident. Thus, the low coefficient values indicated in the early storm period are presumed erroneously to prevail at the beginning of the concentration period. Reference to Figure 9 and consideration of previous statements regarding the chronological position of the time of concentration during which average design rainfall occurs will indicate the fallacy of this practice. Experience has shown the peak rainfall rate of the great majority of significant storms occurs appreciably after beginning of rainfall. Thus, a substantial period of rainfall usually will have occurred before the beginning of the concentration period, and the low coefficients indicated at the beginning of the rainfall are in no way representative of conditions during the storm when the average design intensity occurs.

Proper use of Figure 11, therefore, requires knowledge of the storm pattern which is not provided by elemental application of the rational method. Figure 11 may be applied to the storm hydrograph, or, if coefficient modification by the rational method is to be used, the beginning of the time of concentration must be taken after the proper interval following the start of the storm. If this procedure is followed, it will be found that the coefficient tends to increase slightly as the rainfall continues. For example, if 30 min of rainfall precedes the start of a 20-min time of concentration, the coefficient for pervious surfaces will change from 0.40 to 0.47 during the 20-min time of concentration.

Application of the zone principle requires development of a composite coefficient which depends on shape of the drainage area and time

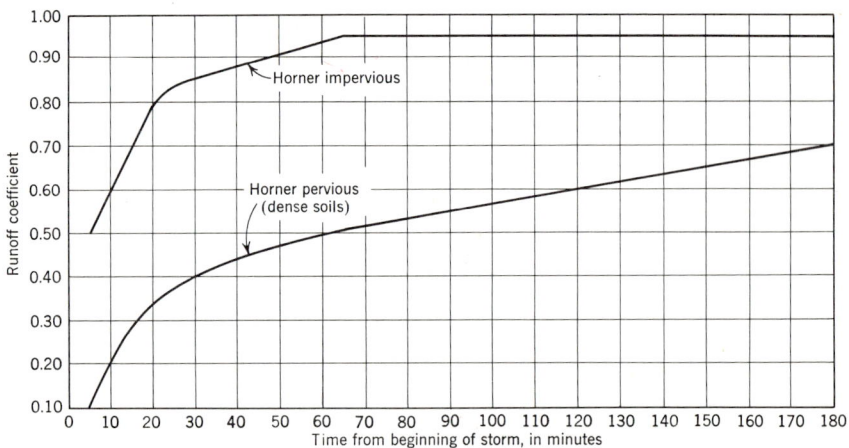

FIGURE 11.—Variation in runoff coefficient with period of rainfall.

QUANTITY OF STORMWATER 53

of concentration. In view of the small change in coefficient encountered in many applications, the decision must be made for the particular study area as to whether use of the zone principle is warranted.

4. Application of Rational Method

After the items discussed in the preceding sections have been determined or estimated, a tentative arrangement of the proposed system including the location of inlets is made to permit division of the whole drainage area into subdistricts tributary to sections of the storm sewer system.

Coefficients are selected or estimated which are appropriate for the land use and development expected at the end of the period of design.

Rainfall frequency consistent with degree of protection desired is selected and intensity-duration curves for the locality are developed from available rainfall records or from such other data as may be available (6).

Many designers find it convenient to reduce the rainfall-intensity relationship to a family of curves of runoff for rainfall intensities of selected frequencies. Such curves indicate the product of the runoff coefficient and average rainfall intensity (18). Other designers suggest weighting the size of areas to reflect the variation in imperviousness of individual subareas from the average imperviousness of the project area. The weighted area size then is applied to the runoff curve developed for the average degree of imperviousness.

Figure 12 is a typical plan for design of a small storm sewer project. Table XIII is a summary of computations illustrating the application of the rational method to the storm drainage system shown in that figure. Figure 13 is an illustrative storm drain profile of the trunk drain of Figure 12. The computations and the profile also illustrate many of the hydraulic principles discussed in Chapter 5. The form of the tabulation is simplified from the more comprehensive forms developed by some designers for day-to-day use.

The example is based on the following conditions:

(a) Runoff Coefficients.—
  1. Residential area, 0.3
  2. Business area, 0.6
  3. Average coefficient weighted according to amount of each type of area tributary to given inlet.
(b) 5- yr frequency rainfall as shown in Figure 8.
(c) 20-min inlet time assumed.
(d) Sewer capacity by Manning formula, $n=0.013$.
(e) Outlet unsubmerged with free outfall.
(f) Flow profile analysis was made by the methods outlined in Chapter 5. Manhole losses were computed as sudden expansions and contractions in a 3.5-ft (1.05-m) diam manhole with inverts developed for the lower one-half of the pipes.

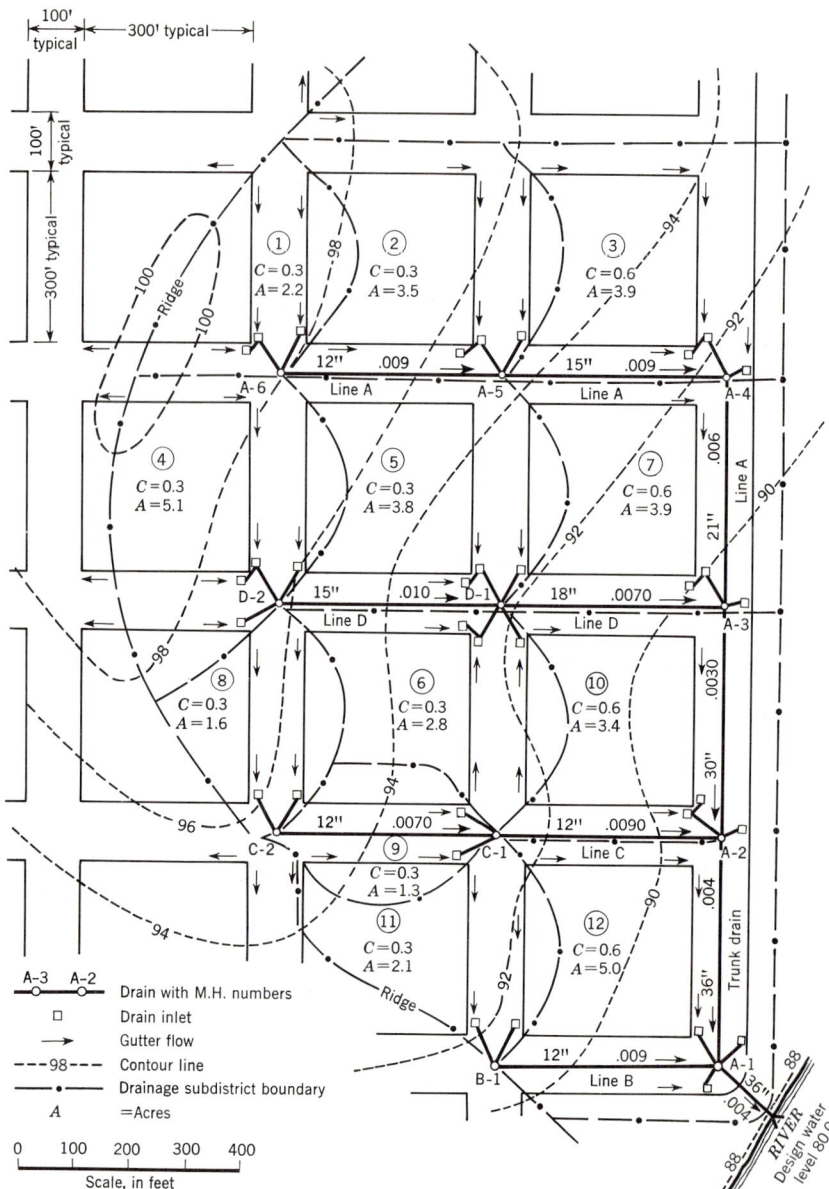

FIGURE 12.—Illustrative storm drain design sketch.

5. Modifications of the Rational Method

In 1932, Gregory and Arnold (19) developed a modification of the rational formula to recognize such factors as watershed shape and slope, stream pattern, and elements of channel flow. In 1938, Bernard (20) developed similar modifications more clearly representing the many variables of runoff, with charts and nomographs to facilitate use of the more complex formulas. Application of these modifications generally has been limited to areas larger than those encountered in most urban drainage projects.

Date: June 17, 1968
Date: June 22, 1968

Flow in Conduit

Design Conditions

... Line Elev.

							Piezometric or Water Surface (ft) (29)
—	—	—	—	—	—	—	—
—	—	—	—	—	—	94.25	—
4.06	0.26	—	—	En. 0.11	0.03	94.22	93.96‡
—	—	—	2.38‖	—	—	—	—
2.55	0.10	—	—	Ex. 0.40	0.04	91.94	91.84
—	—	—	—	—	—	91.90	—
—	—	—	—	En. 0.10	0.03	91.87	91.60
4.1	0.27	0.0062	2.48	—	—	—	—
				En. 0.36	0.10		
—	—	—	—	Ex. 0.32	0.09	89.39	89.12
—	—	—	—	—	—	89.20	—
—	—	—	—	En. 0.07	0.03	89.17	88.80
4.9	0.37	0.0055	2.20	—	—	—	—
—	—	—	—	Ex. 0.19	0.07	86.97	86.60
—	—	—	—	—	—	86.90	—
—	—	—	—	En. 0.04	0.02	86.88	86.38
5.64	0.50	0.0046±	1.82±	—	—	—	—
—	—	—	—	Ex. 0.08	0.04	85.06	84.56
—	—	—	—	—	—	85.02	—
5.98*	0.56	—	—	En. 0.02	0.01	85.01#	84.45#
—	—	—	1.11‖	—	—	—	—
5.03	0.39	—	—	Ex. 0.04	0.01	83.90	83.54
—	—	—	—	—	—	83.89	—
6.78	0.71	—	—	En. 0.02	0.01	83.88	83.17**
—	—	—	0.64‖	—	—	—	—
8.10	1.02	—	—	Ex. 1.00	1.02	83.24	82.22
—	—	—	—	—	2.17	80.0	80.0
Total			10.53		3.67		

× 0.405 = ha; cfs/acre × 4.2 = cu m/min/ha; cfs × 1.7 = cu m/min.

III.—Storm-Drainage Design Summary—Rational Method

ency: 5 yr—Boston (see Figure 8) Computed by: R.E.V.
0.013 Checked by: H.S.
an: Figure 12
ofile: Figure 13

Cumulative		Time		Rainfall-Runoff			Full at Invert Slope		
rea, Σ (acre)	Weighted Runoff Coef., C	Increment $\Delta = \frac{L}{V}$ (min)	Cumulative, Σ (min)	Avg. Rainfall Intensity, i (in./hr)	Unit Runoff, Ci (cfs/acre)	Runoff, Design Flow, Q (cfs)	Cap., Q_t (cfs)	Vel., $V_t = \frac{Q_t}{A_t}$ (fps)	Depth of Flow, d (ft)
(12)	(13)	(14)	(15)	(16)	(17)	(18)	(19)	(20)	(21)
—	—	—	20.0	3.1	0.93	—	—	—	—
2.2	0.3	—	—	—	—	2.0	—	—	0.61† Hyd. jump
—	—	1.6	—	—	—	—	3.4	4.3	Full and surcharged
—	—	—	21.6	3.0	0.90	—	—	—	—
5.7	0.3	—	—	—	—	5.1	—	—	—
—	—	1.6	—	—	—	—	6.1	5.0	Full and surcharged
—	—	—	23.2	2.9	1.22	—	—	—	—
9.6	0.43	—	—	—	—	11.7	—	—	—
—	—	1.4	—	—	—	—	12.3	5.1	Full and surcharged
—	—	—	24.6	2.8	1.10	—	—	—	—
25.2	0.39	—	—	—	—	27.7	—	—	Full most of length
—	—	1.2	—	—	—	—	26.0	5.3	
—	—	—	25.8	2.75	1.12	—	—	—	—
31.5	0.41	—	—	—	—	35.3	—	—	Back-water
—	—	1.2¶	—	—	—	—	42.1	6.0	
—	—	—	27.0	3.7	1.15	—	—	—	—
38.6	0.43	—	—	—	—	44.4	—	—	2.62** Draw-down
—	—	—	—	—	—	—	42.1	6.0	
—	—	—	—	—	—	—	—	—	2.17†
—	—	—	—	—	—	—	—	—	—

s explained in Chapter 5.
nin.

** Normal depth.
Note: In. $\times 2.54 =$ cm; ft $\times 0.3048 =$ m; acre

TABLE X

Project: Lexingford Storm Sewers
District: Riverside
Subdistricts: 1-12
Street(s): Misc.

Storm Frequ
Manning n:
Reference P
Hydraulic P

		Conduit						Drainage Area		
									Subdistricts	
Line	Location	Type Structure	Pipe Diam (in.)	Length (ft)	Invert Slope, S_0	Invert Elev. (ft)	Approx. Surface Elev. (ft)	Designation	Area (acre)	Runoff Coef., C
(1)	(2)	(3)	(4)	(5)	(6)	(7)	(8)	(9)	(10)	(11)
1	Trunk Line A	—	—	—	—	—	—	—	—	—
2	21+25	MH A-6	—	—	—	93.35	98.4	①	2.2	0.3
3	—	—	—	—	—	—	—	—	—	—
4	—	Pipe	12§	400	0.0090	—	—	—	—	—
5	—	—	—	—	—	—	—	—	—	—
6	17+25	MH A-5	—	—	—	89.75	94.9	②	3.5	0.3
7	—	—	—	—	—	—	—	—	—	—
8	—	Pipe	15	400	0.0090	—	—	—	—	—
9	—	—	—	—	—	—	—	—	—	—
10	13+25	MH A-4	—	—	—	86.15	91.8	③	3.9	0.6
11	—	—	—	—	—	—	—	—	—	—
12	—	Pipe	21	400	0.0060	—	—	—	—	—
13	—	—	—	—	—	—	—	—	—	—
14	9+25	MH A-3	—	—	—	83.75	89.7	④⑤⑥⑦	5.1, 3.8, 2.8, 3.9	0.3, 0.3, 0.3, 0.6
15	—	—	—	—	—	—	—	—	—	—
16	—	Pipe	30	400	0.0040	—	—	—	—	—
17	—	—	—	—	—	—	—	—	—	—
18	5+25	MH A-2	—	—	—	82.15	89.5	⑧⑨⑩	1.6, 1.3, 3.4	0.3, 0.3, 0.6
19	—	—	—	—	—	—	—	—	—	—
20	—	Pipe	36	400	0.0040	—	—	—	—	—
21	—	—	—	—	—	—	—	—	—	—
22	1+25	MH A-1	—	—	—	80.55	88.5	⑪⑫	2.1, 5.0	0.3, 0.6
23	—	—	—	—	—	—	—	—	—	—
24	—	Pipe	36	125	0.0040	—	—	—	—	—
25	0+00	Outlet	—	—	—	80.05	85.0	—	—	—
26	River	—	—	—	—	—	—	—	—	—
27										

* Computation of head losses made as explained in Chapter 5.
† Flow at critical depth.
‡ 93.96 = 93.35 + 0.61.
§ Selected as min diam pipe.

‖ By subtraction in Col. 28
\# Backwater computation
¶ $\Delta t = \dfrac{400 \times 1/60}{\frac{1}{2}(5.98 + 5.03)} = 12$

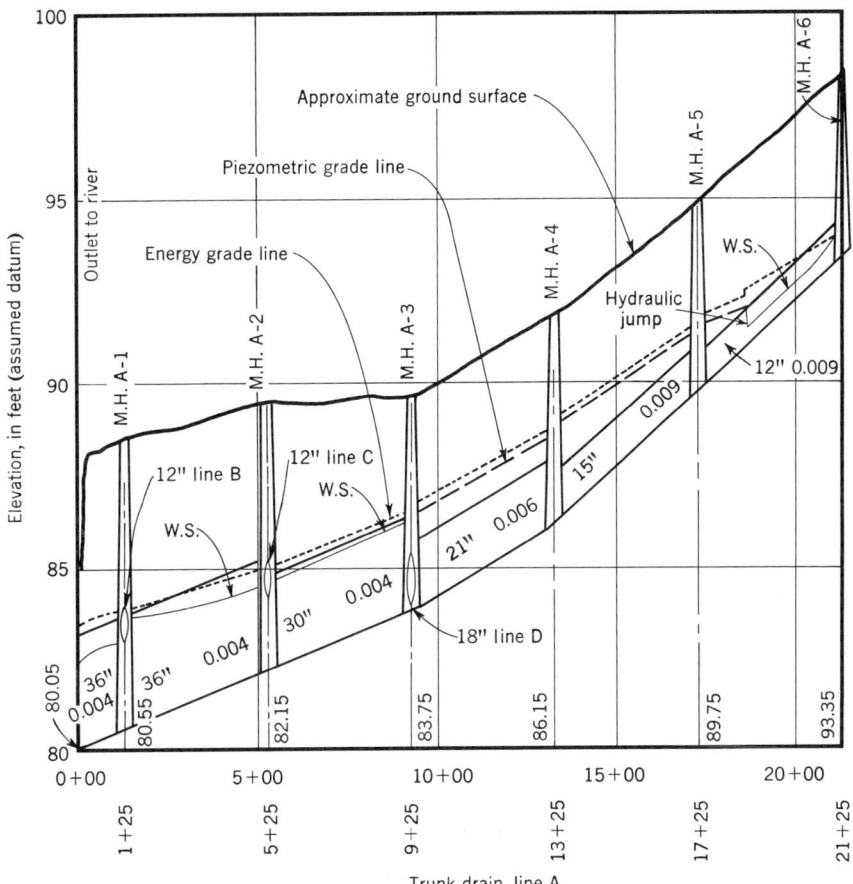

FIGURE 13.—Illustrative storm drain design profile.

The rational method with modification in the form of a reduction constant to allow for differences between point and areal average rainfall intensity and for the effect of channel storage has been used for areas less than about 5 sq miles (13 sq km) (21). A modified rational method which separates rainfall-runoff relationships from storage effects is explained by Chow (22) (23).

C. OTHER METHODS OF RUNOFF DETERMINATION

Other approaches to the determination of runoff include the hydrograph (overland-flow) method, the inlet method, the unit hydrograph method, and methods based on statistical analysis of flooding occurrences (24) and flood frequencies (25).

1. Hydrograph (Overland-Flow) Method

Measurements of rainfall and runoff (actual sewer gagings) for virtually all intense rainfalls from 1914 to 1933 for several city inlet areas in

St. Louis, Mo., were presented and analyzed in 1934 by Horner and Flynt (26). The ratio of runoff to rainfall, defined in several ways, was found to vary widely. Rainfall and runoff rates were studied for frequency determinations as independent phenomena with the two sets of curves considered comparably representative of equivalent probabilities of occurrence. It was suggested that ratios between corresponding values be applied to appropriate rainfall-frequency curves for other localities to give approximate runoff values for similar conditions of surface.

Horton (27) focused attention on the concept of infiltration and introduced the idea of a limiting infiltration capacity through his definition of it as ". . . the maximum rate at which the soil, when in a given condition, can absorb falling rain." This stimulated interest in the belief on the part of some engineers that runoff should be considered to be a residual, after deduction of certain abstractions or "losses" from rainfall, the principal such abstraction being infiltration.

Several reports have been made of attempts to use this approach in the determination of runoff. Horner and Jens (28) suggested a method for applying infiltration data to storm rainfall patterns to obtain net rainfall or supply (also recognizing the other abstractions). They then processed this net rainfall through overland flow detention and gutter storage to obtain inflow rates at the inlet serving part of a typical city block. The example was elaborated to obtain the runoff for design at the 20-, 40-, and 60-min points in a simplified network of several typical city blocks. Comparison of the results of assuming a uniform average rainfall rate and an advanced, intermediate, or delayed storm pattern also was presented. Runoff hydrographs for urban watersheds are presented by Willeke (29), and synthetic unit hydrographs by Morgan and Johnson (30).

Hicks (31) developed a method of computing urban runoffs for the Los Angeles area based on experimental work for the determination of the principal abstractions from rainfall and actual gagings of local drainage areas in the metropolitan region. The Los Angeles method includes use of a design storm between the medium and delayed pattern, infiltration values developed from tests on small plots, and antecedent precipitation evaluated through a composite moisture factor. Times of concentration were determined from experimentally investigated overland and gutter flows. The Hicks methodology has been in use for more than 25 yr in the city of Los Angeles, and is used also by a considerable number of satellite communities.

Extensive studies of a hydrograph method have been made by the city of Chicago. A detailed explanation of the hydrograph type of analysis used there is presented by Tholin and Keifer (4), and of the synthetic storm pattern by Keifer and Chu (32). A summary of the method and a comparison of measured and computed runoffs are given by Jens and McPherson (5).

The procedure in the studies was to develop a methodology based on a design storm pattern statistically derived from rainfall records and then isolate and evaluate all the influences which transform the hyetograph, or time-intensity pattern of rainfall, into the sewer flow hydrograph for a typical unit drainage area.

Large areas of the Chicago street system are arranged in 5-acre (2-ha) blocks; and two blocks, 10 acres (4.1 ha), were taken as the typical unit drainage area for the hydrograph analysis. A detailed plan of land use and surface drainage of a typical Chicago residential area is shown in Figure 14.

Various types of uniform land use, ranging from suburban residential to industrial and commercial, were analyzed and their significant characteristics determined. The individual influences evaluated include infiltration capacity of pervious areas, depression storage, overland-flow detention, and detention in gutters, house drains, catchbasins, and lateral sewers.

After the hydrograph of sewer flow from a 10-acre (4.1-ha) unit area had been determined, the hydrograph was routed through the branch and trunk storm sewers, using a simple time-offset method based on time of flow between junction points in the sewer system.

An important value in the use of the hydrograph method as contrasted with the use of empirical formulas or the rational method is the possiblity of graphical visualization because, on the hydrograph volume equals the rate multiplied by time. Thus, the areas shown under the various rate curves represent volumes of water measured in inches of depth.

The typical Chicago unit of 10 acres (4.1 ha) may be typical of some other cities and the computed hydrograph from a 10-acre (4.1-ha) drainage area, with the rate of runoff expressed in in./hr, would be directly comparable with similar hydrographs for areas of approximately the same size and land use in other cities having similar rainfalls, topography, and soil conditions.

In the design of urban drainage for a large metropolitan area such as that of Chicago, the digital computer has proved useful. A program for urban runoff computation such as prepared by Keifer and Khan (33) will determine the design storm pattern from a selected intensity-duration-frequency curve for given maximum duration and time location of storm peak; compute hydrographs of flow at two gutter inlets for appropriate constants to describe infiltration, depression storage, and overland and gutter flow; and calculate the routing of flows through lateral and branch storm drains to obtain hydrographs at selected critical points. Figure 15 shows an intensity-duration curve and hyetograph of 5-yr frequency storm rainfall from the Tholin and Keifer paper (4). Figure 16 shows a typical family of curves of runoff rates in the storm-sewer system vs. time of travel and percent of directly connected impervious areas as computed by the hydrograph method.

2. Inlet Method

Another approach studied at The Johns Hopkins University (34) (35) (36) (37) and at New Mexico State University (38) (39) is based on rainfall measurements and inlet and sewer gagings in urban areas at Baltimore, Md., and other municipalities, including one in England. The

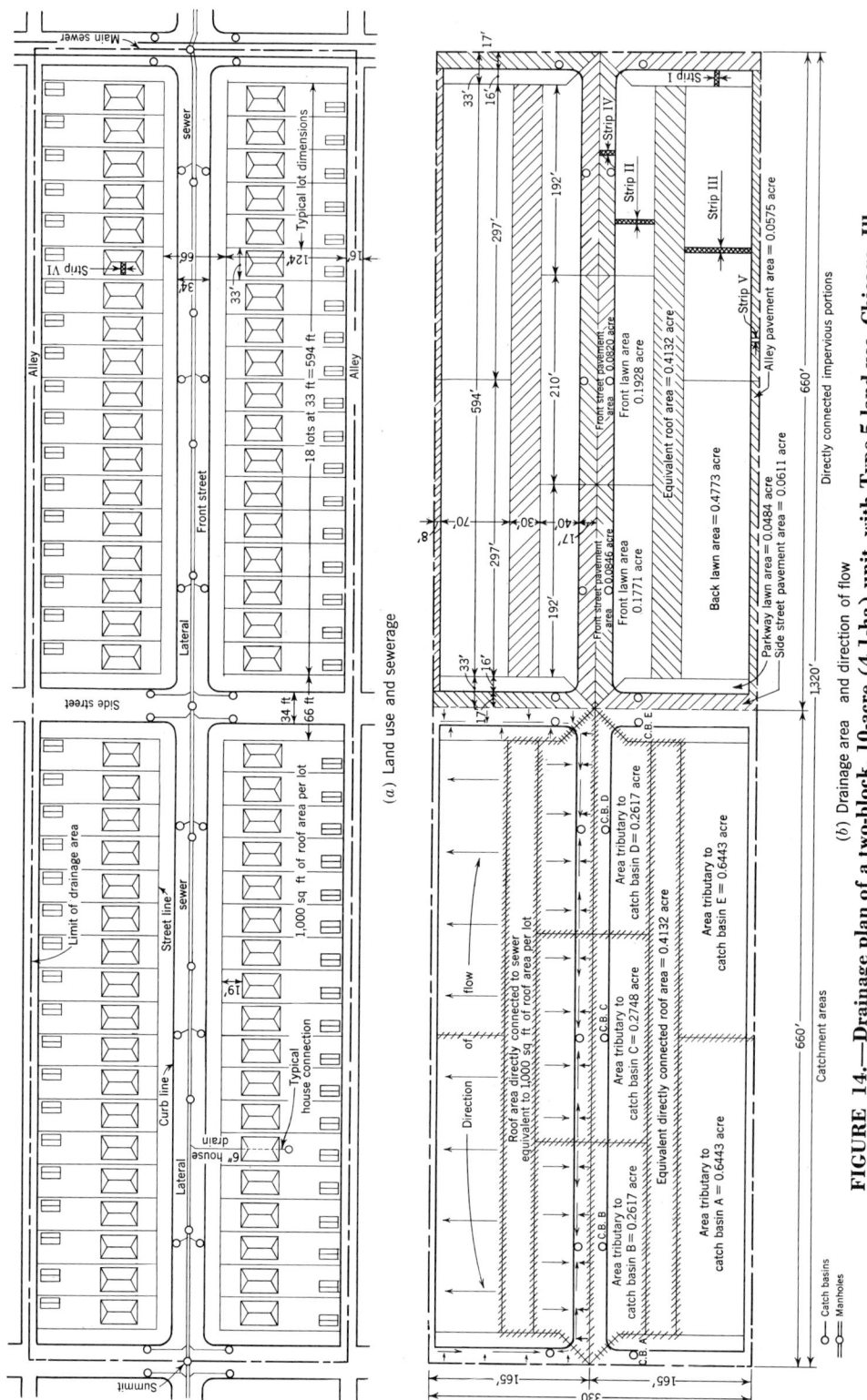

FIGURE 14.—Drainage plan of a two-block, 10-acre (4.1-ha) unit, with Type 5 land use, Chicago, Ill.

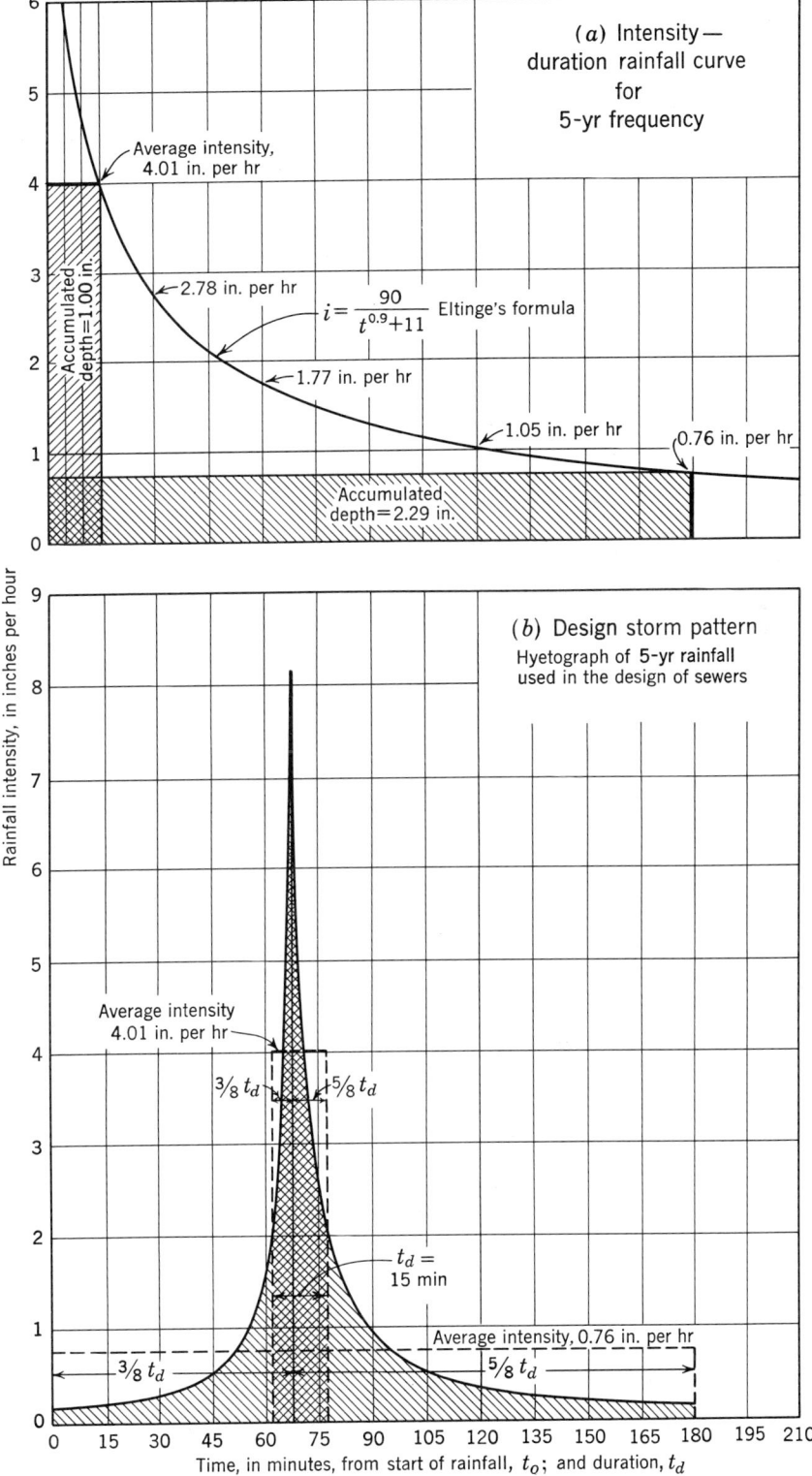

FIGURE 15.—Intensity-duration curve and concomitant storm pattern, Chicago, Ill. (Hyetograph of 5-yr rainfall used in design of sewers based on statistical analysis of location of peak and amount of antecedent rain of recorded heavy rainfalls.)

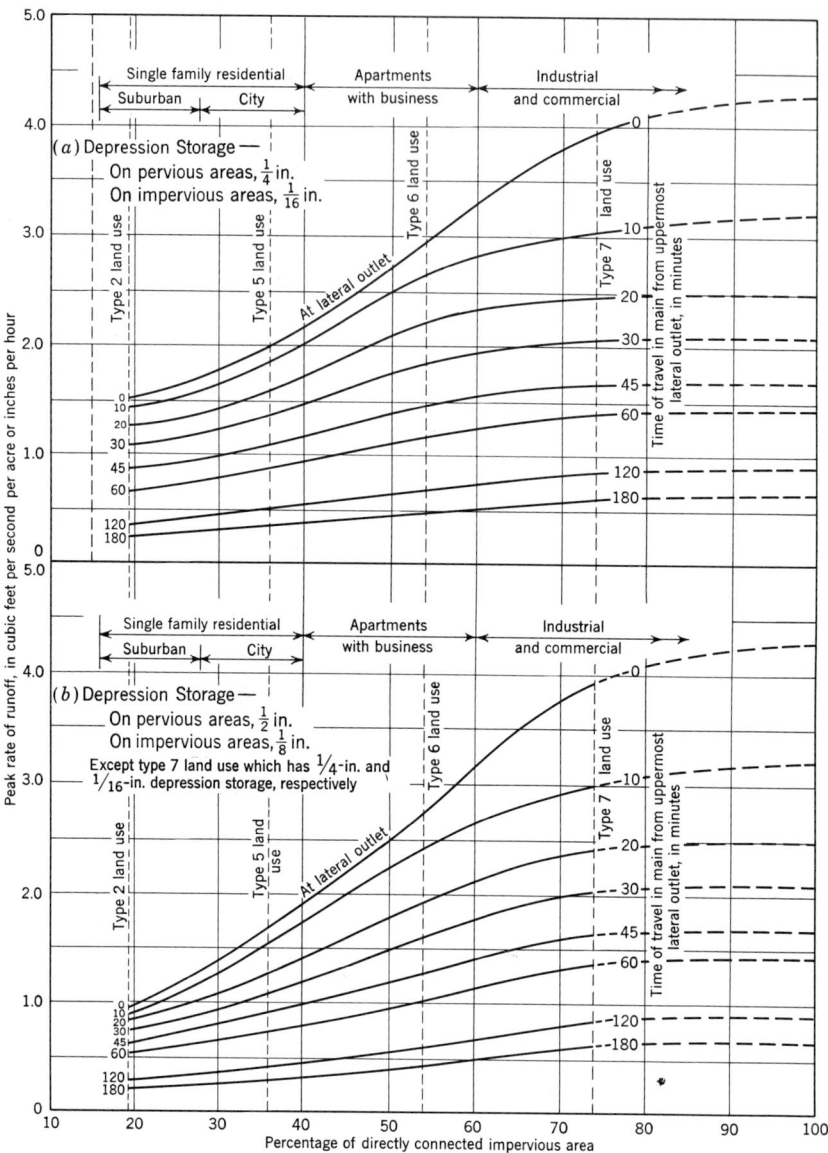

FIGURE 16.—Runoff rates in sewer system vs. travel time and percentage of directly connected impervious areas, average ground slope, 0.02, Chicago, Ill.

method comprises three parts: (a) determining the flow to each inlet; (b) attenuating the peak flow from each subarea (group of inlets) as it moves down the pipe; and (c) summing the attenuated peaks to determine the total peak at the design point. Using this method, computed peak flows for 4 yr of data agree closely with major peaks. The limits of applicability regarding size and type of area are being investigated.

The inlet method is summarized by Jens and McPherson (5) and by Kaltenbach (40), the latter with the suggestion that the frequency of occurrence of peak rates of runoff may be greater than the frequency of maximum rainfall intensities.

3. Unit Hydrograph Method

The unit hydrograph method depends on the correlation of characteristics of measured sewer outflow hydrographs from urban areas of varying types, to permit construction of synthetic unit hydrographs for areas under design, as discussed by Eagleson (41). Outflow hydrographs developed by this method have particular application to the sizing of impounding basins and drainage pumping stations, for which the rational method provides no sound basis for design. The unit hydrograph method is reported to have been used for runoff rate determination for Philadelphia in 1947.

D. OVERLAND FLOW

The specialized problem of the hydraulic phenomena involved in overland flow, both laminar and turbulent, is important in estimating the volume detained in the sheet of water as it is accumulated on the surface of the ground to sufficient depth to cause flow, as well as the rate of flow to gutters. This complex subject is discussed by Tholin and Keifer (4), Horton (42), Izzard (43) (44), Yu and McNown (45), Grace and Eagleson (46), Woo and Brater (47), and Morgali and Linsley (48), among others.

E. RETENTION BASINS FOR REDUCING STORM SEWER INFLOW

The foregoing discussion has referred to quantities of stormwater discharged directly from the drainage basin to the storm sewer. This condition requires that the storm sewer system be capable of receiving runoff at whatever rate it may be discharged from the drainage basin to the sewer system through the storm sewer inlets, the only modifying effects between rate of rainfall and rate of runoff being those which are characteristic of the particular drainage basin.

Consideration should be given to shallow ponding in park areas and on streets where such occasional accumulation would not be harmful, and to the preservation of natural channels with adjacent strips of land reserved for overflow.

Where topography permits, retention basins may be utilized to reduce the peak rate of discharge to the storm sewer system. Such basins are most feasible where a substantial part of the drainage area lies above the developed area to be served by the storm sewer system and where the topography and land use will permit relatively inexpensive impoundment of surplus waters prior to their discharge to the storm drainage system. Greatest reduction in peak flow rates will be achieved by constructing the detention basin or impoundment as far toward the lower part of the drainage basin as possible in order to control the runoff from the greatest possible part of the drainage basin.

The basic relationship expressing the functioning of a stormwater retention basin is:

Rate of inflow − rate of outflow = rate of change of storage.

Analysis of a stormwater retention basin requires knowledge of the rate of inflow with respect to time which, as has been emphasized previously, is not given directly by the rational method. Hence, a hydrograph method of analysis is required.

Storage or impoundment will be governed by the chacteristics of the outlet structure, usually a conduit through the dam in the bottom of the valley through which normal discharge from the drainage area is taken without impoundment. Storage commences when the rate of inflow exceeds the capacity of the outlet conduit and continues as long as this condition exists, the reservoir draining when the rate of inflow becomes less than the rate of outflow. The size of the outlet structure will depend on both the rate and the duration of inflow as shown by the inflow hydrograph and the storage which can be developed economically at the location under consideration.

The effect of retention basins on reducing peak flow in storm sewers is illustrated by the drainage system in a small community in an extremely hilly region. About three-quarters of the developed area of the town lies in a system of valleys, consisting of a main valley along which the main street is located and several smaller tributary valleys. The steep, rocky slopes of the drainage area result in a short time of concentration and a high runoff coefficient. The simultaneous contributions from the branch valleys to the main valley caused frequent flooding of an inadequate drainage system and high property damage. A series of retention reservoirs in the tributary valleys, controlling approximately 46 percent of the 2,600-acre (1,050-ha) drainage area, reduces the estimated 50-yr frequency flow in the main valley drainage conduit from 7,500 to 4,750 cfs (12,750 to 8,100 cu m/min) or to 63 percent of the uncontrolled rate of discharge. In this instance, retention reservoirs proved to be an economical means of reducing the required capacity and, hence, the cost of stormwater conduits.

The feasibility of reserving the stormwater storage capacity of natural flood plains in areas such as park lands, and of constructing deep tunnels of large volume for purposes of detention is discussed by Bauer (49).

References

1. Williams, G. R., "Hydrology." In "Engineering Hydraulics." Ch. 4, p. 309, John Wiley & Sons, Inc., New York (1950).
2. Rousculp, J. A., "Relation of Rainfall and Runoff to Cost of Sewers." *Trans. Amer. Soc. Civil Engr.*, **104**, 1473 (1939).
3. "Hydrology." Manual of Practice No. 28, Amer. Soc. Civil Engr., New York (1949).

4. Tholin, A. L., and Keifer, C. J., "The Hydrology of Urban Runoff." *Trans. Amer. Soc. Civil Engr.*, **125,** 1308 (1960).
5. Jens, S. W., and McPherson, M. B., "Hydrology of Urban Areas." In "Handbook of Applied Hydrology." Sec. 20, McGraw-Hill Book Co., Inc., New York (1965).
6. "Rainfall Frequency Atlas of the United States for Durations from 30-Min to 24-Hr and Return Periods from 1 to 100 Yr." Weather Bureau, Technical Paper No. 40, U.S. Govt. Printing Office, Washington, D.C. (1961).
7. Sherman, C. W., "Frequency and Intensity of Excessive Rainfalls at Boston, Mass." *Trans. Amer. Soc. Civil Engr.* **95,** 951 (1931).
8. Yarnell, D. L., "Rainfall Intensity-Frequency Data." Dept. of Agriculture, Misc. Publication No. 204, U.S. Govt. Printing Office, Washington, D.C. (1935).
9. "Rainfall Intensity-Frequency Regime." Weather Bureau, Technical Paper No. 29, U.S. Govt. Printing Office, Washington, D.C.: Part 1—Ohio Valley (1957); Part 2—Southeastern U.S. (1958); Part 3—Middle Atlantic Region (1958); Part 4—Northeastern U.S. (1959); Part 5—Great Lakes Region (1960).
10. "Generalized Estimates of Probable Maximum Precipitation for the United States West of the 105th Meridian for Areas to 400 Sq Miles and Durations to 24 Hr." Weather Bureau, Technical Paper No. 38, U.S. Govt. Printing Office, Washington, D.C. (1960).
11. "Rainfall Intensities for Local Drainage Design in Western United States, for Durations of 20-Min to 24-Hr and 1 to 100-Yr Return Periods." Weather Bureau, Technical Paper No. 28, U.S. Govt. Printing Office, Washington, D.C. (1956).
12. Bleich, S. D., "Rainfall Studies for New York, N.Y." *Trans. Amer. Soc. Civil Engr.,* **100,** 609 (1935).
13. Chow, V. T., "A General Formula for Hydrologic Frequency Analysis." *Trans. Amer. Geophys. Union,* **32,** 231 (1951); Discussion, **33,** 277 (1952).
14. Chow, V. T., "Design Charts for Finding Rainfall Intensity Frequency." *Water and Sew. Works,* **99,** 2, 87 (1952).
15. Marston, F. A., "The Distribution of Intense Rainfall and Some Other Factors in the Design of Storm-Water Drains." *Trans. Amer. Soc. Civil Engr.,* **87,** 535 (1924).
16. Huff, F. A., and Neill, J. C., "Rainfall Relations on Small Areas in Illinois." *Illinois State Water Survey Bull.* **44** (1957).
17. Horner, W. W., "Modern Procedure in District Sewer Design." *Eng. News,* 64, 326 (1910).
18. Horner, W. W., "Hydrology;" Metcalf, L., and Eddy, H. P., "Sewerage Systems." In "American Civil Engineering Practice." Sec. 13 and 19, John Wiley & Sons, Inc., New York (1956).
19. Gregory, R. L., and Arnold, C. E., "Runoff—Rational Runoff Formulas." *Trans. Amer. Soc. Civil Engr.,* **96,** 1038 (1932).
20. Bernard, M., "A Modified Rational Method of Estimating Flood Flows." In "Low Dams—A Manual of Design for Small Water Storage Projects." Natl. Res. Comm., Washington, D.C. (1938).
21. "Drainage Manual." 3rd Ed., Santa Clara Co., Calif. (1954).
22. Chow, V. T., "Hydrologic Determination of Waterway Areas for the Design of Drainage Structures in Small Drainage Basins." *Univ. of Illinois, Eng. Exp. Sta. Bull.* **462** (1962).
23. Chow, V. T., "Hydrologic Design of Culverts." *Jour. Hydr. Div., Proc. Amer. Soc. Civil Engr.,* HY 4, **88** (1962).
24. Dalrymple, T., "Flood Frequency Analysis." U.S. Geological Survey, Water Supply Paper No. 1543-A, U.S. Govt. Printing Office, Washington, D.C. (1960).
25. Snyder, F. F., "Synthetic Flood Frequencies." *Jour. Hydr. Div., Proc. Amer. Soc. Civil Engr.,* HY5, **84** (1958).
26. Horner, W. W., and Flynt, F. L., "Relation between Rainfall and Runoff from Small Urban Areas." *Trans. Amer. Soc. Civil Engr.,* **101,** 140 (1936).
27. Horton, R. E., "The Roll of Infiltration in the Hydrologic Cycle." *Trans. Amer. Geophys. Union,* **14,** 446 (1933).

28. Horner, W. W., and Jens, S. W., "Surface Runoff Determination from Rainfall without Using Coefficients." *Trans. Amer. Soc. Civil Engr.,* **107,** 1039 (1942).
29. Willeke, G. E., "The Prediction of Runoff Hydrographs for Urban Watersheds from Precipitation Data and Watershed Characteristics." *Jour. Geophys. Research,* **67,** 3610 (1962).
30. Morgan, P. E., and Johnson, S. M., "Analysis of Synthetic Unit—Graph Methods." *Trans. Amer. Soc. Civil Engr.,* **129,** 727 (1964).
31. Hicks, W I., "A Method of Computing Urban Runoff." *Trans. Amer. Soc. Civil Engr.,* **109,** 1217 (1944).
32. Keifer, C. J., and Chu, H. H., "Synthetic Storm Pattern for Drainage Design." *Proc. Amer. Soc. Civil Eng.,* (1957).
33. Keifer, C. J. and Kahn.
34. Bock, P., Viessman, W., Jr., McKay, J. H., Jr., and Schaake, J. C., "Storm Drainage Research Project Reports, 1955–1963." The Johns Hopkins Univ., Dept. San. Eng. and Water Resources, Baltimore, Md. For summary of report on instrumentation see *Jour. Hydr. Div., Proc. Amer. Soc. Civil Engr.,* **89,** HY5, 99 (1963).
35. Viessman, W., and Geyer, J. C., "Characteristics of the Inlet Hydrograph." *Jour. Hydr. Div., Proc. Amer. Soc. Civil Engr.,* HY5, **88,** 245 (1962).
36. Schaake, J. C., Jr., Geyer, J. C., and Knapp, J. W., "Runoff Coefficients in the Rational Method." The Johns Hopkins Univ., Dept. San. Eng. and Water Resources, Storm Drainage Project, Baltimore, Md. Tech. Rept. No. 1 (1964).
37. Schaake, J. C., Jr., "Synthesis of the Inlet Hydrograph." The Johns Hopkins Univ., Dept. San. Eng. and Water Resources, Storm Drainage Project, Baltimore, Md., Tech. Rept. No. 3 (1965).
38. Viessman, W., Jr., and Abdel-Razaq, A. Y., "Time Lag for Urban Inlet Areas." *New Mexico State Univ., Eng. Exp. Sta. Tech. Rept. No. 10* (1964).
39. Viessman, W., Jr., "The Hydrology of Small Impervious Areas," New Mexico State Univ., Eng. Exp. Sta. Tech. Rept. No. 24 (1965).
40. Kaltenbach, A. B., "Storm Sewer Design by the Inlet Method." *Pub. Works,* **94,** 1, 86 (1963).
41. Eagleson, P. S., "Unit Hydrograph Characteristics for Sewered Areas." *Trans. Amer. Soc. Civil Engr.,* **129,** 37 (1964).
42. Horton, R. E., "Surface Runoff Phenomena." Horton Hydrologic Lab. Bull. No. 101 (1935).
43. Izzard, C. F., "Hydraulics of Runoff from Developed Surfaces." *Proc. Highway Res. Bd.,* **26,** 129 (1946).
44. Izzard, C. F., "Surface Profile of Overland Flow." *Trans. Amer. Geophys. Union,* **25,** Part VI, 959 (1944).
45. Yu, Y. S., and McNown, J. S., "Runoff from Impervious Surfaces." *Jour. Hydr. Res.,* **2,** 1 (1964).
46. Grace, R. A., and Eagleson, P. S., "The Modelling of Overland Flow on Impervious Surfaces," Mass. Inst. Tech., Dept. Civil Eng. (1965).
47. Woo, D. C., and Brater, E. F., "Spatially Varied Flow from Controlled Rainfall," *Jour Hydr. Div., Proc. Amer. Soc. Civil Engr.,* HY6, **88,** 31 (1962).
48. Morgali, J. R., and Linsley, R. K., Jr., "Computer Analysis of Overland Flow." *Jour. Hydr. Div., Proc. Amer. Soc. Civil Engr.,* HY3, **91,** 81 (1965).
49. Bauer, W. J., "Economics of Urban Drainage Design." *Trans. Amer. Soc. Civil Engr.,* **129,** 335 (1964).

General References

"Airport Drainage." Federal Aviation Agency, AC 150/5320-5A, U.S. Govt. Printing Office, Washington, D.C. (1965).

"California Culvert Practice." 2nd Ed., Div. Highways, Dept. Pub. Works, State of Calif. (1960).

"Design of Small Dams—Appendix A—Runoff." Bureau Reclamation, U.S. Govt. Printing Office, Washington, D.C. (1960).

Fair, G. M., Geyer, J. C., and Okun, D. A., "Water Supply and Wastewater Removal." John Wiley & Sons., Inc., New York (1966).

Johnstone, D., and Gross, W. P., "Elements of Applied Hydrology." The Ronald Press, New York (1949).

Kinnison, H. B., and Colby, B. R., "Flood Formulas Based on Drainage Basin Characteristics," *Trans. Amer. Soc. Civil Engr.,* **110** (1945).

Linsley, R. K., Jr., Kohler, M. A., and Paulhaus, J. L. H., "Applied Hydrology." McGraw-Hill Book Co., Inc., New York (1949).

Linsley, R. K., Jr., Kohler, M. A., and Paulhaus, J. L. H., "Hydrology for Engineers." McGraw-Hill Book Co., Inc., New York (1958)

CHAPTER 5. HYDRAULICS OF SEWERS

A. INTRODUCTION

A properly functioning sewer must carry the maximum fluid discharge for which it is designed, and transport suspended solids in such a manner that deposits in the sewer and odor nuisances therefrom are kept to a minimum. It follows, then, that a properly designed sewer must take into account peak flow which the sewer must carry, the expected fluctuation of discharge during periods of low flow, and the character of the suspended matter to be transported.

In Chapter 1 sewers were classified according to the type of fluid carried or their use. Hydraulically, sewers are classified in another manner. The flow of wastewater in a conduit may be either open-channel or pressure-conduit (closed-conduit) flow. In open-channel flow the water surface is exposed to the atmosphere. This type of flow occurs in either an open conduit or a partly full closed conduit. Pressure-conduit flow, by definition, totally fills the conduit. If the pressure conduit is circular, the flow may be called pipe flow.

Sanitary sewers often are designed as open channels, that is, with a free surface for ventilation reasons or to provide an additional factor of safety. However, some engineers design them to flow just full at peak design flow, since there will be a free surface for ventilation at all lesser flows. In either case, the invert slope must be sufficient to insure adequate cleansing velocities at a reasonable minimum flow.

It is sometimes an advantage to design storm sewers for surcharge conditions at peak design flow. It also may be economical to design very deep sanitary sewers such as tunnels on this basis. But this type of design should not be used for sanitary sewers which have service connections.

Occasionally there is no other choice and sewers must be constructed with the conduit depressed below the hydraulic grade line for all conditions of flow (inverted siphons, deep tunnels, submerged outlets). This is usually undesirable because of the difficulty of maintaining adequate cleansing or scouring velocities over the wide range of discharge encountered.

Estimated design flow data depend in large measure on assumptions, the accuracy of which is variable. In spite of this, care must be exercised in hydraulic designs to avoid compounding either errors or waste.

The organization of hydraulic computations is discussed in Chapter 6, with examples. A detailed tabular computation sheet for the design of a storm drain is presented in Chapter 4, illustrating many of the principles of flow described hereinafter.

B. TERMINOLOGY AND SYMBOLS

Insofar as possible the terminology used in this chapter conforms to that given in *Nomenclature for Hydraulics* (1). Further reference to terminology may be found in Chow (2), King (3), and Rouse (4) (5) (6).

The symbols and units * used in this chapter are:

A	= cross-sectional area, sq ft
A_f	= total area of a closed conduit, sq ft
B	= dimensionless constant (sediment-scouring characteristic)
b_i	= width of intercepted flow, ft
b_w	= width of water surface, ft
C	= coefficient (dimensionless); Hazen-Williams resistance coefficient (dimensional)
C_c	= coefficient of contraction, dimensionless
D	= diameter of circular conduit or height of other closed conduit, ft
D_g	= diameter of sediment grain, ft
d	= depth of flow above invert, ft
d_c	= critical depth, ft
d_L	= lower stage depth, ft
d_m	= hydraulic mean depth $\left(d_m = \dfrac{A}{b_w}\right)$, ft
d_{m_c}	= hydraulic mean depth at critical flow, ft
d_n	= normal depth, ft
d_U	= upper stage depth, ft
d_1, d_2	= depths before and after hydraulic jump, ft
F	= Froude number, dimensionless
ΣF	= summation of external forces acting on a fluid body, lb
F_1, F_2, F_3	= dimensionless channel cross-section functions
f	= friction factor, Darcy-Weisbach, dimensionless
f_f	= friction factor for conduit flowing full, dimensionless
g	= acceleration of gravity, ft/sec/sec
H	= total head, ft
H_a	= head added, ft
H_L	= head loss, ft
H_o	= specific energy head, ft
h	= head on weir, ft
h_1, h_2	= head on upper and lower ends of side channel weir, ft
h_b	= head loss in bends, ft
h_f	= friction head loss, ft
h_p	= pressure head, ft
h_v	= velocity head $(V^2/2g)$, ft
K, K_1, K_2	= transition loss coefficient, dimensionless
K_b	= coefficient for head loss in bends, dimensionless
k	= effective absolute roughness coefficient, ft
l	= length, ft

Δl	= increment of length, ft
n	= roughness factor (Manning and Kutter), ft$^{\frac{1}{6}}$
n_f	= roughness factor (Manning) for conduit flowing full, ft$^{\frac{1}{6}}$
P	= wetted perimeter, ft
Q	= discharge, cfs
Q_c	= discharge at critical flow, cfs
Q_f	= discharge, conduit flowing full, cfs
Q_i	= intercepted discharge, cfs
Q_w	= total gutter discharge, cfs
Q_n	= discharge at normal flow, cfs
R	= Reynolds number, dimensionless
R	= hydraulic radius (A/P), ft
R_f	= hydraulic radius, full section, ft
r_c	= centerline radius of curvature, ft
S	= slope, dimensionless
S_c	= critical slope, dimensionless
S_e	= slope of energy grade line, dimensionless
S_f	= slope of energy grade line for conduit flowing full, dimensionless
S_o	= slope of invert or bed, dimensionless
S_1, S_2	= energy slope upstream and downstream, dimensionless
s	= specific gravity, dimensionless
V	= velocity (mean, Q/A), fps
V_c	= critical velocity, fps
V_f	= mean velocity of closed conduit flowing full, fps
V_L	= supercritical velocity, fps
V_U	= subcritical velocity, fps
V_1, V_2	= velocity, upper and lower reaches or different conduit sections
v	= local velocity, fps
y	= height above invert, ft
z	= height of invert above datum, ft
a	= energy correction factor, dimensionless
β	= momentum correction factor, dimensionless
γ	= specific weight, lb/cu ft
ν	= kinematic viscosity, sq ft/sec

* Sq ft × 0.0929 = sq m; ft × 0.3048 = m; lb × 0.454 = kg; cfs × 1.7 = cu m/min.

C. HYDRAULIC PRINCIPLES

1. Type of Flow

Flow is said to be steady if the rate of discharge at a point in a conduit remains constant with time and unsteady if it varies. Although the flow in sewers is unsteady, in many instances the hydraulic analysis may be simplified by assuming steady-flow conditions. In other instances, such as the case of a large storm channel designed for flood waves, or where surge or water hammer may be important, as in force mains, the fact that the flow is unsteady should not be ignored. For these conditions, which are beyond the scope of this Manual, the reader is referred to Chow (2), King (3), Rouse (5), Parmakian (7), and Rich (8).

Steady open-channel flow is said to be uniform flow if the velocity and depth are the same from point to point along the conduit and non-uniform if either the velocity, the depth, or both are changing. Steady, pressure-conduit flow is uniform only if the cross-sectional area is uniform. Figure 17 illustrates steady, uniform flow for both open-channel and pressure-conduit flow.

Flow also may be classified as laminar or turbulent and subcritical or supercritical. In laminar flow, the fluid moves along in smooth layers, while in turbulent flow the fluid particles move in irregular paths. Subcritical and supercritical flow, which are types of open-channel flow, are discussed later in this chapter.

2. One-Dimensional Method of Flow Analysis

Fluid flow of the most general type may be non-uniform and thus present a three-dimensional problem. In many instances the condition of uniform flow across a given cross section may be assumed with knowledge of details being unnecessary to the bulk solution of the problem, or else the detail solution is so complex that a gross approximation is all that can be expected. In the one-dimensional method of analysis, variation of flow characteristics across any section is ignored. Only changes of mean values in the direction of flow are considered. Many of the hydraulic designs for sewers may be computed on the basis of this method of analysis. Basic to the one-dimensional method are the continuity, momentum, and energy principles.

(a) **Continuity Principle.**—For steady incompressible flow the continuity equation is expressed as

$$Q = A_1 V_1 = A_2 V_2 \quad \ldots \ldots \ldots \ldots \ldots \ldots \ldots \ldots 1$$

in which the subscripts designate different conduit sections, Q is the discharge, A is the cross-sectional area, and V is the mean velocity. There is

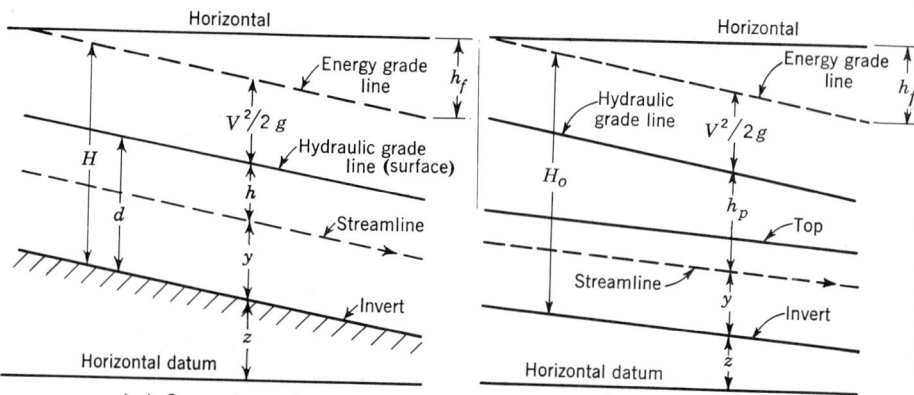

(a) Open channel (b) Closed conduit flow

FIGURE 17.—Comparison of uniform flow in an open-channel and a pressure conduit.

no inexactness in the continuity equation for steady flow if written in terms of the mean velocity of the flow section, since there is no change in the volume of the fluid between Sections 1 and 2.

(b) **Momentum Principle.**—The momentum principle may be written

$$\Sigma F_x = \frac{\gamma}{g} Q [(\beta V_x)_2 - (\beta V_x)_1] \quad \dots \dots \dots \dots \dots \dots 2$$

in which x is any direction, ΣF_x is the summation of the x components of the external forces acting on the fluid body under consideration, γ is the specific weight, g is the acceleration due to gravity, β is the momentum correction factor, V_x is the x component of the mean velocity and the Subscripts 1 and 2 designate different conduit sections. The external force components to be considered are the pressure forces acting at each cross-section, pressure force components acting on any boundaries between the cross sections, and frictional forces and weight forces in the direction of flow. The momentum correction factor arises from the use of the mean velocity in the momentum formula, when in fact the velocity varies across the cross section. Most sewer design work deals with turbulent flow with streamlines following the boundaries so that the assumption of β equal to unity gives sufficiently accurate results.

(c) **Energy Principle.**—The specific energy per pound of fluid at any point along a stream line in a conduit, based on the invert as the datum, is given by the equation

$$H_o = y + h_p + \frac{v^2}{2g} \quad \dots \dots \dots \dots \dots \dots 3$$

in which H_o is the specific energy head, y is the height above the invert or potential energy, h_p is the pressure head above the stream line, and $v^2/2g$ is the kinetic energy per unit weight, also called velocity head of the stream line.

It is important to note from Figure 17 and Equation 3 that the energy grade line will be at the same location regardless of where the datum is taken and that for uniform flow, H_o is equal to a constant along the length of the channel. Since the velocity varies over a cross section from zero at the walls to a maximum near the middle of the channel, the total energy is not the same for each stream line. For hydraulic computations it is convenient to assume that the velocity is the same for each stream line and is equal to the mean velocity or the rate of discharge divided by the cross-sectional area. For this assumption Equation 3 would apply to any and all stream lines in the conduit and therefore to the conduit as a whole. In order to accomplish this, the mean velocity head is multiplied by a, the energy correction factor. For turbulent flows, a is approximately unity, thus for open-channel flow

$$H_o = d + V^2/2g = d + \frac{Q^2}{2gA^2} \quad \dots \dots \dots \dots \dots \dots 4$$

in which d is the depth.

The energy equation also may be referenced to a horizontal datum and the more general form of the energy equation may be written

$$\left[a\frac{V^2}{2g}+h_p+z+y\right]_1+H_a=\left[a\frac{V^2}{2g}+h_p+z+y\right]_2+H_L\ldots\ldots\ldots 5$$

in which $z+y$ is the elevation above the datum of the point where the pressure head is equal to h_p, the Subscripts 1 and 2 refer to different conduit sections, and H_L is the energy loss (due to the conduit friction and conduit form losses) while H_a is the energy gain (as, say, from a pump in a force main) between Sections 1 and 2. For an open channel, if no energy is added, this may be rewritten as

$$[z+H_o]_1=[z+H_o]_2+H_L\ldots\ldots\ldots\ldots\ldots\ldots 6$$

3. Energy Losses

The resistance to flow of fluids may be classified into conduit surface, or frictional, and conduit form, or drag and separation losses. Although there is not a clear demarcation between the types of losses, conduit surface losses are caused primarily by viscous and turbulent shears along the boundary of the conduit, while form losses may be caused by shear as well as pressure differentials caused by separation of the flow from the boundary of the conduit, drag on obstructions placed in the flow, changes in conduit alignment, etc. For pressure-conduit flow, form losses often are called "minor losses." Yet there are times when they are relatively large compared to the friction loss.

For friction, the rate of energy loss in laminar flow may be evaluated as a function of the first power of the velocity; whereas in turbulent flow, the loss will be a function of approximately the second power. Form losses too usually are evaluated as a function of the second power of the velocity. Discussions of both types of losses are found in a later section of this chapter.

4. Pressure-Conduit Flow Analysis

The continuity and energy principles are the primary tools used in pressure-conduit analysis. If Equation 1 is substituted in Equation 5,

$$\left(\frac{aQ^2}{2gA^2}+h_p+z+y\right)_1+H_a=\left(\frac{aQ^2}{2gA^2}+h_p+z+y\right)_2+H_L\ldots\ldots\ldots 7$$

For a straight pipe discharging from a reservoir, Point 1 may be chosen at the reservoir surface and Point 2 at the center of the emerging jet. Let the datum be through the center of the emerging jet, whose pressure is assumed to be atmospheric. If H is the distance of the reservoir surface above the center of the jet,

$$(0+0+H)_1+0=\left[\frac{1\times Q^2}{2gA^2}+0+0\right]_2+H_L$$

or

$$Q/A=[2g(H-H_L)]^{\frac{1}{2}}\ldots\ldots\ldots\ldots\ldots\ldots 8$$

The head loss, H_L, includes the pipe friction loss, h_f, and an entrance loss. Pressure-conduit "minor losses", and further discussion of pressure conduit analysis will be found in References (3) and (5) and in Chapter 7.

5. Open-Channel Profile Analysis

For uniform flow in an open channel, the slope of the energy and hydraulic grade lines will be the same as the slope of the invert, and the depth of flow will adjust itself to produce a velocity commensurate with the friction losses in the channel. Flow at this condition will be at the so-called normal depth of flow, d_n. Since for uniform flow the hydraulic and energy grade lines are parallel, the energy loss may be associated with either grade line. In non-uniform flow problems, however, the energy and the hydraulic grade lines are not parallel. Friction losses, therefore, must be applied to the energy grade line or substantial errors may result in the computations. The factors affecting the difference in elevation of the energy grade line between any two cross sections are the conduit friction loss and form losses caused by change in alignment or cross-sectional area.

It is rare in the design of sewers that uniform flow will prevail in all reaches. In general, there will be interconnected regions of uniform and non-uniform flow, the latter possibly taking the form of backwater curves, drawdown curves, or hydraulic jumps. Except for minimum-sized sewers and preliminary design of larger sewers, it is good practice to calculate the hydraulic profile for the various reaches of conduit. Profile calculations must proceed from a point for which the depth and velocity are known. Sometimes this point may be taken as a river or lake stage with a particular frequency of occurrence or, in the case of an ocean outfall, as a specific tide and swell condition. In other cases, the starting point may be a uniform-flow condition in one reach of conduit. When drawdown or backwater curves are encountered, the selection of a starting point is not always a simple procedure. In many cases of non-uniform flow the most common starting point for hydraulic profile computations is a control section for which the total energy above the invert is a minimum for a given discharge or, conversely, the rate of flow is maximum for a given total energy. This condition is defined as critical flow, or flow at critical depth.

Critical flow is characterized by a Froude number, **F**, equal to unity. If **F** is less than unity, the flow is subcritical, and if it is greater, the flow is supercritical. The Froude number is defined as:

$$\mathbf{F} = \frac{V}{\sqrt{gd_m}} \dotfill 9$$

The hydraulic mean depth, $d_m = A/b_w$, in which b_w, is the width of the water surface. For a rectangular channel, $d_m = d$.

The primary significance of the Froude number is that it represents the ratio of the mean flow velocity to the velocity of propagation of a small gravity wave. A gravity wave can be propagated upstream in subcritical flow since the wave velocity is greater than the flow velocity, but can be propagated only downstream in supercritical flow. Consequently, for water surface profile analysis, the analysis begins at the control point and proceeds upstream when the upstream flow is subcritical, and works downstream from the control point when the downstream flow is supercritical. At a control section the depth of flow is equal to the critical depth.

It may be seen from Equation 4 that for a given channel section and discharge, the specific energy head is a function of a depth of flow only. If the depth of flow is plotted against the specific energy head, a specific energy curve, Figure 18, is obtained. The curve shows that for all flows except critical flow, there are two possible alternate stages or depths at which flow may occur for any value of the specific energy head and discharge. The alternate depth which occurs depends on the channel slope and friction, and location of control sections.

If flow takes place in an open channel at the upper of the two alternate depths, it is known as upper-stage or subcritical flow; if at the lower depth, it is known as lower-stage or supercritical flow. Depths of flow within 10 to 15 percent of the critical depth are likely to be unstable because slight changes in energy may cause great changes in the depth to either upper- or lower-alternate stage. The designer should be able to identify the alternate stages of flow as well as the case of flow at critical depth. Illustrative examples for this purpose are presented hereinafter.

It was mentioned previously that at critical flow the Froude number is equal to unity. It should be noted further, that for critical flow $V_c^2/2g = d_{m_c}/2$, and that $H = H_{o_{min}} = d_c + d_{m_c}/2$. For a rectangular channel $d_c = d_{m_c}$, therefore $H = H_{o_{min}} = \dfrac{3d_c}{2}$.

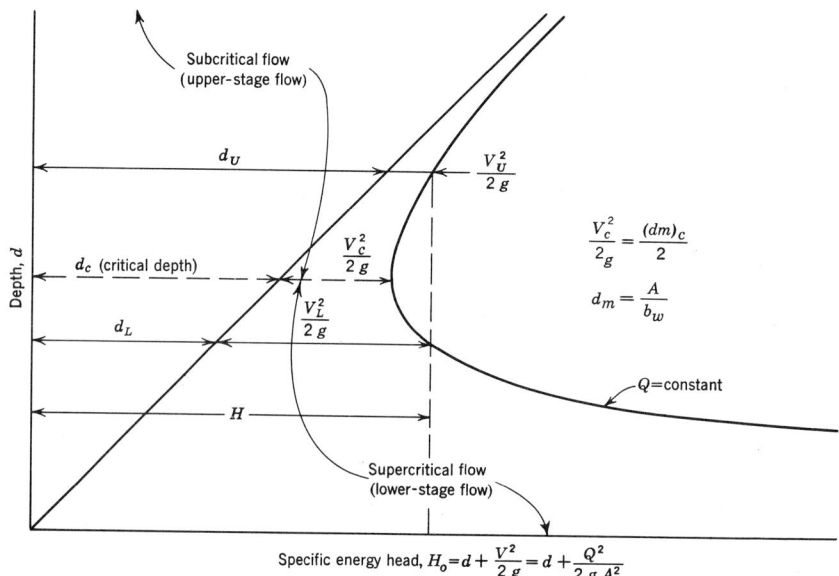

FIGURE 18.—Specific energy curve.

If the flow passes from a subcritical stage on a gently sloping channel to a supercritical stage in a steeply sloping channel, as shown in Figure 19a, it must pass through a control section. The control section is located in the vicinity of the break in grade and critical flow occurs there. It is therefore a starting point for hydraulic profile computations in a problem of this nature. The drawdown curves produced for this condition of flow also are shown in Figure 19a. The upstream slope, which is less than the critical slope, is called a subcritical slope, or more commonly, a mild slope. The downstream slope, which is greater than the critical slope, is called a supercritical slope, or more commonly, a steep slope.

If the flow in an upstream channel is subcritical and is controlled downstream by discharging into a relatively quiet body of water, or by encountering an obstruction such as a weir, a backwater curve is produced in the upstream channel as shown in Figure 19b.

Figure 20 illustrates one of the possible water surface profiles for discharge from a steep to a mild slope channel, with supercritical flow on the upper channel and subcritical flow on the lower one. To pass from supercritical to subcritical flow, the flow must pass through a hydraulic jump with a consequent loss in head. In the illustration, the energy conditions are such that the jump must take place on the mild slope; but, if the required downstream total energy necessary to transport the flow is greater than that which would result if the jump occurred on the mild slope, the jump must take place on the steep slope. In either case, there is a backwater or drawdown curve from the jump to the break in grade. The

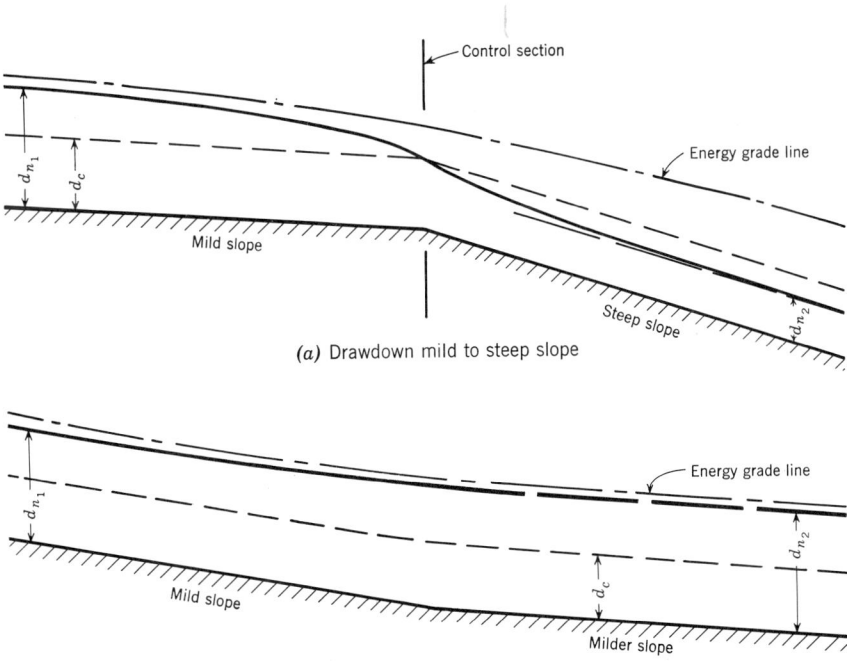

(a) Drawdown mild to steep slope

(b) Backwater mild to milder slope

FIGURE 19.—Non-uniform flow hydraulic profiles.

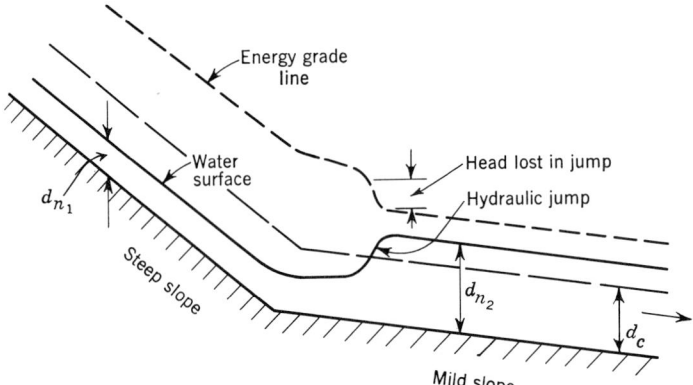
Figure 20.—Hydraulic jump profile.

loss in head in the jump may be computed by means of the momentum principle referred to previously. To compute the hydraulic profile for a case such as illustrated in Figure 20, the total energy must be known at some point on the mild slope downstream from the break in grade. The usual case is for uniform flow to exist at some point both upstream and downstream from the jump.

More detailed discussion of open-channel profile analysis may be found elsewhere (2) (3) (5) (9).

D. FLOW FRICTION FORMULAS

1. General

The basis for the first flow friction formula was developed circa 1768 by Antoine Chezy. Chezy reasoned that V^2P/AS or V^2/RS would be constant for any one channel, and would be the same for any similar channel. The square root of V^2/RS became known as the Chezy coefficient and numerous investigators undertook the task of determining how this coefficient varied under different conditions. An interesting summary of the various major investigations and formulas as well as an excellent bibliography is given in Reference (10). A critical analysis of open-channel resistance has been presented by Rouse (11). Present design practice tends to the use of the Manning formula for open-channel flow and the Manning, Hazen-Williams, or Darcy-Weisbach formulas for closed-conduit or pressure flow.

2. Darcy-Weisbach Equation

Of the various formulas in current use, the Darcy-Weisbach equation represents a basic approach which is dimensionally correct, may be used for any fluid, and applied over the widest range of flow conditions. For open channels the formula may be expressed

$$S_e = \frac{f}{4R} \frac{V^2}{2g} \quad \ldots \ldots \ldots \ldots \ldots \ldots 10$$

in which S_e is the slope of the energy grade line, R is the hydraulic radius, and f is the dimensionless friction factor. For pipe flow the formula may be written

$$h_f = \frac{fl}{D} \frac{V_f^2}{2g} \quad \dots \dots \dots \dots \dots \dots \dots \dots \dots \dots 11$$

in which h_f is the friction head loss, l is the length of conduit, D is the pipe diameter, and V_f is the mean velocity of the pressure conduit.

The coefficient f has been found to be primarily a function of the Reynolds number and the relative roughness of the conduit. The Reynolds number used to determine frictional resistance is defined normally as

$$\mathbf{R} = \frac{4RV}{\nu} \quad \dots \dots \dots \dots \dots \dots \dots \dots \dots \dots 12$$

in which ν is the kinematic viscosity. For a circular pressure conduit this reduces to

$$\mathbf{R} = \frac{DV}{\nu} \quad \dots \dots \dots \dots \dots \dots \dots \dots \dots \dots 12a$$

The Reynolds number is dimensionless and used to compare the importance of viscous effects with relation to roughness effects in determining frictional resistance. The larger the Reynolds number, the greater the influence of the relative roughness in determining the frictional resistance.

The relative roughness is defined normally as the dimensionless ratio of effective absolute roughness, k, to four times the hydraulic radius. For a circular pressure conduit, the relative roughness reduces to k/D. The effective absolute roughness is related to the size of the conduit surface irregularities that project into the flow. The roughness of the conduit has been found to depend not only on the size but on the shape and spacing of the surface irregularities. For most commercially available pipe, and job-formed conduits of materials similar to those used in commercial pipe, k is found to be relatively constant for a given material, and the relative roughness adequate to define the conduit roughness. It is difficult to develop roughness values for natural channels where sand waves may form in the bottom, or where the surface is composed of large, irregularly shaped material. Summaries of usual k values for various materials are given in Table XIV and elsewhere (2) (3) (5) (10) (12).

Various semi-empirical equations have been developed to relate the coefficient f to the relative roughness and Reynolds number. For circular pressure-conduit flow the relationships are rather well defined, but for open-channel flow they are somewhat complicated by the influence of the shape of the channel and the Froude number. In general, for fixed-bed open channels which are not very narrow and are rectangular or trapezoidal in cross section, reasonably good predictions of the friction factor result for turbulent, subcritical flow when the Reynolds number and relative roughness are utilized as defined above. For other shapes the predictions are less certain, although in general, the more turbulent the flow, the less the deviation.

For design purposes a graphical representation of the various relationships is usually more convenient than an equation. Such graphs for pipe flow have been published by Moody (3) (13) and Rouse (4) (5) (14); Chow (2) has published similar graphs for various open channels; and the Corps of Engineers has published graphs for both pipes and open channels (12).

One form of the graph has been included as Figure 21, in which it is seen that for transitional flow, f varies with both **R** and $k/4R$, whereas for completely turbulent flow f varies with $k/4R$ only. For pressure-conduit flow (fixed flow area), f is determinate for completely turbulent flow. But for transitional flow it has to be determined by trial and error methods since it varies with V. For open-channel flow (variable flow area), f requires a trial and error solution for both transitional and completely turbulent flow because it varies with both R and V. Determination of the normal depth for a given discharge can best be done by construction of a stage-discharge diagram.

The Darcy-Weisbach formula has a design shortcoming in that f is quite variable with size of conduit even for the same material. In addition, more experimental work should be done on k values, thus the possibilities for greater accuracy inherent in the Darcy-Weisbach formula compared to the more empirical Kutter, Manning, and Hazen-Williams formulas have yet to be realized. While the coefficients for the Kutter, Manning, and Hazen-Williams formulas tend to be more nearly constant with size of conduit, they do vary with size, and in addition, are subject to variations caused by the shape of channel, type of roughness, etc. Judgment is required in choosing a coefficient no matter what formula is used. Further discussion of friction coefficients will be found in another section of this chapter.

3. Kutter and Manning Formulas

Kutter's formula, published about 1869, received wide acceptance and use in estimating open-channel flows. The formula is rather unwieldy, but since it was accepted so widely, many tables and graphs for its solution were prepared. Through use of the formula, early designers became familiar with values of the Kutter roughness coefficient, n, applicable to sewers. Kutter's formula (English units) is

$$V = \left[\frac{\frac{1.81}{n} + 41.67 + \frac{0.0028}{S_e}}{1 + \frac{n}{\sqrt{R}}\left(41.67 + \frac{0.0028}{S_e}\right)} \right] \sqrt{RS_e} \quad \ldots \ldots \ldots \ldots 13$$

in which V is the mean velocity of flow, R is the hydraulic radius, S_e is the slope of energy grade line, and n is the coefficient of roughness. Some authors have suggested that the term $0.0028/S_e$ be omitted since it originally was added to make the formula agree with data now known to be inaccurate (2).

The Manning formula, because of its greater simplicity, the fact that the n value is substantially equal to Kutter's n for types of pipe com-

HYDRAULICS OF SEWERS

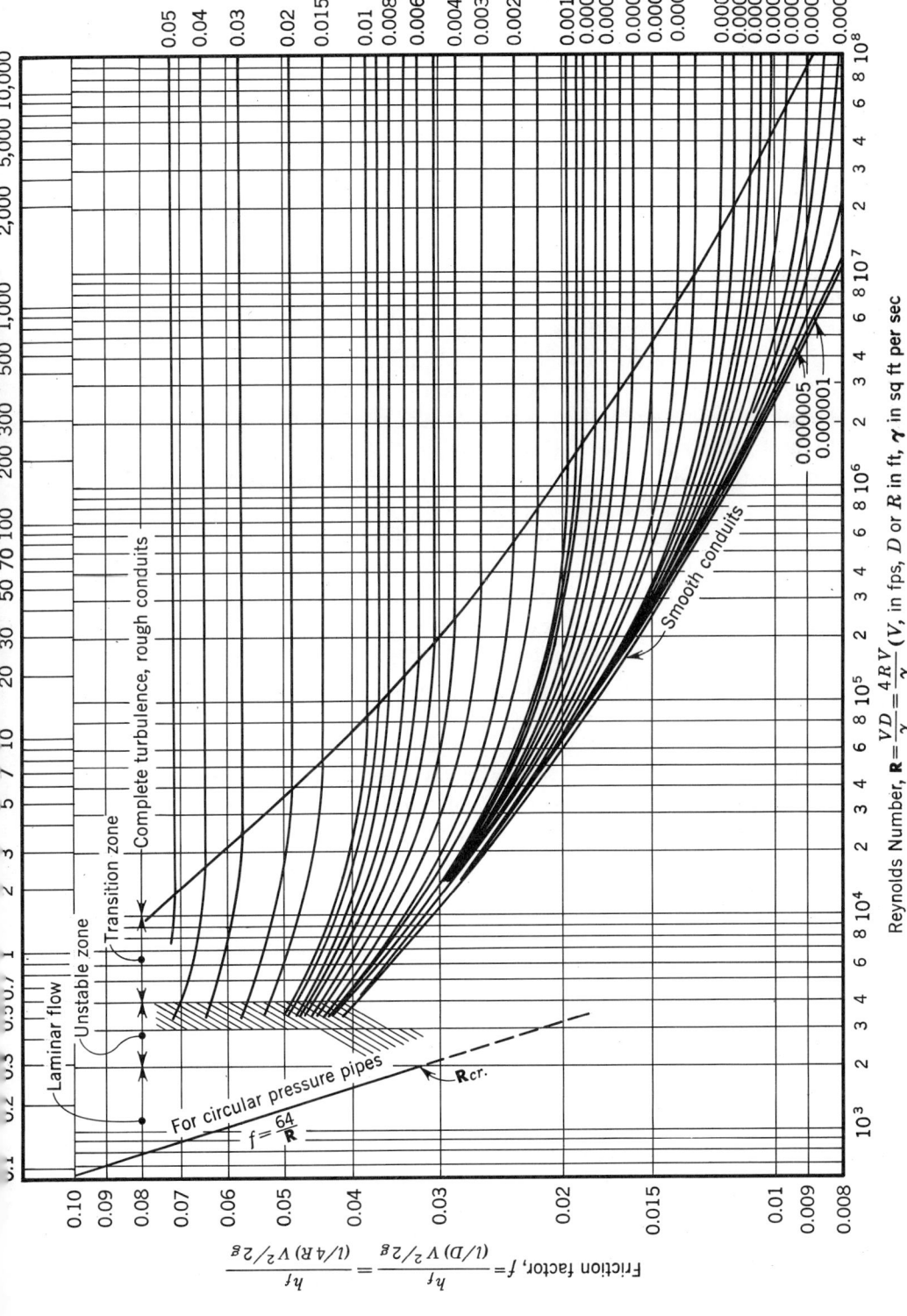

FIGURE 21.—Friction factors as a function of Reynolds number and relative roughness.

monly used in sewer construction, and the recommendations of various authors and authorities, has come into more general use for sewer design and largely has replaced the Kutter formula in engineering practice. The Manning equation is

$$V = \frac{1}{n}(R^{\frac{2}{3}}S^{\frac{1}{2}}) \quad \text{(metric units)} \quad \ldots\ldots\ldots\ldots\ldots\ldots 14$$

$$V = \frac{1.49}{n}(R^{\frac{2}{3}}S^{\frac{1}{2}}) \quad \text{(English units)} \quad \ldots\ldots\ldots\ldots\ldots 15$$

in which the terms have been defined previously.

The Kutter and the Manning equations are used for pipes and conduits of all shapes flowing either full or partly full. The graphs commonly available for their solution usually are compiled for pressure-conduit flow only, and the so-called hydraulic-elements graphs described hereinafter or tabular values thereof are used to determine the flow characteristics at other than full flow. Figure 22 is an alignment chart prepared by Camp (15) for the solution of the Manning equation for circular pipes flowing full. It may be used with values of n up to 0.10. This chart, as do most other charts for the same purpose, also gives a solution of the equation $Q_f = A_f V_f$ for the full pipe. The use of the chart is illustrated by the broken lines.

It also is possible to use this chart for other shapes of closed conduits and open channels if the discharge scale is ignored and the diameter scale is taken to represent values equal to four times the hydraulic radius of the actual cross section. If, for example, Q is given, the choice of S and n will give a point on the transfer line. By pivoting from this point, sets of values of $4R$ and V may be obtained. By trial and error a compatible set of values for the given Q and shape of cross section may be found which furnishes the correct depth of flow, d.

For frequent use, graphs of slope vs. discharge plotted on log-log scales with lines of constant conduit size and velocity are probably more convenient than the alignment chart of Figure 22. Individual graphs usually are prepared for each n value. Such graphs may be found in some texts or in other source material.

Many engineers prefer to use $Q/S^{\frac{1}{2}}$ values (16) (17) for closed-conduit flow; these values are constant for any given conduit size and n value. For open-channel flow tabular values of $Qn/(l^{8/3}S^{\frac{1}{2}})$ often are used for design to determine depth of flow or channel size (3) (5) (9). Special slide rules also are available for the solution of the Manning equation. Usual ranges of Manning's n for particular materials will be found in Table XIV.

4. Hazen-Williams Formula

The Hazen-Williams formula is

$$V = 0.85\,C\,R^{0.63}S^{0.54} \quad \text{(metric units)} \ldots\ldots\ldots\ldots\ldots\ldots 16$$

$$V = 1.32\,C\,R^{0.63}S^{0.54} \quad \text{(English units)} \ldots\ldots\ldots\ldots\ldots\ldots 17$$

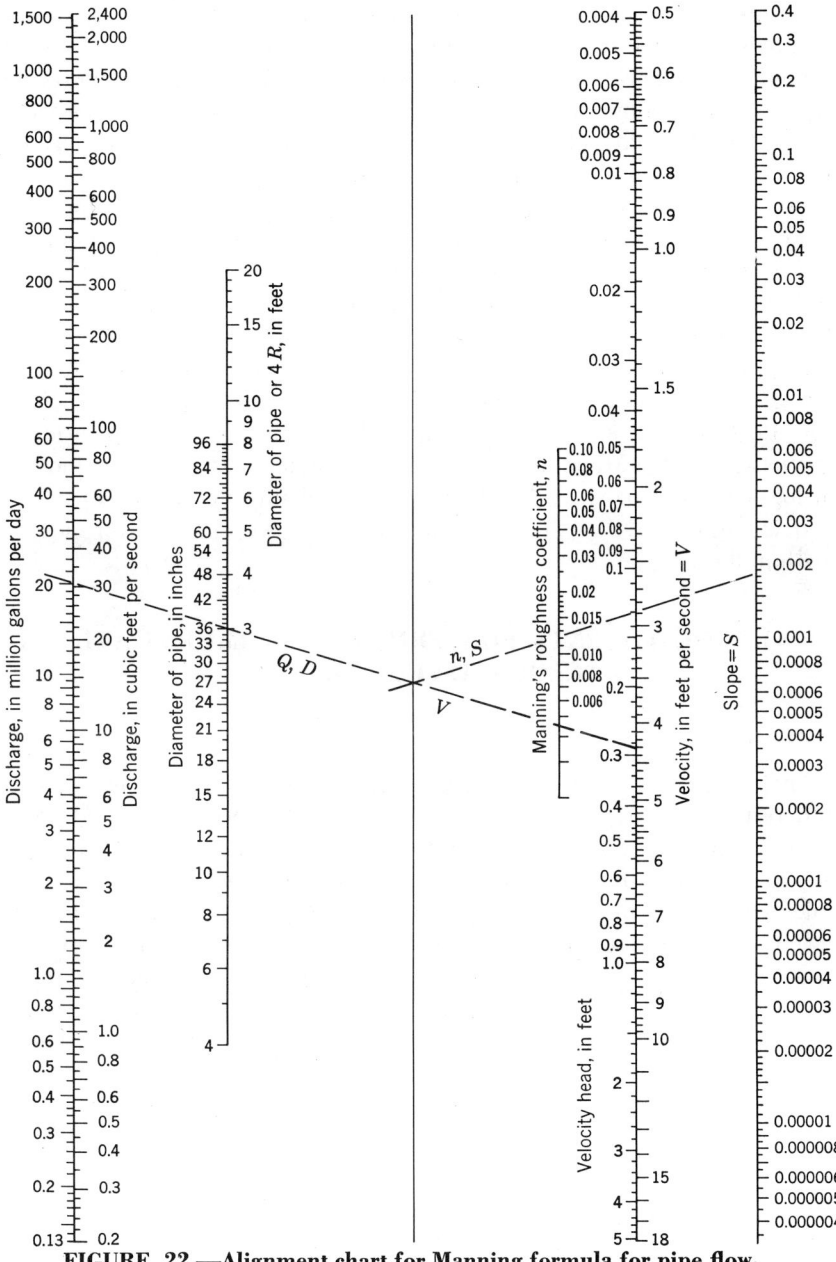

FIGURE 22.—Alignment chart for Manning formula for pipe flow.

in which the nomenclature is basically the same as that used in Equation 13 and C is a coefficient related to roughness.

The formula is used widely for pressure-conduit or pipe flow, although it is equally applicable to open-channel conditions. Published values for C have come largely from pipe-flow experiments, while many of the reported n values are from open-channel flow tests. The Hazen-Williams

C is primarily dependent on the roughness of the conduit. A wide variety of values published in table form are available (18).

The following are values of C suggested by King (3) for pipes carrying water. The C value for new, smooth pipes generally is taken to be from 130 to 140. For old, tuberculated cast-iron pipes, the C value may be 60 to 80; and, for small pipes badly tuberculated, the value may be as low as 40. Low C values resulting from tuberculation of cast iron pipe may be avoided by the use of cement-lined pipe, preferably with a bituminous seal coat. To estimate friction losses for future conditions, lower values of C are used to allow for reductions in carrying capacity resulting from films and slimes on the pipe interior. For smooth concrete and cement-lined pipes, a C value of 100 to 120 commonly is used for future conditions.

Figure 23 is an alignment chart prepared by Camp for the solution of the Hazen-Williams formula for pipes flowing full. As was mentioned previously for the Manning formula, graphs of slope vs. discharge plotted on log-log scales or special slide rules are probably more convenient for frequent use. $Q/S^{0.54}$ values also can be computed and used for closed-conduit flow.

E. COEFFICIENTS FOR FRICTION FORMULAS AND FACTORS AFFECTING THEM

1. General

There have been many experiments in both the laboratory and the field to determine the friction coefficients for various materials and conditions. In the laboratory accurate measurements can be obtained, but it is difficult to duplicate conditions of flow equivalent to those in a sewer. On the other hand, field measurements in existing sewers may reflect unknown variables peculiar to the particular sewer being investigated, as well as errors in measurement and an inability to control identifiable variables.

Factors which affect the choice of a coefficient are conduit material, Reynolds number, size and shape of conduit, and depth of flow. In addition to these interrelated factors the following should be considered:

(a) Rough, opened, or offset joints,
(b) Poor alignment and grade due to settlement or lateral soil movement,
(c) Deposits in sewers,
(d) Amount and size of solids being transported,
(e) Coatings of grease or other matter on interior of sewer,
(f) Tree roots, joint compounds, and mortar dams resulting from poor or deteriorated jointing and other protrusions, and
(g) Flow from laterals disrupting flow in the sewer.

Where deposition will not be a problem, the designer may be justified in using values of coefficients which will result in somewhat lower predicted friction losses for storm sewers over those used for sanitary or combined sewers. For conduits of minimal slope and velocity, deposition

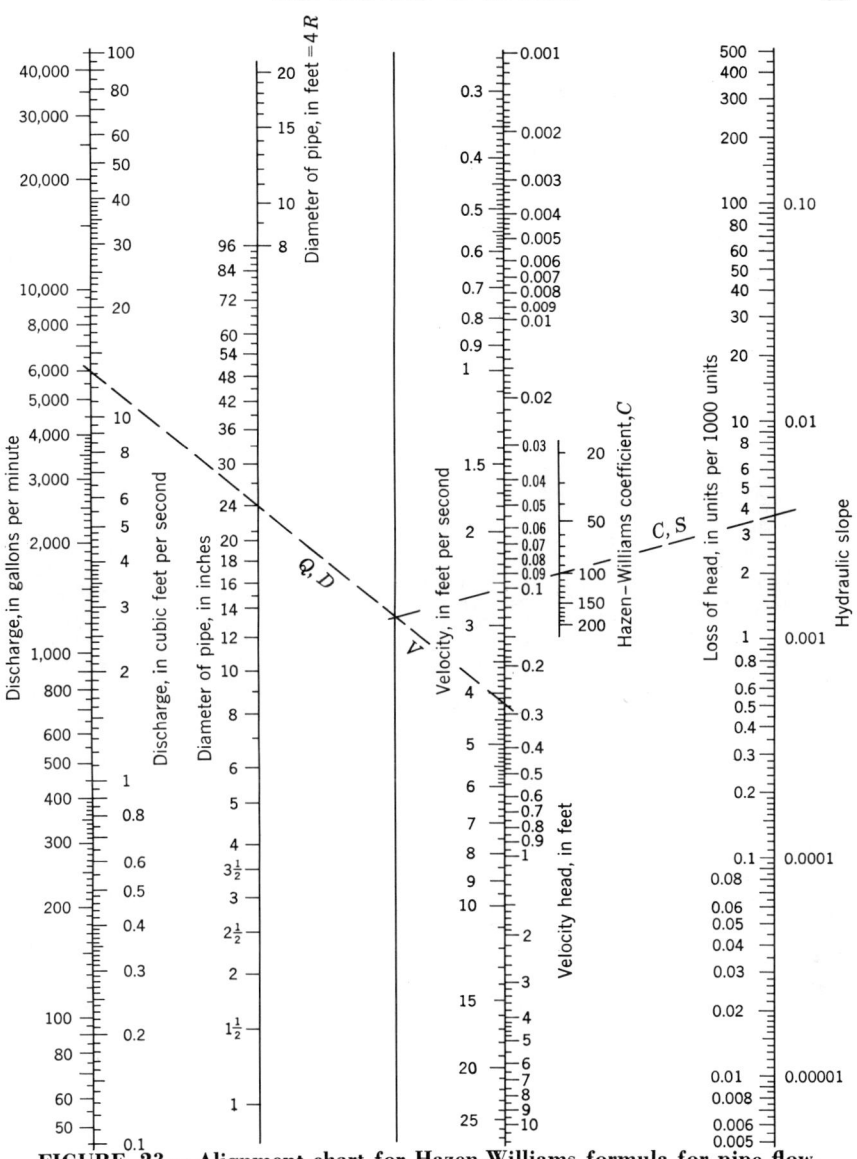

FIGURE 23.—Alignment chart for Hazen-Williams formula for pipe flow.

may be aggravated so that a higher than normal friction loss allowance may be justified.

2. Conduit Material

The type of joint used and the material of which the conduit is constructed are basic to the selection of a coefficient. A friction coefficient may be selected directly for the Manning and Hazen-Williams formulas, but usually is obtained indirectly for the Darcy-Weisbach formula by means of a value of k and a Moody-type diagram. Table XIV is a summary of commonly used coefficients and k values compiled from the various sources. The values are for sewer design and hence are higher than

TABLE XIV.—Values of Effective Absolute Roughness and Friction Formula Coefficients

Conduit Material	Effective Absolute Roughness (Darcy-Weisbach) k (ft)	Manning n (ft$^{\frac{1}{6}}$)	Hazen-Williams $C*$
Closed conduits			
Asbestos-cement pipe	0.001–0.01	0.011–0.015	100–140
Brick	0.005–0.02	0.013–0.017	—
Cast iron pipe			
Uncoated (new)	0.00085	—	—
Asphalt dipped (new)	0.0004	—	—
Cement-lined & seal coated	0.001–0.01	0.011–0.015	100–140
Concrete (monolithic)			
Smooth forms	0.001–0.005	0.012–0.014	—
Rough forms	0.005–0.02	0.015–0.017	—
Concrete pipe	0.001–0.01	0.011–0.015	100–140
Corrugated-metal pipe (½-in. × 2⅔-in. corrugations)			
Plain	0.1 –0.2	0.022–0.026	—
Paved invert	0.03 –0.1	0.018–0.022	—
Spun asphalt lined	0.001–0.01	0.011–0.015	100–140
Plastic pipe (smooth)	0.01	0.011–0.015	100–140
Vitrified clay			
Pipes	0.001–0.01	0.011–0.015	100–140
Liner plates	0.005–0.01	0.013–0.017	—
Open channels			
Lined channels			
a. Asphalt	—	0.013–0.017	—
b. Brick	—	0.012–0.018	—
c. Concrete	0.001–0.03	0.011–0.020	—
d. Rubble or riprap	0.02	0.020–0.035	—
e. Vegetal	—	0.030–0.40 *	—
Excavated or dredged			
Earth, straight and uniform	0.01	0.020–0.030	—
Earth, winding, fairly uniform	—	0.025–0.040	—
Rock	—	0.030–0.045	—
Unmaintained	—	0.050–0.14	—
Natural channels (minor streams, top width at flood stage < 100 ft)	0.1 –3.0	—	—
Fairly regular section	—	0.03 –0.07	—
Irregular section with pools	—	0.04 –0.10	—

* Assume dimensional units contained in 1.32 term in formula. See References (2) (19)(20). (Varies with depth and velocity.)

the values obtained in laboratory tests with clear water and clean conduits.

The range in coefficients for a given pipe material is explained partially by the disturbing influences mentioned previously in the general discussion of coefficients. A coefficient which will yield higher friction losses should be selected for sewers with high disturbing influences.

Because of the physical and hydraulic conditions which may influence a friction formula coefficient, the values given in Table XIV for one friction formula are not necessarily equivalent to the values for another formula. They do, however, reflect ranges commonly used in design.

3. Size of Conduit and Reynolds Number

For a given conduit material and disturbing influences the size of the conduit will determine the relative roughness. From an examination of Figure 21 the following generalizations may be made (exclusive of the unstable zone between laminar and transitional flow):

(a) For laminar flow and for smooth conduits for all flow, f is independent of the relative roughness and decreases with an increase in Reynolds number (increase in conduit size for constant V and ν).

(b) For rough conduits with transitional flow, f decreases with a decrease in relative roughness (increase in conduit size for constant k) and with an increase in Reynolds number (increase in conduit size for constant V and ν).

(c) For rough conduits with fully turbulent flow f is independent of Reynolds number and decreases with a decrease in relative roughness (increase in conduit size for constant k).

Thus, for the usual range of flows in sewers (fully turbulent or transitional), f will decrease with an increase in conduit size.

The following remarks and those in the next section are concerned primarily with factors affecting Manning's n. The Manning formula is assumed independent of Reynolds number so that it should be applicable only to fully turbulent flow. For this flow condition, the effect of conduit size alone may be examined by a comparison of n and f. (If the designer considers the Hazen-Williams equation more reliable than the Darcy-Weisbach equation, then n and C values might be related.) These values are related as follows for circular conduits:

$$f = \frac{185 n^2}{D^{1/3}} \quad \dots \dots \dots \dots \dots \dots \dots \dots \dots \dots \dots \dots \dots 18$$

in which the terms have been defined previously. Thus, for a given value of n and conduit diameter, f can be calculated. From the assumption of fully turbulent flow, the effective absolute roughness, k, can be ascertained for the type of conduit material by reference to Figure 21. The n value for a different size pipe of the same material can be calculated by determining the new value of f from Figure 21 and then using the ratio

$$\frac{n_1}{n_2} = \left(\frac{D_1}{D_2}\right)^{1/6} \left(\frac{f_1}{f_2}\right)^{1/2} \quad \dots \dots \dots \dots \dots \dots \dots 19$$

in which the subscripts denote the two conduit sizes. Substitution of the empirical equation for f for fully turbulent flow into Equation 19 gives

$$\frac{n_1}{n_2} = \left(\frac{D_1}{D_2}\right)^{1/6} \left[\frac{2 \log \frac{D_2}{k} + 1.14}{2 \log \frac{D_1}{k} + 1.14}\right] \quad \dots \dots \dots \dots \dots 20$$

An examination of this equation will indicate that except for extremely rough pipes n must increase as the diameter of the pipe increases. This effect is more pronounced, the smoother the pipe. Thus, from a hydraulic standpoint, the practice of using somewhat smaller values of n for large pipe than for small pipe is generally not correct. (A similar deduction is obtained if n and C are related.)

For example, suppose that tests of a 12-in. (30.5-cm) pipe, with a D/k value of 1,000, indicate an n value of 0.010; then, the value for a 60-in. (152.5-cm) pipe with the same k value (Equation 20) would be $n = 0.01094$, say 0.011. If an n value is developed for large pipe, it may be prudent to use the same value for small pipes, since they may be subject to greater disturbing influences.

For geometrically similar channels of non-circular cross section roughly similar conclusions may be reached if $4R$ is substituted for the pipe diameter.

There is no paradox in that, on the one hand, f will decrease with conduit size while on the other hand n generally must increase. This results from the fact that the Manning coefficient, assumed as a constant, does not fully account for all of the variables and conditions of flow.

4. Shape of Conduit and Hydraulic-Elements Graphs

For both pressure-conduit and open-channel flow the hydraulic radius is used in the various friction formulas with satisfactory accuracy even though many of the experimental data were developed from circular pressure conduits. For roughnesses of the type often found in pipe work (such as provided by concrete surfaces), pipe resistance diagrams of the Moody type (Figure 21) may be used for estimating f for open channels. As mentioned previously, $4R$ should be used in place of D in computing the Reynolds number and relative roughness. It should be realized that the shape of the conduit changes the frictional resistance from that predicted by the Moody chart. Chow (2) has discussed the shape effect on open-channel frictional resistance.

The effects of the channel shape on Manning's n have not been defined explicitly, but if shape affects f it must also affect n.

The effects of depth of flow on f and n have not been defined completely. The indirect effect of depth of flow or the height of vegetal cover on the resistance coefficient has been investigated in grass-lined channels (2) (19) (20), and also for natural channels and flood plains (2). A variation of f or n with depth of flow without an apparent change in boundary roughness has been observed in circular conduits and other sewer shapes flowing partly full. This variation is due to the effects of both size and shape.

A hydraulic-elements graph may be used for the solution of problems involving open-channel flow in closed conduits. It also is useful to illustrate the effects of variation of n with depth of flow. Figure 24 is a hydraulic-elements graph for sewers of circular cross section and uniform smoothness throughout the surface area, prepared by Camp (21). Computations for the construction of such a graph are set forth by Fair

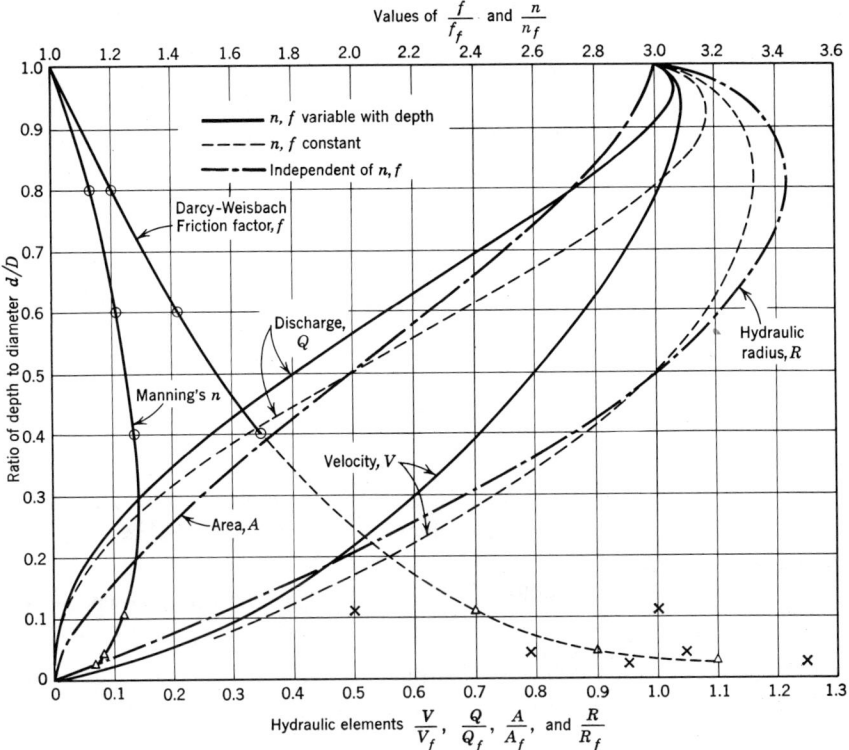

FIGURE 24.—Hydraulic-elements graph for circular sewers.

et al. (22). Graphs for sewers of other than circular cross section may be developed by the same general method.

Most of the hydraulic-elements graphs in common use have been prepared on the assumption that the Manning n does not change with the depth of flow for the particular conduit shape. Nonetheless, many experimenters have observed a variation of n with depth of flow. The experiments of Wilcox (23) and of Yarnell and Woodward (24) show that the value of n for a pipe flowing partly full is greater than for the full pipe; and the average n values for 824 experiments are as indicated by the curve through the points marked by circles in Figure 24. A similar curve for the Darcy-Weisbach fraction factor f also is shown in the same figure.

The relation between the two friction coefficients is

$$\frac{n}{n_f} = \left(\frac{R}{R_f}\right)^{1/6} \left(\frac{f}{f_f}\right)^{1/2} \qquad \qquad \text{.21}$$

which is similar to Equation 19.

The points in Figure 24 marked by triangles and x's were estimated from the measurements made by Johnson (25) in large Louisville, Ky., sewers flowing partly full. Since individual values of f/f_f in the experiments of Wilcox and of Yarnell and Woodward varied widely from the average for a particular value of d/D, the reliability of the averages used in Figure 24 may be questioned. Tests by Schmidt (26) on a large concrete arch-section sewer in Kansas City, Mo., indicated a variation in n with depth, with n increasing with depth up to about one-half of the section height. Limited tests by Bloodgood and Bell (27) on asbestos-cement, cast iron, and vitrified clay pipes apparently indicate agreement with the variation in n shown in Figure 24. Recent tests of existing sewers by Pomeroy (28) also indicate agreement with the variation in n for the larger depths of flow. At very low depths of flow the results of this experimentation indicate quite low n values, provided the sewers were clean and well constructed. It is interesting to note, however, that tests by Neale and Price (29) on hydraulically smooth polyvinyl chloride pipe flowing at relative depths of 0.2 to full do not indicate a significant variation in n with depth. Their data for discharge and velocity approximate the curves which are obtained for the assumption of constant n.

Until more information and better analyses are available, the decision to use a constant or variable n must be left to the individual designer.

In Figure 24 the curves for discharge and velocity obtained from the assumption of a constant n are shown by broken lines; those for the assumption of variable n are shown by solid lines.

The method of using the hydraulic-elements graph is illustrated below:

Example 1. A 12-in. (30.5-cm) sewer is laid on a grade of 3.0 units per 1,000 units. Find the velocity and discharge if the sewer is flowing 0.4 full [4.8-in. (12.2-cm) depth]. Assume Manning's $n=0.013$.

(a) From Manning's chart, Figure 22, $Q_f=2.0$ cfs (3.4 cu m/min) and $V_f=2.5$ fps (0.76 mps)

(b) From hydraulic-elements chart, Figure 24, for $d/D=0.4$,
  1. n constant

    $Q/Q_f=0.33$ and $V/V_f=0.90$

    $Q=0.33\times2.0=0.66$ cfs (1.12 cu m/min); $V=0.90\times2.5=2.25$ fps (0.68 m/sec)

  2. n variable

    $Q/Q_f=0.27$ and $V/V_f=0.71$

    $Q=0.27\times2.0=0.54$ cfs (0.92 cu m/min); $V=0.71\times2.5=1.78$ fps (0.54 m/sec)

Many engineers prefer to use tabular values of $Qn/D^{8/3}S^{1/2}$ for open-channel flow in circular conduits. Values based on constant n are given in Table XVI. These values could be modified to reflect a variation in n with depth, and be based on $Qn_f/D^{8/3}S^{1/2}$ to avoid a trial-and-error solution.

F. SCOURING VELOCITIES

1. Self-Cleaning Velocities

A theoretical development by Shields (30), based on experiments by himself and others on bed-load movement of unigranular materials, indicated that the critical tractive force required to produce motion of the particles along the bottom is approximately proportional to the diameter of the particles and their submerged weight per unit of volume. From findings of Shields, Camp (15) derived the following equation for the velocity required to transport sediment:

$$V = \sqrt{\frac{8B}{f} g (s-1) D_g} = \frac{1.486}{n} R^{1/6} \sqrt{B (s-1) D_g} \quad \ldots \ldots \ldots 22$$

in which s is the specific gravity of the particle, D_g is the diameter of the particle, B is a dimensionless constant with a value of about 0.04 to start motion of granular particles and of about 0.8 for adequate self-cleaning of sewers and drains, and the other terms are as defined before.

Camp's formula indicates that the velocity required to transport material in sewers is only slightly dependent on conduit shape and depth of flow (as reflected by the hydraulic radius to the one-sixth power) and primarily dependent on the particle size and specific weight. Consider a friction factor, f, of 0.025 for a pipe flowing full. Then, from Equation 22, the size of sand (specific gravity of 2.65), which will be transported effectively ($B=0.8$) at 2 fps (0.61 mps) velocity is 0.000295 ft or 0.09 mm, and the diameter of organic material (specific gravity taken arbitrarily at 1.2) is 0.00242 ft or 0.74 mm. Material nearly 20 times these sizes, if not sticky, barely will be moved along the sewer invert ($B=0.04$).

Grit chambers frequently are designed to permit the settling of sand particles 0.2 mm or larger in size. If B is taken at 0.04 for impending scour and f is assumed to be 0.025, the velocity for impending scour is, from Equation 22, approximately 0.69 fps (0.26 m/sec).

Velocities commonly used in the design of grit chambers range from 0.5 to about 1.0 fps (0.15 to 0.30 m/sec). The critical size of organic matter with a specific gravity of 1.2 is approximately 8 times as large as the sand size, or about 1.6 mm.

However, if the specific gravity of organics is taken as 1.01, which may be closer to the actual figure, the theoretical velocity required to transport a 0.00242-ft or 0.74-mm diam organic particle becomes 0.45 fps (0.137 m/sec) or conversely, a velocity of 2.0 fps (0.61 m/sec) will be sufficient to transport a 0.0491-ft or 15.0-mm diam organic particle. The low velocities actually required to transport organics may explain why many sewers laid at extremely flat grades do not cause excessive trouble due to the deposition of these materials.

2. Velocity for Equal Cleansing

Since the flow in a sewer varies with time, so must the depth of flow and mean velocity. If the velocity in a sewer is adequate for self-cleansing (transport of organics and grit up to a certain size) at one particular rate of flow, it does not follow that it will be adequate at lower rates. In the design of a sanitary sewer, therefore, an attempt should be made to obtain adequate scouring velocities at the average or at least at the maximum flow at the beginning of the design period. It is, of course, also necessary to size the sewer to have adequate capacity for the peak rate of flow at the end of the design period.

It may not be possible to have scouring velocities in a storm sewer at minimum flows since that flow approaches zero. Nonetheless, the storm sewer should be self-cleansing during moderate storms when it flows at considerably less than design capacity.

It rarely is possible to design combined sewers with adequate self-cleansing velocities at minimum dry-weather flow if the capacity of the sewer also must be adequate for the stormwater runoff. Hence, combined sewers often are subject to deposition during dry weather and are dependent on frequent rainfall for flushing.

Fair et al. (22) extended Camp's work to develop, for flows at less than full depth, the slope required to transport the same size particles that would be transported by the sewer flowing full at a given velocity. Useful relationships are:

$$\frac{S}{S_f} = \frac{R_f}{R}, \frac{V_s}{V_f} = \frac{n_f}{n}\left(\frac{R}{R_f}\right)^{1/6}, \frac{Q_s}{Q_f} = \frac{n_f}{n}\frac{A}{A_f}\left(\frac{R}{R_f}\right)^{1/6}$$

in which V_s is the velocity for self-cleansing and Q_s is the discharge at self-cleansing velocity. Figure 25 shows the required slopes and hydraulic elements. This graph indicates that no change in the slope is required

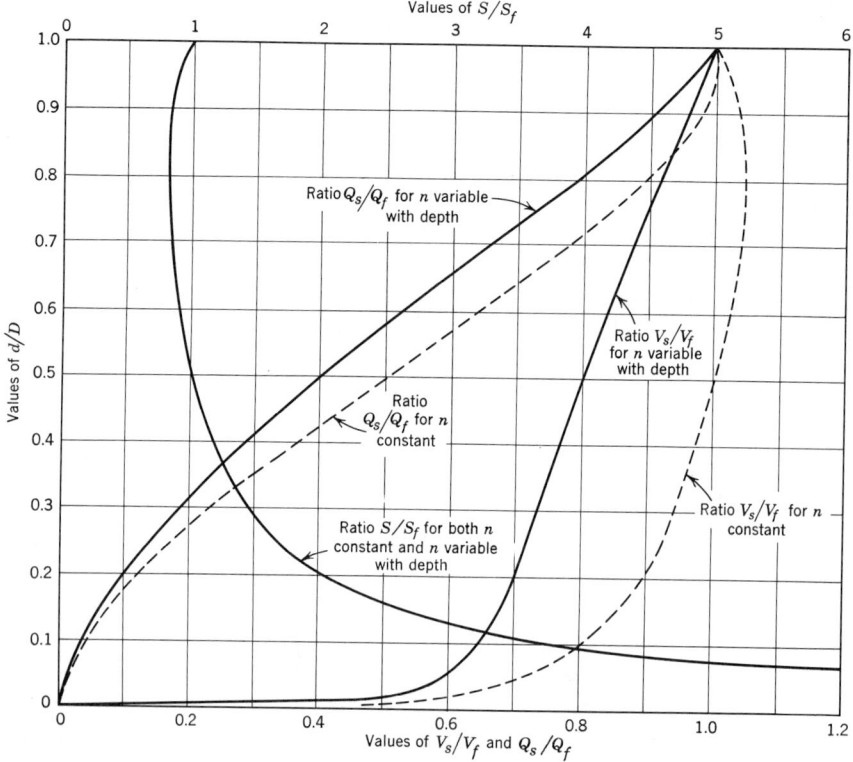

FIGURE 25.—Hydraulic elements of circular sewers that possess equal self-cleansing properties at all depths.

when the sewer flows more than half full, but that the slope must be doubled when the depth of flow drops to 0.2 full and quadrupled at 0.1 depth.

Example 2. A 10-in. (25.4-cm) sewer is to discharge 0.4 cfs (6.8 cu m/min) at a velocity equivalent in self-cleansing action to that of the same size sewer flowing full at 2.0 fps (0.61 mps). Find the velocity of flow and the required slope for $n_f = 0.013$.

(a) From Equation 22 with $B = 0.8$, grit of up to 0.12 mm will be transported effectively in a 10-in. (25.4-cm) sewer flowing full at 2 fps (0.6 m/sec). From Manning's chart, Figure 22, $Q_f = 1.1$ cfs (1.87 cu m/min) and $S_f = 2.5$ units per 1,000 units; hence, $Q_s/Q_f = 0.4/1.1 = 0.364$.

(b) From Figure 25 for constant n and $Q_s/Q_f = 0.364$, $V_s/V_f = 0.97$ and $S/S_f = 1.20$; hence $V_s = 0.97 \times 2.0 = 1.94$ fps (0.59 m/sec) and $S = 1.20 \times 2.5 = 3.0$ per 1,000. Thus, to effectively transport grit of up to 0.12 mm, the sewer must be sloped at 0.003 if n is assumed constant.

(c) From Figure 25 for variable n and $Q_s/Q_f = 0.364$, $V_s/V_f = 0.79$ and $S/S_f = 1.05$; hence, $V_s = 0.79 \times 2.0 = 1.58$ fps (0.48 m/sec) and $S = 1.05 \times 2.5 = 2.63$ per 1,000.

The increase in the coefficient of roughness is seen to be an aid to the cleansing action of flow and to allow lower velocities and slopes.

G. NON-UNIFORM FLOW PROBLEMS

1. Normal and Critical Depths

In calculations involving non-uniform flow, the normal and critical depths need to be known. Even for uniform flow it often is necessary to compute critical depth so that a comparison may be made with the normal depth. Various tables, graphs, and alignment charts have been prepared for many conduit shapes to facilitate these computations and to avoid the necessity of trial-and-error solutions.

For similar geometrical shapes the following relationships may be obtained:

$$A = F_1 l^2 \quad \text{.........................23}$$

$$R = F_2 l \quad \text{..........................24}$$

$$d_m = F_3 l \quad \text{..........................25}$$

in which F_1, F_2, and F_3 are functions of the channel shape which may be constants or expressible in terms of one or more dimensionless parameters and l is a length characteristic of the channel or the flow. With these relationships the Manning formula may be written

$$\frac{Qn}{l^{8/3} S^{1/2}} = 1.49 \, F_1 \, F_2^{2/3} \quad \text{...................26}$$

and the relationship for critical flow $V_c^2/2g = d_{m_c}/2$ becomes

$$\frac{Q_c}{l^{5/2}} = \sqrt{g}\, F_1 F_3^{1/2} \quad\ldots\ldots\ldots\ldots\ldots\ldots\ldots\ldots .27$$

Values of the right hand sides of Equations 26 and 27 may be computed and tabulated or plotted in terms of convenient dimensionless parameters to render unnecessary the development of specific relationships for each combination of channel shape, size, n value, and slope. Tabular values for various channels are given in (3) (5) (9) (16) (17).

Table XVI lists the hydraulic elements of circular pipes including the parameters of Equations 26 and 27. The table is explained further in the discussion of drawdown and backwater curves.

For infrequent normal depth calculations a trial-and-error solution of the Manning equation in the form $Q = \dfrac{1.49}{n} AR^{2/3} S^{1/2}$ may be accomplished for a compatible set of d_n, A, and R values. Critical depth determinations for circular and rectangular conduits may be facilitated by the alignment chart, Figure 26; for other channels a trial-and-error solution of the equation $\dfrac{Q_c^2}{A^2 g} = d_{m_c}$ may be used. This use of Figure 26 is illustrated in the following example:

Example 3. Given a flow of 400 cfs (680 cu m/min) in a rectangular conduit 8 ft (2.54 m) wide by 6 ft (1.83 m) high, find (*a*) the critical depth and (*b*) the corresponding specific head. Do the same for an 8-ft diam sewer, (*c* and *d*). Given a flow of 400 cfs in an 8-ft diam sewer having a slope of 3 ft per 1,000 ft, $n=0.013$, find (*e*) the stage of flow and (*f*) determine whether the flow is stable.

(a) $b_w = 8$ ft and $Q = 400$ cfs; from Figure 26, $d_c/b_w = 0.53$ and $d_c = 0.53 \times 8$ ft $= 4.24$ ft (1.29 m)

(b) $d_c/b_w = 0.53$; from Figure 26, $H_c/b_w = 0.80$ and $H_c = 0.80 \times 8$ ft $= 6.40$ ft (1.95 m)

(c) $D = 96$ in. and $Q = 400$ cfs; from Figure 26, $d_c/D = 0.63$ and $d_c = 0.63 \times 96$ in. $= 60.5$ in. (1.54 m)

(d) $d_c/D = 0.63$; from Figure 26, $H_c/D = 0.91$ and $H_c = 0.91 \times 96$ in. $= 87.4$ in.

(e) From Figure 22, $Q_f = 500$ cfs (850 cu m/min) and $Q/Q_f = 400/500 = 0.80$; from Figure 24, $d/D = 0.76$ and, as $0.76 > 0.63$, the flow is subcritical.

(f) As $[(0.76-0.63)/0.63] \times 100\% = 20.6\% > 10\%$, the flow will be stable.

2. Drawdown and Backwater Curves

Prediction of the shape of drawdown and backwater curves is essential for the design of some sewers. In the case of drawdown curves, it sometimes is possible to make a saving in cost by reducing the size of the conduit or lowering the roof, thus possibly avoiding overhead structures.

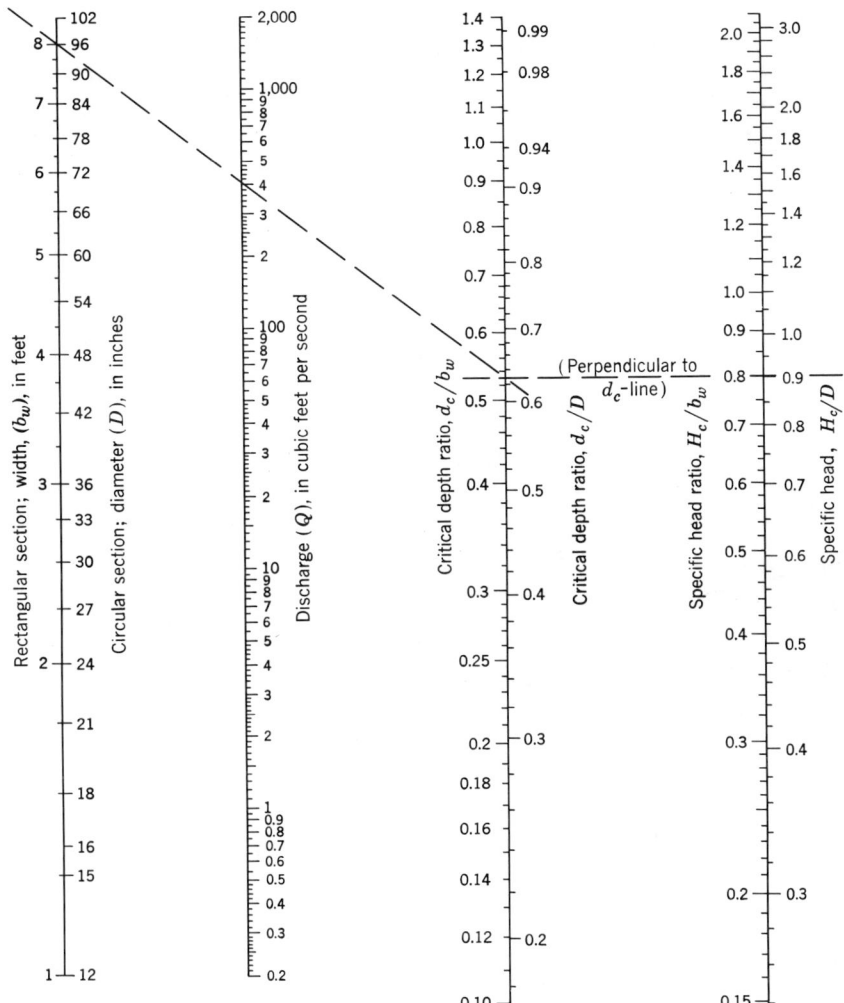

FIGURE 26.—Critical depth of flow and specific head in rectangular and circular conduits.

In the case of backwater curves, it is frequently desirable to know the amount by which the depth is increased at various points along the curve, and the distance upstream to which the backwater curve extends. Various curves are possible depending on the flow conditions, control, and the channel slope. Most frequently encountered in sewer design are the six curves for mild and steep slopes. These are illustrated in Figure 27.

Backwater and drawdown profiles may be developed by stepwise calculations or by integration, analytical (2) (31) (32) (33) or graphical (2) (34).

For stepwise calculations of the reach of conduit between cross sections of given depth

$$\Delta l = \frac{\Delta(d+h_v)}{S_e - S_o} \quad \ldots\ldots\ldots\ldots\ldots\ldots\ldots 28$$

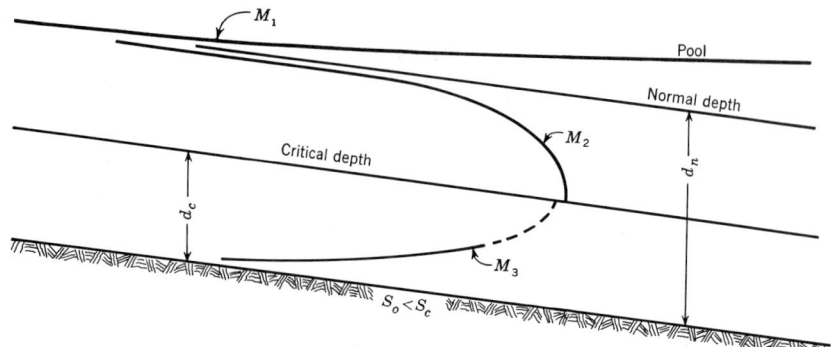

M_1—Backwater from reservoir or from channel of milder slope $(d>d_n)$
M_2—Drawdown, as from change of channel of mild slope to steep slope $(d_n>d>d_c)$
M_3—Flow under gate on mild slope, or upstream profile before hydraulic jump on mild slope $(d<d_c)$

(a) Mild slope $(d_n>d_c)$

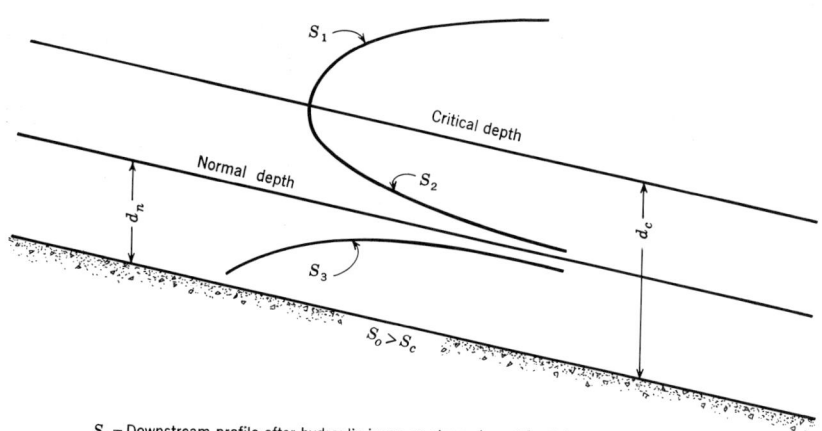

S_1—Downstream profile after hydraulic jump on steep slope $(d<d_c)$
S_2—Drawdown, as from mild to steep slope or steep slope to steeper slope $(d_c>d>d_n)$
S_3—Flow under gate on steep slope, or change from steep slope to less steep slope $(d<d_n)$

(b) Steep slope $(d_n<d_c)$

FIGURE 27.—Open-channel flow classifications.

in which Δl is a portion of the reach of conduit, d is the depth of flow, h_v is the velocity head, S_e is the average slope of the energy grade line, S_o is the slope of the invert, and $\Delta(d+h_v)$ is the change in specific energy between the cross sections. This equation is used in the so-called direct-step method, which is applicable to uniform channels. Another system of analysis, which is applicable to both uniform and non-uniform channels, is the so-called standard-step method. In this procedure,

$$h_f = S_e \Delta l = \tfrac{1}{2}(S_1+S_2)\Delta l \quad \ldots \ldots \ldots \ldots \ldots \ldots \ldots 29$$

in which S_1 and S_2 are the slopes of the energy grade line at the two end sections of the reach under consideration.

The standard-step method has more flexibility and Δl is an independent

HYDRAULICS OF SEWERS

variable, but it requires a trial-and-error solution. For either method the step computation should be carried upstream for subcritical flow and downstream for supercritical flow. Step computations carried in the wrong direction tend inevitably to diverge from the correct profile.

Examples of both step methods for backwater and drawdown profile follow:

Example 4. Direct-Step Method—Calculation of Backwater M-1 Curve.—An 8-ft (2.54-m) diam circular sewer laid at a slope of 1.0 ft per 1,000 ft discharges 80 cfs (136 cu m/min) into a pump well. The wastewater level in the well rises to a maximum of 8 ft (2.54 m) above the invert of the incoming sewer. Trace the profile of the liquid surface in the sewer. Assume a coefficient of roughness of $n=0.013$ for the sewer when flowing full.

(a) An 8-ft (2.54-m) sewer on a grade of 1 ft per 1,000 ft has a capacity of 290 cfs (493 cu m/m), as shown in Figure 22. The value of Q/Q_f, therefore, equals 80/290 or 0.276; and d/D, from Figure 24, equals 0.405 for variable n. Hence, the initial depth of flow is $0.405 \times 8 = 3.24$ ft (1 m), and the terminal depth is 8 ft (2.54 m).

(b) From Figure 26, $d_c/D=0.275$, or the critical depth of flow is 2.2 ft (0.7 m). Since the critical depth is below the normal depth of 3.24 ft (1 m), the flow is subcritical and a smooth backwater curve of the M-1 type will result. Had the initial depth been below critical depth, a hydraulic jump would have had to occur.

(c) The length of reach in which the depth changes by a chosen amount is given by Equation 28. The calculations are shown in Table XV. The length of run in which the transition from a depth of 3.24 to 8.00 ft (1 to 2.54 m) takes place is about 7,800 ft (2,380 m).

In addition to the standard-step method, the following example will illustrate the use of Table XVI, values of hydraulic elements of pipes (n constant).

Example 5. Standard-Step Method—Calculation of Drawdown S-2 Curve.—Flow from a channel enters a 36-in. (94.6-cm) diam circular pipe which is laid on a slope of 0.02. If the discharge is 40 cfs (68 cu m/m), trace the surface profile downstream from the entrance. Assume a coefficient of roughness of $n=0.013$, which is constant with depth of flow.

(a) The normal depth of flow is obtained indirectly from the value $Qn/D^{8/3}S^{1/2}=0.196$. This value is entered in Table XVI to obtain, by interpolation, the corresponding $d_n/D=0.454$, or $d_n=1.36$ ft (0.42 m).

(b) From the value $Q_c/D^{5/2}=40/15.59=2.57$ ft$^{1/2}$/sec and Table XVI, the value $d_c/D=0.687$ is determined by interpolation. Thence, $d_c=2.06$ ft (0.63 m) is calculated.

(c) Since $d_c>d_n$, the flow is supercritical so calculations should start at d_c and proceed downstream. For the conditions stated, critical

TABLE XV.—Calculation of M-1 Backwater Curve for Example 4—Direct-Step Method

d (ft) (1)	d/D (2)	Q/Qf (3)	Qf (cfs) (4)	Vf (fps) (5)	V/Vf (6)	V (fps) (7)	hv (ft) (8)	d + hv (ft) (9)	S (10)	Se (11)	Se − So (12)	Δ(d + hv) (ft) (13)	Δl (ft) (14)
8.00	1.00	1.00	80	1.59	1.00	1.59	0.04	8.04	0.85×10⁻⁴	1.1×10⁻⁴	−8.9×10⁻⁴	−1.98	2,220
6.00	0.75	0.79	101	2.01	0.98	1.97	0.06	6.06	1.35×10⁻⁴	1.7×10⁻⁴	−8.3×10⁻⁴	−0.97	1,170
5.00	0.62	0.60	133	2.64	0.90	2.38	0.09	5.09	2.1×10⁻⁴	3.4×10⁻⁴	−6.6×10⁻⁴	−0.93	1,410
4.00	0.50	0.40	200	3.97	0.80	3.18	0.16	4.16	4.8×10⁻⁴	6.0×10⁻⁴	−4.0×10⁻⁴	−0.36	900
3.60	0.45	0.33	242	4.80	0.75	3.60	0.20	3.80	7.2×10⁻⁴	8.6×10⁻⁴	−1.4×10⁻⁴	−0.29	2,080
3.24	0.40	0.28	290	5.77	0.72	4.16	0.27	3.51	10.0×10⁻⁴				$l = \overline{7{,}780}$

Explanation:
Col. 1: Assumed depths between initial depth of 3.24 ft and terminal depth of 8.00 ft.
Col. 2: Col. 1 ÷ D (diameter of sewer).
Col. 3: From Fig. 24 for d/D in Col. 2.
Col. 4: 80 cfs ÷ Col. 3.
Col. 5: Col. 4 ÷ 50.3 sq ft.
Col. 6: From Fig. 24 for d/D in Col. 2.

Col. 7: Col. 5 × Col. 6.
Col. 8: Velocity head for Col. 7.
Col. 9: Col. 1 + Col. 8.
Col. 10: S from Fig. 22 for $D = 8$ ft, Q_f of Col. 4, and $n = 0.013$.
Col. 11: Arithmetic mean of successive pairs in Col. 10.
Col. 12: Col. 11 − S_o.
Col. 13: Difference between successive amounts in Col. 9.
Col. 14: Col. 13 ÷ Col. 12.

Note 1: To obtain accuracy, the assumed depths should be chosen such that differences between successive values of velocities shown in Col. 7 will be less than 10 per cent; to make Table XV concise, this limit was not met.
Note 2. Ft × 0.3048 = m; cfs × 1.7 = cu m/min.

TABLE XVI.—Tabular Values of Hydraulic Elements of Pipes (n constant)

$\dfrac{d}{D}$	$\dfrac{A}{D^2}$	$\dfrac{Qn}{D^{8/3}S^{1/2}}$	$\dfrac{Qc}{D^{5/2}}$	$\dfrac{d}{D}$	$\dfrac{A}{D^2}$	$\dfrac{Qn}{D^{8/3}S^{1/2}}$	$\dfrac{Qc}{D^{5/2}}$
0.01	0.0013	0.00007	0.0006	0.51	0.4027	0.239	1.4494
0.02	0.0037	0.00031	0.0025	0.52	0.4127	0.247	1.5041
0.03	0.0069	0.00074	0.0055	0.53	0.4227	0.255	1.5598
0.04	0.0105	0.00138	0.0098	0.54	0.4327	0.263	1.6166
0.05	0.0147	0.00222	0.0153	0.55	0.4426	0.271	1.6741
0.06	0.0192	0.00328	0.0220	0.56	0.4526	0.279	1.7328
0.07	0.0242	0.00455	0.0298	0.57	0.4625	0.287	1.7924
0.08	0.0294	0.00604	0.0389	0.58	0.4724	0.295	1.8531
0.09	0.0350	0.00775	0.0491	0.59	0.4822	0.303	1.9147
0.10	0.0409	0.00967	0.0605	0.60	0.4920	0.311	1.9773
0.11	0.0470	0.01181	0.0731	0.61	0.5018	0.319	2.0410
0.12	0.0534	0.01417	0.0868	0.62	0.5115	0.327	2.1058
0.13	0.0600	0.01674	0.1016	0.63	0.5212	0.335	2.1717
0.14	0.0668	0.01952	0.1176	0.64	0.5308	0.343	2.2886
0.15	0.0739	0.0225	0.1347	0.65	0.5404	0.350	2.3068
0.16	0.0811	0.0257	0.1530	0.66	0.5499	0.358	2.3760
0.17	0.0885	0.0291	0.1724	0.67	0.5594	0.366	2.4465
0.18	0.0961	0.0327	0.1928	0.68	0.5687	0.373	2.5182
0.19	0.1039	0.0365	0.2144	0.69	0.5780	0.380	2.5912
0.20	0.1118	0.0406	0.2371	0.70	0.5872	0.388	2.6656
0.21	0.1199	0.0448	0.2609	0.71	0.5964	0.395	2.7416
0.22	0.1281	0.0492	0.2857	0.72	0.6054	0.402	2.8188
0.23	0.1365	0.0537	0.3116	0.73	0.6143	0.409	2.8977
0.24	0.1449	0.0585	0.3386	0.74	0.6231	0.416	2.9783
0.25	0.1535	0.0634	0.3667	0.75	0.6319	0.422	3.0606
0.26	0.1623	0.0686	0.3957	0.76	0.6405	0.429	3.1450
0.27	0.1711	0.0739	0.4259	0.77	0.6489	0.435	3.2314
0.28	0.1800	0.0793	0.4571	0.78	0.6573	0.441	3.3200
0.29	0.1890	0.0849	0.4893	0.79	0.6655	0.447	3.4111
0.30	0.1982	0.0907	0.5226	0.80	0.6736	0.453	3.5051
0.31	0.2074	0.0966	0.5569	0.81	0.6815	0.458	3.6020
0.32	0.2167	0.1027	0.5921	0.82	0.6893	0.463	3.7021
0.33	0.2260	0.1089	0.6284	0.83	0.6969	0.468	3.8062
0.34	0.2355	0.1153	0.6657	0.84	0.7043	0.473	3.9144
0.35	0.2450	0.1218	0.7040	0.85	0.7115	0.477	4.0276
0.36	0.2546	0.1284	0.7433	0.86	0.7186	0.481	4.1466
0.37	0.2642	0.1351	0.7836	0.87	0.7254	0.485	4.2722
0.38	0.2739	0.1420	0.8249	0.88	0.7320	0.488	4.4057
0.39	0.2836	0.1490	0.8672	0.89	0.7384	0.491	4.5486
0.40	0.2934	0.1561	0.9104	0.90	0.7445	0.494	4.7033
0.41	0.3032	0.1633	0.9546	0.91	0.7504	0.496	4.8724
0.42	0.3130	0.1705	0.9997	0.92	0.7560	0.497	5.0602
0.43	0.3229	0.1779	1.0459	0.93	0.7612	0.498	5.2727
0.44	0.3328	0.1854	1.0929	0.94	0.7662	0.498	5.5182
0.45	0.3428	0.1929	1.1410	0.95	0.7707	0.498	5.8119
0.46	0.3527	0.201	1.1900	0.96	0.7749	0.496	6.1785
0.47	0.3627	0.208	1.2400	0.97	0.7785	0.494	6.6695
0.48	0.3727	0.216	1.2908	0.98	0.7817	0.489	7.4063
0.49	0.3827	0.224	1.3427	0.99	0.7841	0.483	8.8261
0.50	0.3927	0.232	1.3956	1.00	0.7854	0.463	—

depth will occur a short distance downstream from the point of entry.

(d) Arbitrary lengths of reach and assumed depths of flow are selected and Equation 29 is used to determine h_f. The friction loss is applied to the previous H and the result is compared with the assumed value. If they are sufficiently close the calculation is continued, if not, a new trial depth of flow is assumed. The calculations are shown in Table XVII. The length of reach in which the transition from a depth of 2.06 to 1.36 ft (6.28 to 4.15 cm) takes place is about 250 ft (70 m).

3. Hydraulic Jump

Hydraulic jump is a local phenomenon whereby flow in a channel abruptly changes from supercritical flow at a relatively shallow depth to subcritical flow at a greater depth. Flow in the channel upstream or downstream from the jump may be either uniform or nonuniform, depending on the characteristics of the channel. The jump is always from a stage less than critical depth to a stage greater than critical depth, there being no design in nature for a reversal of the phenomenon. The upper stage following a jump is called the sequent depth and is always greater than critical depth but less than the depth alternate to the original lower stage depth because of the energy loss in the jump.

The hydraulic jump may be employed as a device for dissipation of excess energy, as where a steep sewer enters a larger sewer at a junction. In stormwater projects, the hydraulic jump may be utilized to use up excess energy and avoid scour of earthen channels. In problems involving the hydraulic jump, the most important consideration is often the location of the jump. It is possible to approximate this location by methods described elsewhere (2) (3) (5) (9).

The basic equations for the solution of a hydraulic jump may be derived from an application of the continuity and momentum principles. For a moderately sloping rectangular channel,

$$\frac{d_2}{d_1} = \tfrac{1}{2}(\sqrt{1+8\mathbf{F}_1^2}-1) \dots \dots \dots \dots \dots \dots 30$$

$$\Delta H = H_1 - H_2 = \frac{(d_2-d_1)^3}{4d_1 d_2} \dots \dots \dots \dots \dots \dots 31$$

in which d_1 and d_2 are the depths before and after the jump, $\mathbf{F}_1$ is the Froude number of the upstream flow, ΔH is the lost head, and H_1 and H_2 are the specific heads of the flow before and after the jump.

For a detailed description of the theory as well as applications for other channel shapes and steeply sloping channels and reader is referred to standard texts such as those by Chow (2), King (3), Rouse (5), Woodward and Posey (9), and Bakhmetoff (35).

4. Culverts

Culvert design commonly is encountered in the design of open ditches or channels to convey storm discharges through a drainage system. In-

HYDRAULICS OF SEWERS 99

TABLE XVII.—Calculation of S–2 Drawdown Curve for Example 5—Standard-Step Method

Q (cfs) (1)	$Qn/D^{8/3}$ (2)	Sta. (3)	Elev Water Surface (ft) (4)	Elev Invert (ft) (5)	d (ft) (6)	d/D (7)	A (sq ft) (8)	V (fps) (9)	$V^2/2g$ (ft) (10)	Assume Elev H (ft) (11)	$Qn/D^{8/3}S^{1/2}$ for d/D (12)	S_e (13)	Avg S_e (14)	H_f (15)	Computed Elev H (16)
40	.02778	0+00	102.06	100.00	2.06	0.687	5.18	7.72	0.93	102.99	0.378	.0054	—	—	102.99
—	—	0+10	101.54	99.80	1.74	0.580	4.25	9.41	1.37	102.91	0.295	.0089	.0072	0.07	102.92
—	—	0+30	100.99	99.40	1.59	0.530	3.80	10.53	1.72	102.71	0.255	.0119	.0104	0.21	102.71
—	—	0+60	100.29	98.80	1.49	0.497	3.50	11.43	2.03	102.32	0.230	.0146	.0133	0.40	102.31
—	—	1+50	98.39	97.00	1.39	0.463	3.20	12.50	2.43	100.82	0.203	.0187	.0167	1.50	100.81
—	—	2+50	96.37	95.00	1.37	0.457	3.14	12.74	2.52	98.89	0.199	.0195	.0191	1.91	98.90

Explanation:
Col. 2: Useful constant for reach calculations.
Col. 3: Stationing arbitrarily established to define backwater curve.
Col. 4: First line of calculation is known elevation at point of control, remaining lines assumed elevations.
Col. 6: Depth of flow, Col. 4 − Col. 5.
Col. 8: Area of channel for depth in Col. 6, from Table XVI, A/D^2 values for d/D in Col. 7.
Col. 9: Q/A, Col. 1 ÷ Col. 8.
Col. 10: $V^2/2g$ from Col. 9.
Col. 11: Col. 4 + Col. 10.
Col. 12: From Table XVI $Qn/D^{8/3}S^{1/2}$ for d/D in Col. 7.
Col. 13: (Col. 2 ÷ Col. 12)^2.
Col. 14: Average friction slope of adjacent stations (no entry on first line of reach).
Col. 15: Length between stations × Col. 14.
Col. 16: First line for actual elevation of energy head at point of control. Remaining lines are values at prior station plus or minus Col. 15. Col. 16 should be in approximate agreement with Col. 11 before proceeding.

adequate attention to design of culverts may result in their being too large and therefore unnecessarily expensive; or too small, causing unanticipated, large headwater depths.

The hydraulic design of any culvert may be affected by its cross-sectional area, shape, entrance geometry, length, slope, construction material, and the depth of ponding at the inlet (headwater) and outlet (tailwater) to the structure. Culvert flows are classified as having either inlet or outlet control (36) (37), that is, whether the discharge capacity is controlled by either the outlet or inlet characteristics. It is possible by hydraulic computations to determine the probable type of flow under which a culvert will operate for a given set of conditions. They may be avoided by computing headwater depths for both inlet and outlet control and then using the higher value to indicate the type of control and the anticipated headwater depth for the culvert under consideration. For a complete discussion of the hydraulic design of culverts the designer is referred to publications available elsewhere (17) (38) (39) (40) (41). Electronic computer programs have been developed for determining the size of several commonly used culvert types (42) (43) (44).

Flow through culverts operating with inlet control often results in only partial use of the barrel size dictated by an acceptable headwater depth. An improvement of the inlet beyond the conventional treatment can increase the hydraulic efficiency and thus, the economy. A research paper on one form of improved inlet is presented by French (45).

Culvert outlet velocities are commonly greater than the flow velocities that would occur in the natural channel exclusive of a culvert, all other factors being equal. If outlet velocities are higher than the downstream channel material can withstand, energy dissipation should be provided to reduce the scour to a tolerable level.

5. Stormwater Inlets

The hydraulic efficiency of stormwater inlets varies with gutter flow, street grade and crown, and with the geometry of the inlet depression. In the design of inlets, freedom from clogging or from interference with traffic often takes precedence over hydraulic considerations.

Inlet capacities usually are determined by consulting rating curves or by solving empirical equations derived from model or prototype tests. Caution should be exercised in extrapolations beyond the sizes tested. The Johns Hopkins report (46), from which this section is abstracted, includes basic equations necessary for determining capacities of various grate, gutter, combination, and deflector inlets in depressed and undepressed settings; also included are rating curves covering a multitude of inlet types and settings. Experimental data for grated inlets have been published by the Highway Research Board (47). These studies considered various styles of grates in undepressed gutters.

If rating curves are not available or lengthy solutions of empirical equations are not warranted, a graphical method is presented (45) which may be used to estimate capacities of single- or closely-spaced inlets. This method is based on the assumption that the velocity throughout the cross section of the gutter flow is uniform. It follows that the ratio of the intercepted width of flow, b_i, to the total width of flow, b_w, geometrically determines the percentage of capture. Another assumption is that there is no carryover across the grate in a single inlet or across the downstream grate in a multiple inlet. The method makes use of diagrams, Figures 28 A, B, C, D, which show flow lines for gutter flows on street grades of various crown slope. Figure 29 is a graph showing the relationship between ratio of intercepted width and intercepted flow.

Example 6. Using Figures 28B and 29, find the capacity of an undepressed combination inlet with grate 3.67 ft (1.12 m) by 1.67 ft (0.51 m) in a street with a 4-percent grade, and 1:18 crown slope. Manning's $n = 0.013$.

(a) Draw outline of inlet on the appropriate flow diagram ($S = 0.04$), Figure 28B.

(b) Determine by inspection the flow for which the outermost flow line intersects the outer downstream corner of the inlet and note its original distance from the curb. In this example, at a flow of 0.8 cfs (1.36 cu m/m), the outer flow line intersects the outer downstream corner of the inlet. The original width of this intercepted flow, b_i, is 2.5 ft (0.75 m)

(c) Also, from Figure 28B, determine by inspection flow widths, b_w, for 1, 2, 4, and 6 cfs. These are listed in Col. 2 of Table XVIII.

(d) Compute interception width ratios, b_i/b_w. In the example, $b_i/b_w =$ 2.5 ÷ values in Col. 2.

(e) From Figure 29 determine interception flow ratios, Q_i/Q_w, for corresponding b_i/b_w ratios. These values are listed in Col. 4.

(f) Determine intercepted flows, Q_i, Col. 5 = Col. 1 × Col. 4.

(g) Data from a rating curve derived by the empirical equations are given in Col. 6. Comparison with Col. 5 shows the simplified method yields results well within the usual engineering accuracy.

This method may be used for closely spaced inlets and for inlets placed perpendicular to each other provided that the downstream grate is not overtopped.

TABLE XVIII.—Capacity of Undepressed Combination Inlet

Gutter Flow, Q_w (cfs) (1)	Flow Width, b_w (ft) (2)	Interception Width Ratio, b_i/b_w (3)	Interception Flow Ratio, Q_i/Q_w (4)	Intercepted Flow, Q_i	
				By Simplified Method (cfs) (5)	By Rating Curve (cfs) (6)
1	2.8	0.89	0.98	0.98	0.99
2	3.5	0.72	0.92	1.84	1.8
4	4.6	0.54	0.79	3.16	3.4
6	5.3	0.47	0.72	4.32	4.5

Note: Cfs × 1.7 = cu m/min; ft × 0.3048 = m.

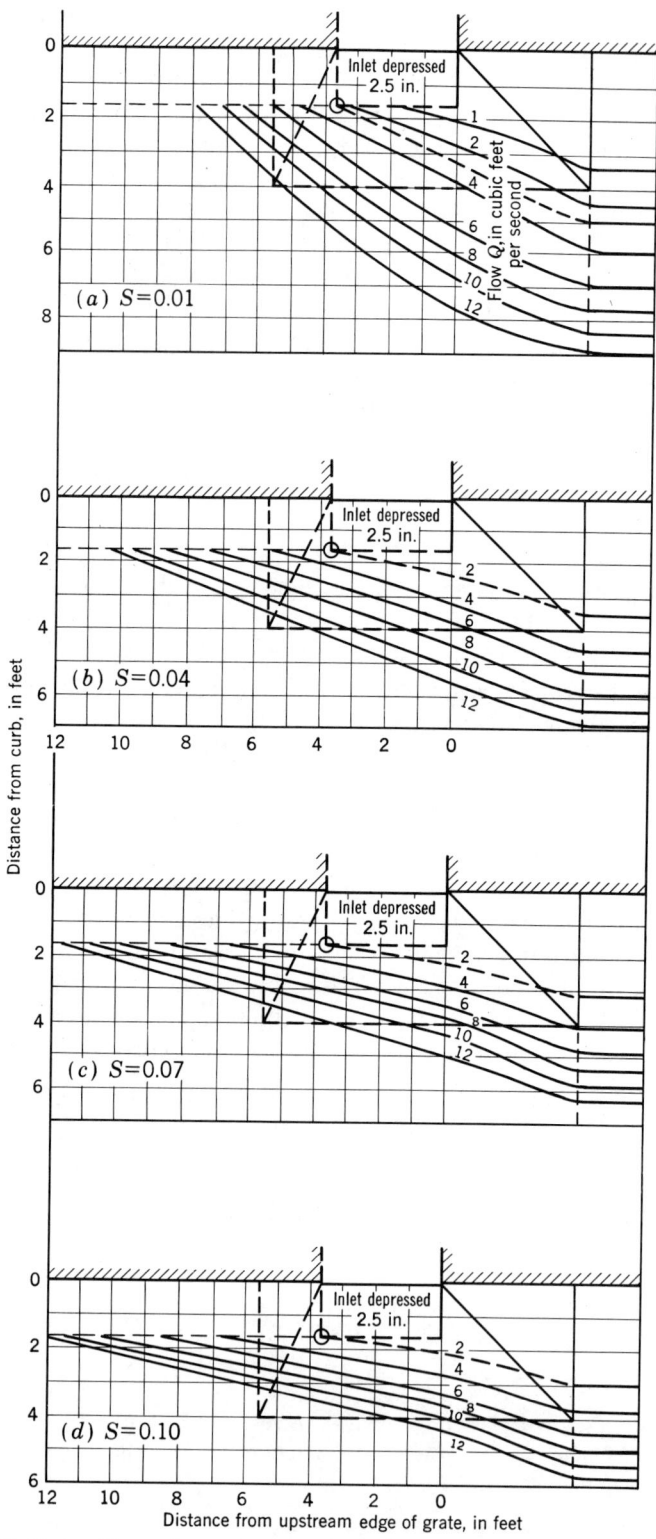

FIGURE 28A.—Flow diagram for simplified method (combination inlets); crown slope, 1:18; $n=0.013$; depression, 2.5 in. (6.4 cm) deep and 4 ft wide (1.2 m).

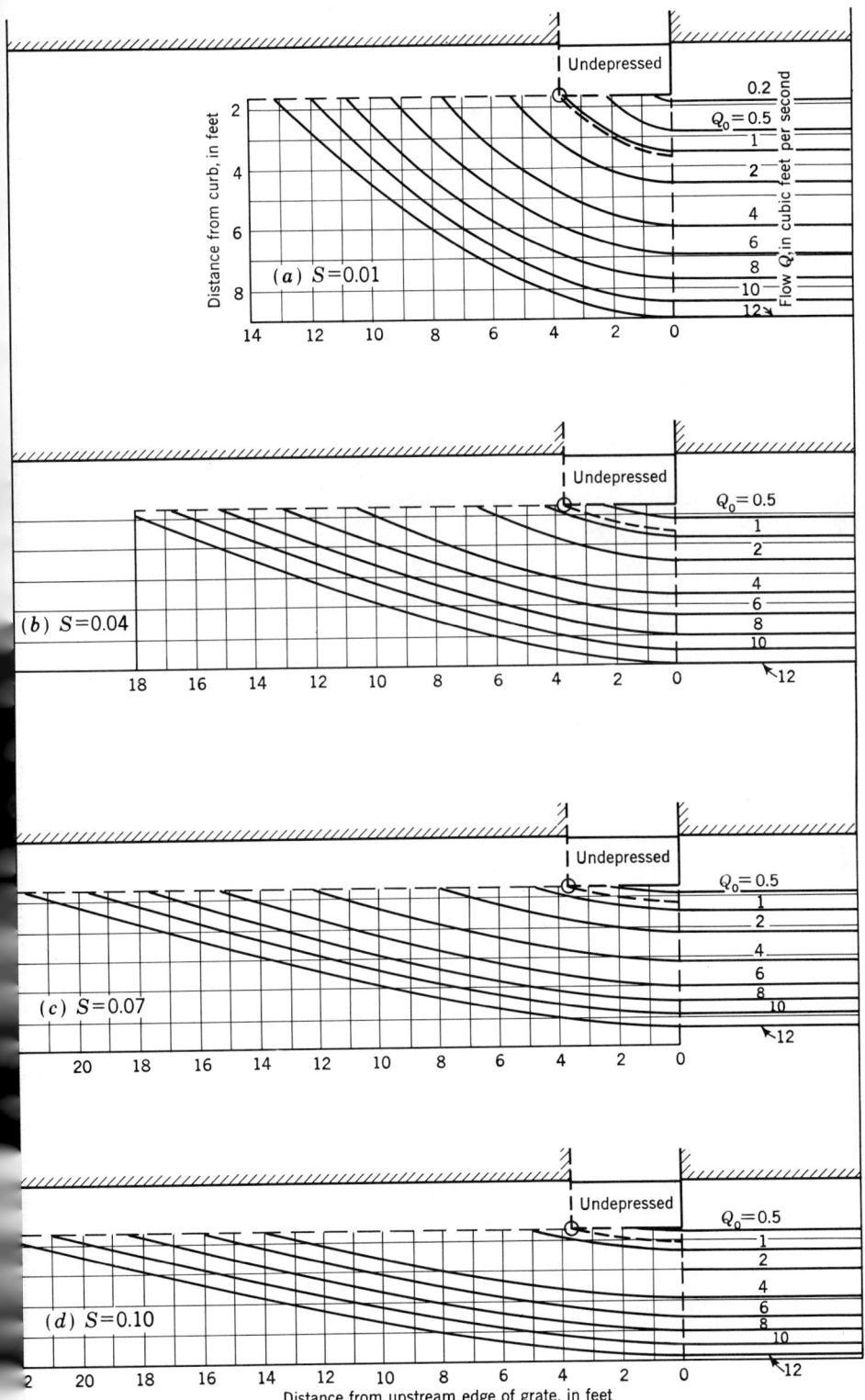

FIGURE 28B.—Flow diagram for simplified method (combination inlets); crown slope, 1:18; $n = 0.013$; undepressed inlet.

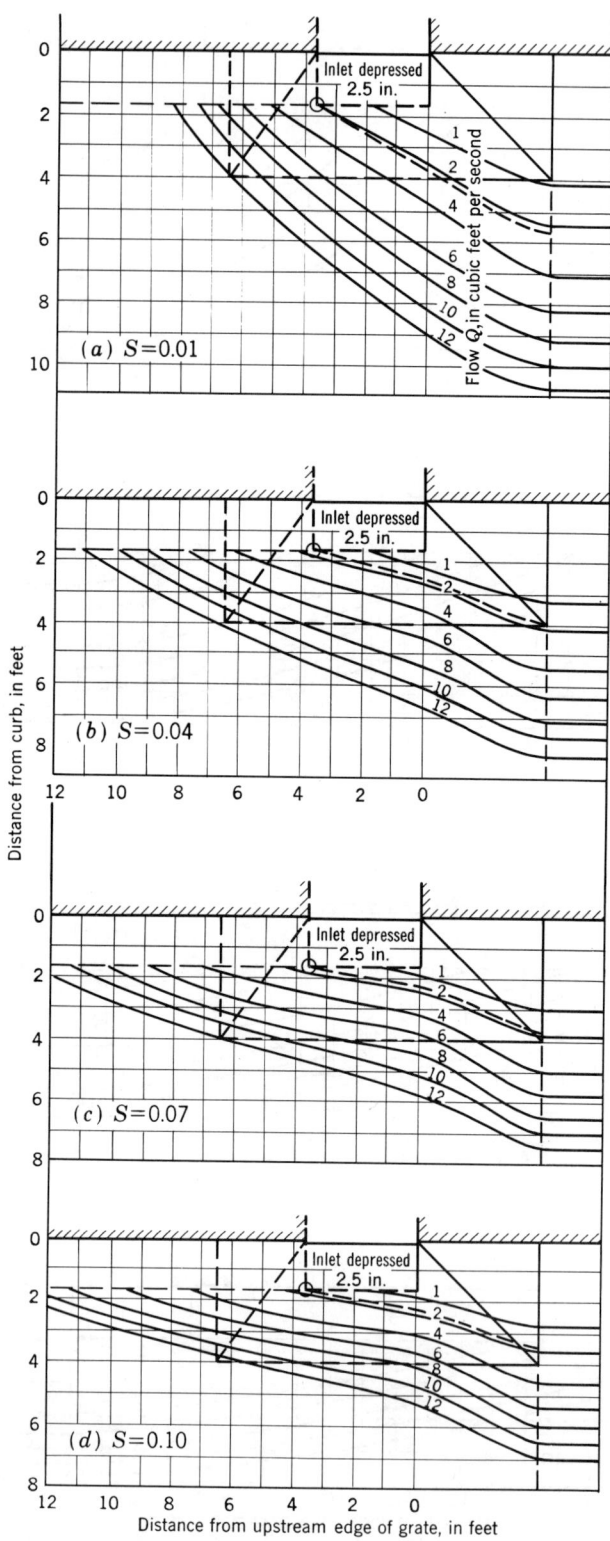

FIGURE 28C.—Flow diagram for simplified method (combination inlets); crown slope, 1:24; $n = 0.013$; depression, 2.5 in. (6.4 cm) deep and 4 ft wide (1.2 m).

HYDRAULICS OF SEWERS 105

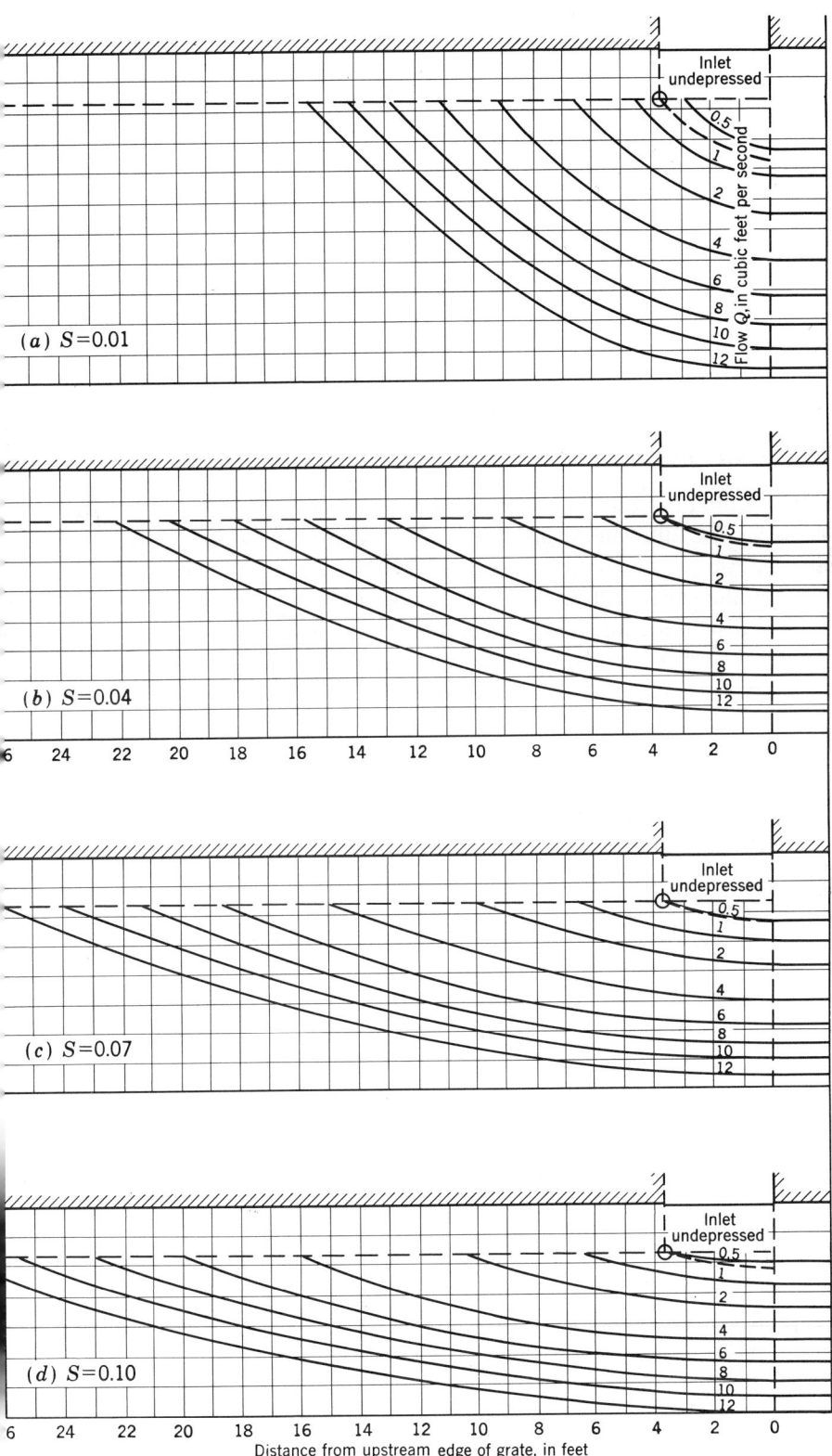

FIGURE 28D.—Flow diagram for simplified method (combination inlets); crown slope, 1:24; $n=0.013$; undepressed inlet.

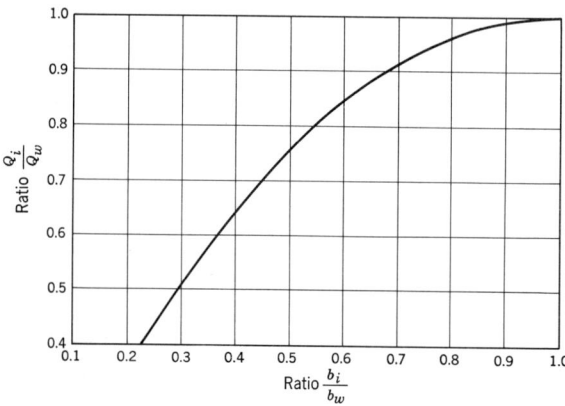

FIGURE 29.—Basic b_i/b_w curve for simplified method.

6. Transitions

In the broadest sense, transitions connect conduits of different characteristics. The differences may be flow area, shape, grade, alignment, conduit material, or a combination of several characteristics. Transitions may be streamlined carefully and gradual or, as a limiting case, sudden.

Head lost in a transition is primarily a function of the velocity head. If the velocity is low, the transition loss cannot be very great. However, even small losses may be significant in flat terrain. And in addition there is always a chance that deposits will impose significant losses. A sewer design is incomplete without a consideration of the necessary transitions and resulting energy losses.

It often is convenient to assume for design purposes that the energy loss and changes in depth, velocity, and invert elevation, if any, occur at the center of the transition. These changes may be distributed throughout the length of the transition in actual detailing. If the designer will carry the energy head, piezometric head (depth in an open channel), and invert as elevations, and work from the energy grade line, he will not need to remember any formulas to determine if the invert should drop and how much. Note that if calculations indicate a rise in invert, it usually is not installed because of the damming effect and consequent deposition of solids. Because of inherent differences in the flow, transitions for pressure conduits will be discussed separately from those for open channels.

(*a*) **Pressure Conduits.**—Transitions in small sewers may be confined within the manhole. But special structures may be required for larger ones. If a sewer is flowing surcharged, the form and friction losses are independent of the invert slope; therefore, for this case, it is usually hydraulically unnecessary and perhaps an unwarranted expense to provide an invert drop at transitions.

If the areas before and after a transition are known, it often is convenient to express the transition loss in terms of the area ratios and either the upstream or downstream velocity. For an expansion,

$$H_L = K \frac{(V_1 - V_2)^2}{2g} = K \frac{V_1^2}{2g} \left[1 - \left(\frac{A_1}{A_2}\right)\right]^2 \quad \ldots \ldots \ldots \ldots 32$$

in which H_L is the energy loss, K is a coefficient equal to about 1.0 for a sudden expansion and about 0.2 for a well-designed transition, and the Subscripts 1 and 2 denote the upstream and downstream sections, respectively. Further information on losses in a gradual pipe expansion will be found in Chapter 7.

For a contraction,

$$H_L = K_1 \frac{V_2^2}{2g} \left(\frac{1}{C_c} - 1\right)^2 \approx K_2 \frac{V_2^2}{2g} \left[1 - \left(\frac{A_2}{A_1}\right)^2\right]^2 \quad \ldots \ldots \ldots 33$$

in which K_1 and K_2 are empirical coefficients, C_c is a coefficient of contraction, and the other terms and subscripts are as used in Equation 32. K_2 is equal to 0.5 for a sudden contraction and about 0.1 for a well designed transition.

These equations may be applied to approximate the energy loss through a manhole for a circular pipe flowing full. If the manhole invert is fully developed, that is, semi-circular on the bottom and vertical on the sides from one-half depth up to the top of the pipe, then for the expansion $A_1/A_2 = 0.88$ (approximately) and for the contraction $A_3/A_2 = 0.88$ (approximately) in which the Subscripts 1, 2, and 3 refer to the upstream pipe, the manhole, and the downstream pipe, respectively. The expansion is sudden; therefore, $K = 1$. The contraction may be somewhat rounded if the downstream pipe has a bell or socket. In this case K may be assumed to be 0.2. The expansion energy loss is 0.014 $V_1^2/2g$ and the contraction energy loss is 0.010 $V_3^2/2g$. Thus, it may be seen that if the invert is developed fully the manhole loss is small, but if the invert is developed only for one-half of the depth, or not at all as is the case in some storm sewers, the losses will approach 1.0 $V_1^2/2g + 0.5\ V_3^2/2g$ if the manhole is large compared to the conduits.

(*b*) **Open Channels.**—The hydraulics of open-channel transitions are further complicated by possible changes in depth. For a more detailed discussion the designer is referred to Chow (2), King (3), Rouse (5), and (48). The Bureau of Public Roads has published design methods for the particular case of transitions at bridge waterways (49).

For an open-channel transition, in subcritical flow, the loss of energy usually is expressed as

$$H_L = K \, \Delta(V^2/2g) \quad \ldots \ldots \ldots \ldots \ldots \ldots 34$$

in which $\Delta(V^2/2g)$ refers to the change in velocity head before and after the transition. For well designed transitions with no wall deflecting more than 12.5 deg, K is approximately 0.1 for channel contractions and 0.2 for channel expansions.

In transitions for supercritical flow, additional factors must be considered. Standing waves of considerable magnitude will be produced in transitions. The height of these waves must be estimated to provide proper channel depth. In addition, in long transitions, air entrainment may cause bulking of the flow with resultant greater depths of the air-water mixture.

Previously mentioned references will aid in the design of many transitions, but for important ones model tests may be necessary.

7. Bends

Reliable information concerning head losses for flow through bends in large conduits is almost entirely lacking. Allowances depend to a considerable extent on the judgment of the individual designer guided by information available on losses in bends in small conduits.

The head losses in bends in excess of that caused by an equivalent length of straight pipe may be expressed by the relation

$$h_b = K_b \frac{V^2}{2g} \quad \ldots \ldots \ldots \ldots \ldots \ldots \ldots \ldots 35$$

in which K_b is the bend coefficient.

The bend coefficient has been found to be a function of (a) the ratio of the radius of curvature of the bend to the width of the conduit, (b) deflection angle of the conduit, (c) geometry of the cross section of flow, and (d) the Reynolds number and relative roughness.

The interrelated effects of these parameters have yet to be fully assessed for either pressure-conduit or open-channel flow, but certain tentative conclusions may be reached. For pipe flow (and probably for open-channel flow), K_b increases with increasing relative roughness and with increasing deflection angle.

Reynolds number effects have not been well delineated and conflicting reports exist concerning the variation of K_b with the radius of curvature. Some investigations indicate there is an optimum value of r_c/D for 90-deg pipe bends; however, this optimum has been ascribed to values of r_c/D ranging from 3 to 10. For gross estimating purposes, a value of K_b of 0.4 probably can be used safely for 90-deg circular pipe bends for which the ratio of center line radius of curvature to pipe diameter exceeds unity, a value of K_b of 0.32 for 45-deg bends, and linear proportioning for other deflection angles.

Shukry (2) (50) has presented information about K_b for open-channel bends. Backwater effects and a description of the open-channel flow pattern have been reported by Muller (51) and Ippen and Drinker (52), respectively.

Information about standard screwed and flanged pipe bends will be found in Reference (53); and head loss coefficients for single- and multi-mitered pipe bends are given in Reference (54). A graphical summary (55) of the results obtained by different investigators and a discussion of some of the factors influencing the loss coefficient (6) are available.

Of perhaps as much significance as the energy loss is the superelevation associated with open-channel bends. For subcritical flow the necessary superelevation between the inside and outside walls is $\frac{b_w}{r_c}\left(\frac{V^2}{g}\right)$. For supercritical flow a pattern of standing cross waves will occur unless special precautions are taken to cancel them. As a first approximation, the maximum height of the waves which occur at both the inside and outside walls is $\dfrac{b_w}{r_c}\left(\dfrac{V^2}{g}\right)$ above and below the surface of the approaching flow rather than above the surface along the inside wall. This phenomenon as well as methods of reducing the waves is discussed in more detail by others (2) (3) (5) (48).

8. Junctions

A junction occurs where one or more branch sewers enter a main sewer. The hydraulic design of a junction is, in effect, the design of two or more transitions, one for each flow path. Allowances should be made for head loss due to curvature of the paths and impact at the converging streams.

In the case of combined sewers, it is necessary to consider the dry-weather conditions as well as those obtaining at maximum flow. In each case the designer is concerned with both velocities and energy losses. There normally will be an excess of energy at junctions during dry-weather periods.

Theoretically, the impact loss at a junction may be computed using the momentum principle as it is applied to the computation of head loss in the hydraulic jump. At a junction of two or more streams, the pressure plus linear momentum below the junction point must equal the sum of the components of the pressure and momentum of each tributary stream. There always will be an energy loss due to impact at the junction.

In practice, however, it is difficult to apply the momentum theory to the design of junctions. To do so it is necessary to make pressure and momentum computations for cross sections just above and below the junction. The pressure components in the direction of flow along the walls and floor of the channels also must be included. These components are a considerable part of the total pressure, but they cannot be determined accurately because of the influence of the impact of the streams and the curvature of the channels.

Tests on storm drain junctions conducted at the University of Missouri (56) for surcharged flow in pipes indicate that if most of the flow passes straight through a junction chamber without appreciable expansion of the flow, the head losses are not appreciably different for chambers with no guiding of the flow and those for which a smooth channel and rounded entrances are provided. However, if most of the flow must be carried around a bend in the junction chamber, a simple deflecting device will keep head losses well below those which would be encountered with no guiding and close to those which would result from a smooth well-rounded junction. Apart from hydraulic considerations, well-rounded junctions are required for separate sanitary and combined systems to prevent deposition.

Chow (2) presents a discussion of open-channel junctions based on the work of Taylor (57) and Bowers (58). As would be expected the effects are considerably more complex because of the variable depths involved.

Recent experimental studies of open-channel junctions for supercritical flow by Behlke and Pritchett (59) indicate that relatively simple means are available to eliminate the cross wave problem. In the proposed method the downstream channel is widened to accommodate the combined flow, and a tapered baffle, which is actually an extension of the main upstream channel wall, is extended downstream in front of the side channel. The effect of the baffle is to maintain the main channel flow essentially uninterrupted except for a gradual spillover into the side channel. The side channel flow is trained in a direction parallel to the downstream channel. With a properly proportioned baffle, which is dependent on the Froude number of the flow, the diagonal wave and pile-up problems are reduced or eliminated.

Because of the difficulty in evaluating hydraulic losses at junctions, general classifications of conditions may be established within which the designer uses judgment tempered by certain rules of thumb. Some of these general classifications are discussed below.

If the available energy at the junction is small, gently sloping transitions may be used. The angle of entry (with respect to the axis of the main sewer) used by some designers is either 30 or 45 deg (if the ratio between the branch diameter and the main diameter is one-half or less); and the junction is sized so that the velocities in the merging streams are approximately equal at maximum flow.

If considerable energy is available in large sewers at a junction, a series of steps may be provided in the branch to produce a cascade, or a junction may be designed to produce a hydraulic jump in the branch to dissipate the excess energy before entering the main sewer. In a combined sewer, a combination of the cascade and jump may be used. In this case the dry-weather flow forms a gentle cascade, whereas under maximum flow conditions the stepped structure is flooded out and a hydraulic jump occurs in the branch. Because of the possible release of foul gases from combined or sanitary sewage in cascades, care should be exercised in their use.

Vertical pipe drops with vitrified clay or cast iron lining are used frequently at junctions for which the main sewer lies well below the branch sewers, particularly if the ratio of the diameter of the branch sewer to the diameter of the main sewer is small. The pipe drops often are constructed with an entrance angle of 30 deg with the main sewer.

At junction points for combining large storm flows, a manhole should be provided, not only for access to each line, but also as a vent for air trapped during rapid filling of the sewer.

9. Vertical Drops and Other Energy Dissipators

Where trunk sewers or storm drains must be constructed through highly developed areas, it is sometimes more economical to build them as tunnels. In such cases the flow to be intercepted from the existing laterals may be dropped vertically through shafts to the tunnel. Some difficult hydraulic problems are encountered with deep vertical drops, the solution of which may be best obtained by means of model studies.

Vertical drops must be designed so as to avoid the entrapment of air in the shaft. Air entrapped in a shaft can result in surges which may reduce seriously the capacity of the intake. Moreover, entrapped air in a shaft may be carried to the tunnel below where it must flow along the tunnel to escape at another shaft, possibly interfering with the flow of sewage in the tunnel or causing damage if its escape is violent. Entrapped air is particularly hazardous if the tunnel is flowing full and under pressure.

The most effective way to minimize air problems is to design the shaft so that the water or sewage will spiral downward with an open vortex in the middle for the full depth of the drop. To accomplish this the flow is directed tangentially into the inlet chamber at the head of the shaft.

The free fall of water or sewage in a deep shaft will result in large impact losses and excessive turbulence at the bottom because the full kinetic energy corresponding with the vertical drop must be absorbed there. If the shaft is designed for vortex flow downward along the walls with an air core in the middle, much of the energy corresponding to the vertical drop will be absorbed by friction on the shaft wall. The velocity at the bottom may be reduced to a value well below that corresponding to the height of the drop.

When excessive turbulence at the bottom of a vertical drop may interfere seriously with the flow in the tunnel, it may be desirable to terminate the drop in a horizontal branch leading downstream to the tunnel in order to quiet the flow before it enters the main flow. If the tunnel flows full, the air core within the spiral may be placed under high vacuum by means of pumps, thus raising the water level at the bottom of the shaft well above the tunnel and creating a considerable volume of water at the bottom of the shaft for dissipation of kinetic energy (60).

Model studies made in connection with the Pittsburgh, Pa., sewerage system led to a design of vertical drops which provides for a circular inlet tank having a diameter four to six times that of the shaft (61).

Another type of tunnel drop incorporates a water cushion or tumbling basin to absorb the impact of the falling jet. A horizontal branch leads downstream from the tumbling basin to the main sewer. The depth required for the tumbling basin has been studied by Dyas (62) and Bakhmeteff (35). Dyas experimented by sending thin glass bottles over falls, observing that they were not broken if the depth of cushion exceeded $h^{1/2}d^{1/3}$, in which h is the height of the fall and d is the depth of the crest. For design this was modified to one-half or one-third of the depth thus found, depending on the material of the floor of the basin.

Chicago practice in the design of vertical drop connections to tumbling basins for combined sewers is to train the dry-weather flow at the top of the shaft into a semicircular cast-iron liner, which projects slightly beyond the face of the concrete so that there is a free-falling trajectory directly into the water cushion below. The trajectory is computed for flows up to 19 percent of maximum without the stream of falling water hitting the opposite concrete wall. Flows which exceed this amount are infrequent and of short duration, thus concrete erosion is not a problem. As the flows increase beyond 50 percent of maximum, the drop shaft begins to flood; and at peak flows the losses in the shaft and lower chamber cause the water level to approach the invert level of the upper-level sewer.

Chutes or steeply inclined sewers with special invert construction to resist erosion of high-velocity flows may be used instead of vertical drops. All types, however, including drops, may cause the release of foul gases as well as maintenance difficulties and, therefore, should be avoided where possible.

In open-channel flow it often is necessary to design energy dissipators. These may be of many forms and types, for which considerable studies of both models and prototypes have been made (2) (5) (12) (17). A particularly complete reference has been published by the Bureau of Reclamation (63).

10. Measuring and Regulating Devices

(*a*) **General.**—Most measuring and regulating devices for sewers utilize the principle of a control section. With such a section a stage-discharge or a pressure differential-discharge relationship will be established. This relationship can be predicted for some devices, while for others it must be found by calibration.

(*b*) **Measuring Devices.**—For pressure-conduit flow the usual measuring device will have a discharge relationship of the form

$$Q = CA\sqrt{2g\Delta h_p} \quad \dots \dots \dots \dots \dots \dots \dots \dots 36$$

in which Q is the discharge, C is a coefficient, A is the cross-sectional area of the device, and Δh_p is the difference in piezometric head upstream and downstream of the device.

The general formula for open-channel measuring devices is

$$Q = Clh^n \dots \dots \dots \dots \dots \dots \dots \dots \dots \dots 37$$

in which Q is the discharge, C is a coefficient, l is a characteristic length, h is a characteristic upstream depth, and n is a characteristic exponent which may range from 1.0 to 2.5.

(*c*) **Leaping Weirs.**—A leaping weir is used as a regulating device to permit the dry-weather flow in a combined sewer to enter an interceptor, but to cause the storm flow to leap across the opening to the interceptor and continue to its point of discharge.

Hydraulic design of leaping weirs for the most part has been based on empirical findings and trial-and-error tests. Adjustable plates have been used so that the openings may be modified. If the opening is constructed of masonry, it is common practice first to undersize it and then enlarge it as necessary, based on actual performance.

A rational approach to the design of leaping weirs has been developed by McClenahan (64). Nevertheless, this discussion states that it is usual practice to construct weirs as movable plates and to provide for adjustments of about 50 percent in each direction. In some cases for combined sewers it may be desirable to design the intercepting structure so that backwater in the collecting sewer will offset the tendency of this sewer to accept more flow than the design value as storm flows become large.

(*d*) **Side-Overflow Weirs.**—A side-overflow weir usually is designed to permit the storm flow in a combined sewer to build up to a certain level and then discharge over the side weir into relief sewers or natural watercourses. (The latter practice is not desirable.)

The earliest tests to determine the discharge capacities of side-overflow weirs were conducted by Engels (65). He found that

$$Q = 3.32 l^{0.83} h^{1.67} \dots\dots\dots\dots\dots\dots\dots 38$$

in which Q is the discharge, l is the length of weir crest, and h is the head on weir at the downstream end. Equation 38 is for use with channels with parallel sides. The discharge is increased by using a contracted channel. In the contracted section the weir may occupy either the contracting side or the straight side opposite the contracting side. For the discharge over a side weir in a contracted channel,

$$Q = 3.32 l^{0.9} h^{1.6} \dots\dots\dots\dots\dots\dots\dots 39$$

The ratio of the channel width to the contracted width does not appear in the formula.

Fair *et al.* (22) point out that the crest length of a side-overflow weir, based on Engels' formulations, might be 2 to 3 or more times greater than that provided on a transverse weir to discharge the same flow at the same head.

Babbitt and Baumann (66), from tests on 18- and 24-in. (45.7- and 60.9-cm) vitrified clay sewer pipe, with cut-in weirs of crest lengths ranging from 16 to 42 in. (40.6 to 106.6 cm) found

$$l = 2.3 \, VD \log \frac{h_1}{h_2} \dots\dots\dots\dots\dots\dots\dots 40$$

in which l is the length of the weir, V is the velocity of approach, and h_1 and h_2 represent the head on the upper and lower ends of the weir. Equation 39 is intended to be used for a weir placed between $D/4$ and $D/2$ from the bottom, depths h_1 and h_2 are determined from the normal depths upstream and downstream of the weir. The weir discharge is taken as the difference in flows in the main sewer. Metcalf and Eddy (67) state that this formula is better adapted for design than Engels' as it takes into account h_1 and h_2, the purpose of the weir being to reduce h_2 to a minimum.

Fair *et al.* (22) analyze the problem of computing the discharge over a side-outlet weir by applying Bernoulli's theorem to points at the upper and lower ends of the weir and include an illustrative example. This method first was presented by Forchheimer (68).

Care should be exercised in the use of formulas for the design of side-overflow weirs because the number of tests have been limited and the results have not been conclusive. Close regulation is difficult to obtain with this type of weir, and frequent maintenance may be required to remove rags, paper, and other debris from the weir crest.

Camp (69), in a theoretical development of the hydraulics of lateral spillway channels, has shown the application of the basic differential equations to a "trial-and-error" solution of the side-overflow weir problem. Chow (2) also presents a discussion of lateral spillways.

(e) Other Sewage Regulating Devices.—Orifices consisting of removable adjustable metal plates sometimes are used to restrict the dry-weather flow from small and medium sewers to an interceptor. The orifice device operates on the principle that the rate of flow discharged through the orifice to the interceptor varies with the square root of the head on the orifice, thus fairly wide variations in depth of flow in the upstream sewer result in only moderate variations in the orifice discharge.

There also are a number of automatic mechanical regulating devices commercially available. They can be designed or adjusted for practically any type of control required, and if properly maintained, generally provide much more accurate control than weirs and orifices.

References

1. "Nomenclature for Hydraulics." Manual of Eng. Practice No. 43, Amer. Soc. Civil Engr., New York (1962).
2. Chow, V. T., "Open Channel Hydraulics." McGraw-Hill Book Co., Inc., New York (1959).
3. King, H. W., and Brater, E. F., "Handbook of Hydraulics." 5th Ed., McGraw-Hill Book Co., Inc., New York (1963).
4. Rouse, H., "Elementary Mechanics of Fluids." John Wiley & Sons., Inc., New York (1946).
5. Rouse, H., "Engineering Hydraulics." John Wiley & Sons, Inc., New York (1950).
6. Rouse, H., "Fluid Mechanics for Hydraulic Engineers." Dover Publications, Inc., New York (1961).
7. Parwakian, J., "Water Hammer Analysis." Prentice-Hall, Inc., New York (1961).
8. Rich, G. R., "Hydraulic Transients." McGraw-Hill Book Co., Inc., New York (1951).
9. Woodward, S. M., and Posey, C. J., "Hydraulics of Steady Flow in Open Channels." John Wiley & Sons, Inc., New York (1941).
10. "Friction Factors in Open Channels." Prog. Rept., Task Force on Friction Factors in Open Channels, *Jour. Hyd. Div., Proc. Amer. Soc. Civil Engr.*, **89**, HY4, (1963).
11. Rouse, H., "Critical Analysis of Open-Channel Resistance." *Jour. Hydr. Div., Proc. Amer. Soc. Civil Engr.* **91**, HY4, (1965).
12. "Hydraulic Design Criteria." U.S. Army Engr., Waterways Exp. Sta., Vicksburg, Miss. (1964).
13. Moody, L. F., "Friction Factors for Pipe Flow." *Trans. Amer. Soc. Mech. Engr.*, **66**, 671 (1944).
14. Rouse, H., "Evaluation of Boundary Roughness." *Proc. 2nd Hyd. Conf.*, State Univ. of Iowa, 112 (1942).
15. "Minimum Velocities for Sewers." Final Rept. Comm. to Study Limiting Velocities of Flow in Sewers, *Jour. Boston Soc. Civil Engr.*, **29**, 286 (1942).
16. Hendrickson, J. G., Jr., "Hydraulics of Culverts." Amer. Concrete Pipe Assn., Chicago (1964).

17. "Design of Small Dams." Bureau of Reclamation, U.S. Govt. Printing Office, Washington, D.C. (1960).
18. Williams, G. S., and Hazen, A., "Hydraulic Tables." 3rd Ed., John Wiley & Sons, Inc., New York (1945).
19. "Design Charts for Open-Channel Flow." Hydraulic Design Series No. 3, Bureau of Pub. Roads, U.S. Govt. Printing Office, Washington, D.C. (1961).
20. "Handbook of Channel Design for Soil and Water Conservation." Soil Conservation Service Publ. No. SCS–TP–61, U.S. Govt. Printing Office, Washington, D.C. (1954).
21. Camp, T. R., "Design of Sewers to Facilitate Flow." *Sew. Works Jour.*, **18**, 3 (1946).
22. Fair, G. M., Geyer, J. C, and Okun, D. A., "Water Supply and Wastewater Removal." John Wiley & Sons, Inc., New York (1966).
23. Wilcox, E. R., "A Comparative Test of the Flow of Water in 8-Inch Concrete and Vitrified Clay Sewer Pipe." *Univ. of Washington, Exp. Sta. Series Bull.* **27**, (1924).
24. Yarnell, D. L., and Woodward, S. M., "The Flow of Water in Drain Tile." Dept. of Agriculture Bull. No. 854, U.S. Govt. Printing Office, Washington, D.C. (1920).
25. Camp, T. R., "Discussion—Determination of Kutter's n for Sewers Partly Filled." *Trans. Amer. Soc. Civil Engr.*, **109**, 240 (1940).
26. Schmidt, O. J., "Measurement of Manning's Coefficient." *Sewage and Industrial Wastes*, **31**, 995 (1959).
27. Bloodgood, D. E., and Bell, J. M., "Manning's Coefficient Calculated from Test Data." *Jour. Water Poll. Control Fed.*, **33**, 176 (1961).
28. Pomeroy, R. D., "Flow Velocities in Small Sewers." *Jour. Water Poll. Control Fed.*, **39**, 1525 (1967).
29. Neale, L. C., and Price, R. E., "Flow Characteristics of PVC Sewer Pipe." *Jour. San. Eng. Div., Proc. Amer. Soc. Civil Engr.*, **90**, SA3, 109 (1964).
30. Shields, A., "Anwendung der Aehnlichkeitsmechanik und der Turbulenz Forschung auf die Geschiebebewegung." Mitteilungen der Preussischen Versuchsanstalt für Wasserbau und Schiffbau, Heft 26, Berlin (1936).
31. Von Seggern, M. E., "Integrating the Equation of Non-Uniform Flow." *Trans. Amer. Soc. Civil Engr.*, **115**, 71 (1950).
32. Keifer, C. J., and Chu, H. H., "Backwater Function by Numerical Integration." *Trans. Amer. Soc. Civil Engr.*, **120**, 429 (1955).
33. Chow, V T., "Integrating the Equation of Gradually Varied Flow." *Jour. Hydr. Div., Proc. Amer. Soc. Civil Engr.*, **81**, HY, 1 (1955).
34. Thomas, H. A., "Hydraulics of Flood Movements in Rivers." Carnegie Inst. of Tech., Pittsburgh (1934).
35. Bakhmeteff, B. A., "Hydraulics of Open Channels." Eng. Soc. Monographs, McGraw-Hill Book Co., Inc., New York (1932).
36. "First Progress Report on Hydraulics of Short Pipes—Hydraulic Characteristics of Commonly Used Pipe Entrances." Natl. Bureau of Standards, Rept. No. 4444, U.S. Govt. Printing Office, Washington, D.C. (1955).
37. "Second Progress Report on Hydraulics of Culverts—Pressure and Resistance Characteristics of a Model Pipe Culvert." Natl. Bureau of Standards, Rept. No. 4911, U.S. Govt. Printing Office, Washington, D.C. (1956).
38. "Hydraulic Charts for the Selection of Highway Culverts." Bureau of Pub. Roads, Hydr. Eng. Circular No. 5, U.S. Govt. Printing Office, Washington, D.C. (1965).
39. "Capacity Charts for the Hydraulic Design of Highway Culverts." Bureau of Pub. Roads, Hydr. Eng. Circular No. 10, U.S. Govt. Printing Office, Washington, D.C. (1965).
40. "Drainage for Areas other than Airfields." U.S. Army, Technical Manual 5–820–4 (1965).

41. "Handbook of Concrete Culvert Pipe Hydraulics." Portland Cement Assn., Chicago (1964).
42. "Electronic Computer Program for Hydraulic Analysis of Circular Culverts." Bureau of Pub. Roads, BPR Program HY–1, U.S. Govt. Printing Office, Washington, D.C. (1965).
43. "Electronic Computer Program for Hydraulic Analysis of Pipe-Arch Culverts." Bureau of Pub. Roads, BPR Program HY–2, U.S. Govt. Printing Office, Washington, D.C. (1965).
44. "Electronic Computer Program for Hydraulic Analysis of Box Culverts." Bureau of Pub. Roads, BPR Program HY–3, U.S. Govt. Printing Office, Washington, D.C. (1965).
45. French, J. L., "Tapered Inlets for Pipe Culverts." *Jour. Hydr. Div., Proc. Amer. Soc. Civil Engr.,* **90,** HY2, 255 (1964).
46. "The Design of Storm Water Inlets." The Johns Hopkins Univ., Rept. of the Storm Drainage Comm. (1956).
47. Cassidy, J. J., "Generalized Hydraulics of Grate Inlets." Highway Res. Bd., *Highway Res. Rec.,* **123,** 36 (1966).
48. "High-Velocity Flow in Open Channels: A Symposium." *Trans. Amer. Soc. Civil Engr.,* **116,** 265 (1951).
49. "Hydraulics of Bridge Waterways." Hydraulic Design Series No. 1, Bureau of Pub. Roads, U.S. Govt. Printing Office, Washington, D.C. (1960).
50. Shukry, A., "Flow around Bends in an Open Flume." *Trans. Amer. Soc. Civil Engr.,* **115,** 751 (1950).
51. Muller, R., "Theoretische Grundlagen der Fluss und Wildbachverbauungen." Eadgenössische Technische Hochschule, Zürich, Mitteilungen der Versuchsanstalt für Wasserbau und Erdbau, No. 4 (1943).
52. Ippen, A. T., and Drinker, P. A., "Boundary Shear Stresses in Curved Trapezoidal Channels." *Jour. Hydr. Div., Proc. Amer. Soc. Civil Engr.,* **88,** HY5, 143 (1962).
53. "Hydraulic Handbook," Fairbanks, Morse & Co., Kansas City, Kans. (1965).
54. "Factors Influencing Flow in Large Conduits." Rept. Task Force on Flow in Large Conduits, *Jour. Hydr. Div., Proc. Amer. Soc. Civil Engr.,* **91,** HY 6, 123 (1965); Discussion, **92,** HY 4, 168 (1966).
55. "Pressure Losses for Fluid Flow in 90-Deg Bends." Natl. Bureau of Standards, *Jour. Res.,* Vol. 21, Paper RP1110 (1938).
56. Sangster, W. M., Wood, H. W., Smerdon, E. T., and Bossy, H. G. "Pressure Changes at Storm Drainage Junctions." *Univ. of Missouri, Eng. Exp. Sta. Bull.* **41** (1958).
57. Taylor, E. H., "Flow Characteristics at Rectangular Open-Channel Junctions." *Trans. Amer. Soc. Civil Engr.,* **109,** 893 (1944).
58. Bowers, C. E., "Studies of Open-Channel Junctions." Part V. Hydr. Model Studies for Whiting Naval Air Sta., Milton, Fla., Univ. of Minnesota, St. Anthony Falls Hydr. Lab. Tech. Paper No. 6, Series B (1950).
59. Behlke, C. E., and Pritchett, H. D., "The Design of Supercritical Flow Channel Junctions." Highway Res. Bd., *Highway Res. Rec.,* **123,** 17 (1966).
60. Kennison, K. R., "Boston Metropolitan Water Supply Extension." *Jour. New England Waterworks Assn.,* **48,** 147 (1934).
61. Laushey, L. M., "Studies Show Pittsburgh How to Drop Sewage 90 Ft. Vertically to Tunnel Interceptors." *Eng. News-Rec.,* **38** (1953).
62. Muckleston, H. B., "The Hydraulic Jump in Open-Channel Flow at High Velocity—Discussion." *Trans. Amer. Soc. Civil Engr.,* **80,** 367 (1916).
63. Peterka, A. J., "Hydraulic Design of Stilling Basins and Energy Dissipators." Bureau of Reclamation, Eng. Monograph No. 25, U.S. Govt. Printing Office, Washington, D.C. (1963).
64. McClenahan. In "Handbook of Applied Hydraulics." 2nd Ed., McGraw-Hill Book Co., Inc., New York (1952).
65. Engels, H. "Handbuch des Wasserbaues." Vol. 1, W. Engelmann, Leipzig (1921).

66. Babbitt, H. E., and Baumann, E. R., "Sewerage and Sewage Treatment." 8th Ed., McGraw-Hill Book Co., Inc., New York (1958).
67. Metcalf, L., and Eddy, H. P., "American Sewerage Practice." Vol. I, 2nd Ed., McGraw-Hill Book Co., Inc., New York (1928).
68. Forchheimer, P., "Hydraulik." B. G. Teubner, Leipzig (1930).
69. Camp, T. R., "Lateral Spillway Channels.—Discussion." *Trans. Amer. Soc. Civil Engr.,* **105,** 636 (1940).

CHAPTER 6. DESIGN OF SEWER SYSTEMS

A. INTRODUCTION

If a sewer is to transport stormwater or wastewater from one location to another, it must be set deep enough to receive these flows. It should be resistant to both corrosion and erosion and its structural strength must be sufficient to carry backfill, impact, and live loads satisfactorily. The size and slope or gradient of a sewer must be adequate for the flow to be carried and be sufficient to avoid deposition of solids. The type of sewer joint must be selected to meet the conditions of use as well as those of the ground. Economy of maintenance, safety to personnel and the public, and public convenience during its life as well as during construction also must be considered.

In the design of a sewer or a system of sewers, decisions must be made regarding location, size, slope, and depth of the sewer and the material of construction. Attention also should be given to streamlining of flow through manholes, junction chambers, and other structures to minimize turbulence and head loss and to prevent deposits. These and other features must be shown on drawings and described in specifications in such detail that the sewer will be built as planned. The preparation of these drawings and specifications is covered in Chapter 10.

Another aim of design is to produce a structure which can be built at the lowest annual cost compatible with its function and durability over the years of its life.

B. ENERGY CONCEPTS OF SEWER SYSTEMS

A sewer system is a means of conveyance which utilizes the energy resulting from the difference in elevation of its upstream and downstream ends. If the available drop is limited, economy requires that energy losses due to free falls, sharp bends, or turbulent junctions be held to a minimum.

Generally the total available energy is utilized to maintain proper flow velocities in the sewers with minimum head loss. However, in hilly terrain it may become necessary to dissipate excess energy by special devices.

Since a sewer system must be self-cleansing there is a minimum limit on velocities and slopes. In storm sewers, moreover, maintenance of self-cleansing velocities is more difficult than in sanitary sewers because of the wide fluctuations in storm flow. A discussion of self-cleansing velocities is given in Chapter 5.

Thus, sewer system design is limited, on one hand, by hydraulic losses which must be kept within the limits of available energy and, on the other, by utilizing available energy to maintain self-cleansing velocities. The wider the variation in rate of flow, the more difficult it becomes to meet both conditions.

Where differences in elevation are insufficient to permit gravity flow, external energy must be added to the system by pumps.

The selection of the number and location of pumping stations generally is determined from economic studies. Normally, the costs of construction, operation, and maintenance of pumping stations are compared with the costs of construction and maintenance of gravity sewers.

In addition to cost comparisons, the consequences of pumping station failure due to mechanical or electrical breakdown must be considered. If pumping station outage would result in pollution of a stream or in any way affect the health and safety of a community, the higher cost of a gravity system may be justified.

C. COMBINED VS. SEPARATE SEWERS

Many existing sewer systems collect both wastewater and stormwater. In such systems, stormwater, up to a design limit, is channeled with sanitary sewage to the treatment plant. When combined flows exceed this value, overflows occur and varying amounts of wastewater along with stormwater are discharged to surface streams. When stormwater is moved with wastewater to the treatment plant, pumping and treatment costs and problems are increased.

A preliminary study by the U.S. Public Health Service (1) shows that stormwater and combined wastewater overflows introduce large quantities of polluting materials into the Nation's waters. Even before the contaminated water reaches streams, it may cause public health hazards, nuisances, and loss of property value by flooding basements and streets. Frequent overflow of sewage with stormwater may preclude the use of downstream waters for recreation. Organic matter from the sanitary wastewater may be sufficient to cause sludge deposits and loss of dissolved oxygen, and otherwise reduce the quality of water.

The only effective way to keep sanitary and storm flows apart is by means of separate systems. The engineer should plan on the basis of combined sewers only after carefully considering the long-range consequences. Combined systems are not considered good modern practice. They may be acceptable under certain special conditions such as short extensions to existing combined systems or when all waste can be treated prior to discharge to a watercourse.

Combined sewers may be less costly to construct than separate sanitary and storm sewers. Indeed, the size of a combined sewer usually is only slightly larger than that required for storm flow alone. Nonetheless, savings due to less costly initial construction must be weighed against a probable increase in future costs for maintenance, separation, pumping, and treatment. Initial savings may very well prove to be false economy.

D. LAYOUT OF SYSTEM

The system layout begins by selecting an outlet, delimiting district and subdistrict boundaries, locating trunk and main sewers, and determining the need for and location of pumping stations.

Preliminary layouts can be made largely from maps, provided they show elevations and other pertinent information. In general the sewers will slope in the same direction as the street or ground surface and will be connected by submain and trunk sewers.

An outlet is located according to the circumstances of the particular project. Thus, a system may discharge to a treatment plant, a pumping station, a watercourse, or to a trunk or intercepting sewer.

District boundaries usually conform to watershed or drainage basin areas. It is desirable to have boundaries follow property lines so that any single lot or property is tributary to a single system. This is particularly important when assessments are made by districts against property served. The boundaries of subdistricts within any assessment district may be fixed on the basis of topography, economy of sewer layout, or other practical considerations.

Trunk and main sewers are located in the valleys. Considerations which may affect the exact location are traffic conditions, type of pavement encountered, availability of rights-of-way, etc.

The need for and location of pumping stations must be considered carefully. Maintenance is always a problem in an isolated pumping station. The possibility of flooding and overflow due to pump failures should be taken into account in site selection.

Due consideration should be given to future needs. A system or part of one should be designed to serve not only the present tributary area, but should be compatible with an overall plan to serve an entire drainage area unless this is impractical for economic or legal reasons.

The most common location of sanitary sewers is at or near the center of the street or alley. A single sewer then serves houses on both sides of the street with approximately the same length of house connection. In an exceptionally wide street it may be more economical to lay a sewer on each side. In such a case, the sewer may be outside the curb, between curb and sidewalk, or under the sidewalk. Normally, sidewalk sewers are used only where other locations are not possible. Cooperation and coordination with other utilities is necessary to eliminate unnecessary interference.

Sometimes a sewer must be located in a right-of-way or easement, as for example, at back property lines to serve parallel rows of houses in residential developments without alleys. Easement agreements for such sewers must provide the right of access for construction, inspection, maintenance, and repair. Despite such provisions, access sometimes becomes difficult or irritating to property owners. Locations in streeets or other public properties are preferred.

Storm sewers commonly are located a short distance back of the curb or in the roadway near the curb where they can easily intercept the flow from stormwater inlets. Inlets are located to protect pedestrian crossings, at low points in the street grade, and at a spacing which will avoid overflowing of gutters. Inlet spacings will in general range from 300 to 600 ft, (90 to 180 m), with closer spacing required for flat terrain and for expressways where vehicular traffic is moving at a relatively high speed.

Sewers as a rule are not located in proximity to public water supplies if avoidable. When such locations cannot be avoided, common practice is to use pressure-type pipe, perhaps also encasing it in concrete. Sewers should not be laid in the same trench with water mains. The designer also should check the appropriate state health department regulations for their requirements as to the separation of water and sewer lines.

Manholes are located at the junctions of sewers and at changes in grade or alignment except in curved sewers as discussed in the next section of this chapter. Street intersections are common location points. Where the manhole does not need to be there for a present or future junction, it is better placed elsewhere in order to avoid the subsurface and surface congestion usually present. A manhole or terminal cleanout is placed at the upper end of a sewer for convenience in flushing and cleaning.

Sanitary sewer manholes should not be located where surface water can drain into them. When this is not possible, a special type of cover should be specified. Manholes not in the pavement, especially across open country, should have the rims set above grade to avoid the inflow of stormwater. This also assists in locating them.

Manhole spacing varies. In general, it is greater now than in earlier years, reflecting improved sewer maintenance methods. Maximum spacings are often in the range of 300 to 400 ft (90 to 120 m). When the size is large enough to permit a man to enter, spacing of 500 ft (150 m) or more may be used.

Tees or wyes should be provided for all house connections. The practice of breaking a hole into the side of a sewer and cementing a branch into it for a house connection should be avoided. Downspouts and footing drains should not be permitted to discharge into the sanitary sewer system.

E. CURVED SEWERS

The design and installation of sewers laid on curves have increased in recent years (2), partly because of an increasing trend toward curved streets in residential areas and partly because improved sewer maintenance equipment lessens the need for straight alignments. The installation of curved sewers may result in economies over straight-run sewers. They have the advantage of eliminating manholes that normally would be needed at each change of direction. The installation of sewers parallel to, or on the centerline of a curved street also makes it easier to avoid other utilities.

The curve usually is made by angling the joints. It is necessary to avoid exceeding the maximum angle at which the joint remains tight. Preferably the angle is limited to three deg. This precludes short radius curves. Curved sewers seldom are laid with radii less than 100 ft (30 m), and many cities consider 200 ft (60 m) as the normal minimum.

Modern sewer maintenance equipment, such as flexible sewer rodding machines, hydraulic cleaning machines, or rubber ball and rubber disc-type cleaning equipment, when used on curves with a radius greater than 100 ft (30 m), has presented no additional problems. However, bucket-type cleaning equipment should not be used since the steel cable may damage the pipe wall on the inside of the curve.

F. TYPE OF CONDUIT

After the layout of the system has been developed in the preliminary studies and the locations and tributary areas of each reach of sewer have been determined, it is necessary to consider the shape and type of conduit to be used.

Sewers in the smaller and medium sizes normally are made of manufactured pipe of circular cross section, available in most areas in diameters up to 12 to 14 ft (4 m). Concrete pipe also is made in elliptical shapes for use where there are either horizontal or vertical space limitations. A variety of shapes are used for poured-in-place sections: rectangular, trapezoidal, horseshoe, circular, and others. For large sewers it often is advisable to prepare alternate designs for precast and cast-in-place sections and to receive alternate bids.

Vitrified clay, concrete, and asbestos-cement pipe are used widely for sanitary sewers in sizes up to about 36 or 42 in. diam (0.9 to 1.1 m). Sanitary sewers of greater size generally are made of concrete; occasionally corrugated metal is used. In storm sewers, where there usually is less concern about corrosive conditions, there is greater freedom in the use of metal or cement-bonded materials. A discussion of the properties of different pipe materials is presented in Chapter 8.

The type of material for small sewers sometimes is dictated by excessive trench loads and superimposed traffic loads. Extra-strength clay, cast iron, concrete, or asbestos-cement pipe may be used to gain added strength where needed and in some cases special bedding or cradles are necessary. Guidance relative to the design of sewers to resist trench loads is given in Chapter 9.

G. VENTILATION

Air normally is drawn down sanitary sewers by the flow of sewage. There also is an exchange of air resulting from the rise and fall of the wastewater. Manholes and building vents are generally adequate to keep sewers sufficiently ventilated. Yet there are places where forced-draft ventilation is necessary:

> (a) To remove for satisfactory disposal any fouled air which may escape from the sewer in densely populated areas.

(b) To maintain a sewer atmosphere not excessively deficient in oxygen. In most sewers the usual flow of air will keep oxygen concentration about 90 percent of normal. In some very long sewers with little natural ventilation or in places where the free flow of air is obstructed, the concentration may drop to less than 50 percent. Excessive reduction of oxygen concentration results in a lethal atmosphere. A deficiency of oxygen also may lead to septic sewage and sulfide production.

(c) To keep walls of structures dry and reduce corrosion where sulfide is present in the sewage. This can be effective for protection of manholes, junction chambers, etc., but only short lengths of sewer can be dried effectively.

(d) To preclude explosive atmospheres. Normally all flammable gases and liquids are excluded by ordinances; and the presence of these materials in sewers is an abnormal or accidental occurrence. This type of event is not predictable and is not common enough to warrant forced ventilation of all the sewers, but is an incidental advantage of ventilation where it is required for other reasons.

Where forced ventilation is required, air is exhausted to a high stack or to some deodorizing process.

The City of Los Angeles, in planning a 300-cfs (510-cu m/min), 8-mile (13-km) long outfall to the Hyperion treatment plant, set up experimental blowers to determine the quantity of air which would be brought by the tributary lines to the head end of the line. It was concluded that an air flow of 17,000 cfm (475 cu m/min) was necessary to avoid at all times a condition of positive pressure upstream from this outfall. Fans now draw this amount of air to the wastewater treatment plant for disposal. An air jumper carries the air across an inverted siphon in the outfall.

These comments on ventilation are specific for sewer layout and design. They should not be interpreted as minimizing in any way the need for gas detection and ventilation before and during maintenance operations to eliminate any possible danger to sewer workers.

H. SEWER DESIGN IN RELATION TO SULFIDE GENERATION

Wastewater out of contact with the air for some period, as in a force main, normally results in sulfide production, principally through the bacterial reduction of sulfates. Depending on the metal content of the waste, a portion of the sulfides may appear as insoluble precipitates of iron, zinc, and copper. However, in wastewater of domestic origin, the amount of insoluble sulfide often does not rise above 1 mg/l. The remainder, therefore, is called dissolved sulfide and is in the form of a mixture of hydrogen sulfide, H_2S, and the hydrosulfide ion, HS^-, in proportions depending on the pH. At pH 5, it is nearly all H_2S; at 9 it is nearly all HS^-.

Hydrogen sulfide, a gas under normal conditions, is soluble in water to the extent of 3,000 to 4,000 mg/l at the temperatures usually prevailing in sewers. Because of this relatively high solubility, it does not come out of sewage as bubbles, yet it tends to pass into the air to some extent from exposed surfaces. At points of high turbulence, and especially if there is a vertical drop, rates of release are far greater than where the stream flows smoothly.

Dissolved sulfide in wastewater reaching a treatment plant increases the chlorine demand and makes aerobic treatment more difficult. These problems are minor compared to the troubles resulting from the hydrogen sulfide that escapes into the air. Even though the amount escaping may be only a small part of that carried by the stream, it causes odor nuisances, blackens lead-base paint, and in sewers, produces lethal atmospheres that have claimed the lives of dozens of sewer workers. Furthermore, on the moist walls of the sewer it is oxidized bacterially to sulfuric acid, where it may destroy the structure if made of concrete, asbestos-cement, or metal.

If wastewater is detained in a force main or siphon only a few minutes, the sulfide produced usually will be inconsequential, but longer detention times may cause severe problems. The possible sulfide buildup in a filled pipe can be estimated roughly (3) (4) as

$$\Delta C_s = 0.0026 t C_{\text{EBOD}} \frac{(1+0.01d)}{d} \quad \ldots \ldots \ldots \ldots \ldots \ldots 1$$

in which ΔC_s is the increase of sulfide concentration in the force main in mg/l; t is detention time in the main in min; C_{EBOD} is effective BOD in mg/l, that is the standard 5-day BOD multiplied by $1.07^{(T-20)}$ where T is the Centigrade temperature of the sewage; and d is the pipe diameter in inches.

Often the sulfide-producing slimes on the pipe wall are not in a state of maximum activity, so the concentrations actually produced may be less than estimated. Nevertheless, the formula serves design purposes well for it shows the maximum effect that must be anticipated if all conditions are favorable for sulfide generation.

Sulfide buildup in a force main usually can be prevented by injection of compressed air into the pump discharge (3). Unfortunately, air cannot be used always because of an irregular profile of the force main. Where it is used, about 1 cfm (0.028 cu m/m) of free air is injected for each inch of pipe diameter (10 l/min for each cm of pipe diameter). The manner of introduction of the air is unimportant; but it must be continuous, not just while the pump is running. Because of this continuous service, the compressor ought to be a slow speed, heavy-duty type.

In some instances where air has been trapped in high places in the main and where the treatment has not been sufficient to keep the main free from sulfide all along its length, damage to the pipe has occurred.

Sulfide also may be produced in partly-filled pipes if the rate of oxygen transfer at the surface of the stream is insufficient to keep up with the demand (5). Temperature, wastewater strength, and area of the wetted pipe wall determine the demand. The rate of oxygen transfer is influenced by the surface width of the wastewater stream, its turbulence, which is related to the slope of the conduit, and less directly, to the velocity. Where BOD is high and temperatures are in the vicinity of 30°C, velocities as high as 3 or 3.5 fps (about 1 m/sec) may be required to prevent sulfide build-up. On the other hand, where the wastewater is weak and temperatures moderate, a velocity of 1 fps (0.3 m/sec) while not a self-cleansing condition, might nevertheless be self-oxidizing. In the northern half of the United States serious sulfide generation in gravity sewers is uncommon since sewage temperatures are moderate and the sewers generally are designed to flow at 2 fps (0.6 m/sec) when half full. However, sulfide production in filled pipes can be a problem in any climate.

Inasmuch as biological activity, concentrated largely in the slime layer, increases with an increase of the wetted perimeter and oxygen uptake is proportional to the surface width of the stream, it follows that deep flow in a pipe is more conducive to sulfide generation than shallow flow. Accordingly, where sulfide generation is a critical consideration, a larger pipe is always better than a smaller one for any given slope and sewage flow.

Prospective sulfide conditions should be considered at the time the system is being designed, and the design should provide, where feasible, for a velocity that is self-oxidizing, as well as self-cleansing. If a self-oxidizing condition can be achieved at the average temperature for the warmest three months of the year and the BOD and flow for the highest six hours of the day, it can be assumed that infrequent extremes producing a small amount of sulfide will be of little consequence.

Other measures to minimize odors and other sulfide nuisances are:

(a) Minimize points of high turbulence within the system.
(b) Design pump station wet wells in a way which precludes the surcharge of tributary lines.
(c) Provide air jumpers across large siphons and around lift stations.
(d) Provide forced-draft ventilation if there is a point where air may be depleted seriously of its oxygen content.

If it is not feasible to construct a system that will maintain self-oxidizing conditions throughout, the engineer should anticipate the results and take the necessary remedial steps. Where substantial concentrations of sulfide will prevail, exposed walls must be protected or be constructed of acid-resistant material. This applies generally to surfaces exposed to the atmosphere above sulfide-containing wastewater. Interestingly, those surfaces below the wasterwater level are not affected adversely because the presence of sulfides does not produce an acid condition on the submerged portions. In addition to this kind of approach, the following control methods have been used:

(a) *Chlorination.*—Good for immediate and complete destruction of sulfide; hence, particularly valuable for treating septic sewage entering a treatment plant. Roughly 10 to 12 parts of chlorine will be required for each part of sulfide.
(b) *Addition of Nitrate.*—Less immediate in its action than chlorine. Principally used to treat overloaded stabilization ponds; some-

times for up-sewer treatment; or for treating polluted natural streams.

(c) Addition of Iron Salts.—Converts dissolved sulfide to insoluble iron sulfide; inexpensive for reducing high concentrations to about one mg/l, but not effective for complete elimination of dissolved sulfide. Useful for treating some industrial wastes where subsequent dilution will reduce the remaining dissolved sulfide to a negligible level.

(d) Addition of Zinc Salts.—Reduces dissolved sulfide to zero. Added at any upstream point, it is effective until all of the zinc has combined with sulfide, giving ZnS. The ratio of Zn:S is 2.04. In some places a zinc solution has been prepared from scrap zinc and waste acids.

(e) Dilution of Wastewater.—Where water is available, this is a particularly good method to take care of the temporary problem of low velocities in the early years of operation of a system when the flow is low in comparison with design capacity. The diluting water has the triple effect of increasing velocity, lowering the BOD, and reducing the sulfide concentration.

(f) Lime Treatment.—Liming the sewer once or twice a week inactivates the sulfide-producing films. A dosage of 5,000 to 10,000 mg/l is used for a period of 30 to 60 min in each treatment.

I. DEPTH OF SEWER

Insofar as feasible, sewers should be at such depth that they can receive the contributed flows by gravity. Deep basements and buildings on land substantially below street level may require individual pumping facilities. Catch basin or street inlet connections of storm sewers present no problems of excessive depth since they normally are shallow. Sufficient cover over sewers must be provided to prevent freezing.

Practice varies in methods for determining the minimum depth for a sanitary sewer. One suggestion is that the top of the sewer should not be less than 3 ft (0.9 m) below the basement floor. Another rule places the invert of the sewer not less than 6 ft (1.8 m) below the top of the house foundation. The latter assumes that it is not necessary for a sanitary sewer to serve basement drains, which results in a considerable saving by reducing sewer depth. It also has the advantage of preventing the connection of exterior basement wall drains to the sanitary sewer. This, however, is acceptable only where few basements have sanitary facilities or if basement sump pumps are utilized.

It is common practice to lay house connections at a slope of 2 percent or $\frac{1}{4}$ in./ft (2 cm/m), with a minimum slope of $\frac{1}{8}$ in./ft (1 cm/m). In some developments in which houses are set well back from the street, the length and slope of the house connection may determine minimum sewer depths. Unjustified costs may preclude the lowering of a whole sewer system to provide service for only a few houses.

Where houses have no basements, sewers may be built at shallower depths; but, in business or commercial districts, it may be necessary to lay sewers as deep as 12 ft (3.6 m) or more to accommodate the underground facilities normally found in such areas.

As sewers usually are laid in public streets, consideration must be given in design to prevent undue interference with other underground structures and utilities. The depth of the sanitary sewer is usually such that it can pass under all other utilities.

The depth of a storm sewer need be sufficient only to receive water from the street inlets and from other tributary drains; it must not interfere with connections to the sanitary sewer. Storm sewers sometimes are laid shallow enough to allow water and other service connections to pass under them. When laying sewers at shallow depths, consideration should be given to live and impact loads since special requirements may be necessary in the selection and installation of pipe.

J. MINIMUM AND MAXIMUM VELOCITIES AND DESIGN DEPTHS OF FLOW

1. Minimum Velocities

Minimum velocities should be sufficient to prevent deposition and prevent or retard sulfide formation. Data on minimum velocities in regard to the latter are presented in Section H of this chapter. Designs for minimum velocities to prevent deposition should be based on the maximum flow anticipated within the initial operating period selected by the designer and the sewer agency. Commonly, slopes are calculated so that when flowing half-full or full, the velocity will be 2.0 fps (0.6 m/sec) for sanitary sewers, or 3.0 fps (0.9 m/sec) for storm sewers. This practice is based on the assumption that these minimum slopes will produce self-cleansing velocities. Inasmuch as this assumption ignores altogether the actual flow which will obtain in a sewer and the corresponding depth, the practice may result in deposits if the actual flow is very much less than the half-pipe capacity. To avoid the danger of deposits in sanitary sewers, it is desirable to estimate beforehand the probable daily maximum flows which will occur in the pipe during the early years of use and to select a slope which will permit self-cleansing velocities at these flows. Sometimes the initial flows may be so low in comparison with design flows that this is not practical. In these cases the owner should be warned that frequent flushing or cleaning will be necessary to remove accumulated sediment.

A minimum full-conduit velocity of 2 fps (0.6 m/sec) is a requirement (or recommendation) of most health departments, while some engineers and authorities have adopted a minimum of 2.5 fps (0.75 m/sec). Table XIX gives values of slopes for full-pipe velocities of 2.0 fps (0.6 m/sec) for various diameter pipes based on the Manning formula with n values of 0.013, 0.014, and 0.015.

It should be apparent that slopes steeper than those shown in Table

DESIGN OF SEWER SYSTEMS

TABLE XIX.—Slopes for Full-Pipe Velocity of 2 fps (0.6 m/sec)

Pipe Diam		Slope (per thousand)		
(in.)	(cm)	n = 0.013	n = 0.014	n = 0.015
6	15	4.9	5.7	6.5
8	20	3.3	3.9	4.5
10	25	2.4	2.9	3.3
12	30	1.9	2.3	2.6
15	38	1.4	1.7	1.9
18	45	1.13	1.32	1.51
21	53	0.92	1.07	1.23
24	60	0.77	0.90	1.03

Note: For a velocity of 2.5 fps (0.75 m/sec) the slopes shown above must be multiplied by 1.56.

XIX may be required to obtain the desired velocity in the upper reaches of lateral sewers for which minimum size pipes will flow only partially full even for the ultimate design flow.

For a given flow and slope, velocity is influenced very little, if at all, by pipe diameter. Calculations from Camp's curve of the hydraulic elements of pipes of circular section show less variation than by use of the classical equations. Table XX shows required slopes for a velocity of 2 fps (0.6 m/sec) for various flows. These values are based on recent studies (6) which indicate the effect of pipe size is negligible. The table is calculated for a full-pipe Hazen-Williams C of 100 (n about 0.015), and is valid for any pipe size if the indicated flow is not less than 0.1 percent nor more than 95 percent of full-pipe capacity.

2. Maximum Velocities

For clear water in hard-surfaced conduits, the limiting velocity is very high. Velocities in excess of 40 fps (12 m/sec) have been found harmless

TABLE XX.—Slopes Necessary for Velocities of 2 fps (0.6 m/sec) *

Q (cfs)	Slope (per thousand)
0.1	9.2
0.2	6.1
0.3	4.8
0.4	4.1
0.6	3.22
0.8	2.73
1.0	2.39
1.5	1.89
2.0	1.59
3.0	1.26
4.0	1.06

* For circular pipes of any size when the indicated flow is not less than 0.1 percent nor more than 95 percent of full-pipe capacity.
Note: Cfs × 1.7 = cu m/min.

to concrete channels. Erosion of inverts may result from much lower velocities when sand or other gritty material is carried.

In the case of sanitary sewers where high velocity flow is continuous and grit erosion is expected to be a problem, the limiting velocity often is taken to be about 10 fps (3 m/sec).

Maximum design velocities in storm sewers, which by their nature occur infrequently even if such conduits are designed for a mean annual storm, may be much greater than those for continuous flow.

3. Design Depth of Flow

Sanitary sewers normally are designed to carry the peak design flow with a depth from one-half to full. Alternatively stated, the full-pipe capacity shall be from 100 to 200 percent of the design peak flow. The smallest sewers usually are designed to flow half full.

The degree of conservatism with which design peak flows are established will affect the selection of design depth of flow. For ventilation reasons, and particularly to avoid sulfide generation, it is undesirable for sanitary sewers to flow full or nearly full.

For storm sewers, the most common design practice is to have the line just full or lightly surcharged at design flow, but some engineers go further and allow the energy grade line to rise to within approximately 1 ft (0.3 m) of the gutter invert.

K. INFILTRATION

In many existing sanitary sewers infiltration is a major cause of hydraulic overloading of both the collection system and treatment plant. To handle this excess flow it may become necessary to construct relief sewers and expand existing treatment facilities. Other expenses also are incurred because of this unwanted flow, such as:

(a) higher pumping costs;
(b) caveins and structural failures in sewers and pavements resulting from soil washing into the sewer; and
(c) higher maintenance costs resulting from soil deposits in sewers, additional root penetration into leaky joints, etc.

Infiltration can enter through faulty joints, cracked pipe, or at manholes. Another source that sometimes is beyond the control of the designer is the house sewer. In many cases these connections are responsible for a major portion of the infiltration in the sanitary sewer. The designer, therefore, should recommend and advise the proper authorities that requirements for house sewers be specified by ordinance.

A more detailed discussion of infiltration and related matters is found in Chapter 3.

1. Infiltration-Exfiltration Test Allowance

Specifications governing sewer design and construction should set forth a maximum infiltration or exfiltration allowance. Infiltration specifications are generally in the range of 250 to 500 gpd/in. diam/mile (0.230 to 0.460 cu m/day/cm diam/km). Tests and allowances should include service connections or stub lines extended from the main or lateral sewer to the curb or property line. However, for lateral sewers with many stubs and wyes, the allowance should be increased 50 gpd/in. diam/mile (0.046 cu m/day/cm diam/km). Specifications should require that all visible or detectable leaks be repaired under any circumstances.

2. Infiltration-Exfiltration Testing

It cannot be over-emphasized that proper engineering inspection and field testing are absolutely necessary if infiltration is to be kept within allowable limits.

A rigorous infiltration or exfiltration test is recommended after completion of construction. Under soil and groundwater conditions that insure a water table above the top of the sewer, an infiltration test is sufficient; where the water table is below the invert, an exfiltration test is required.

Flow can be measured by means of weirs or other devices (see discussion in Chapter 7), but the measurements to be valid must be made with the water table at or near its maximum height to indicate the probable maximum infiltration. It also is important that the pipe walls be saturated thoroughly when the infiltration tests are conducted.

Exfiltration is measured by filling a reach of sewer to provide internal pressure and observing either the drop in head or the quantity of water required to maintain the reach in a full condition. The exfiltration test procedure must specify an elevation head, usually expressed as height of water above the top of the pipe at the upstream manhole.

Exfiltration and infiltration tests are not directly interchangeable. One report (7) covering limited tests suggests that the relation between exfiltration and infiltration varies with the head, and another (8) presents formulas for infiltration based on head and other factors. Type of soil, backfill methods, and pipe embedment materials may cause radical variations in exfiltration rates.

It must be anticipated that tests made shortly after completion of construction of the sewer usually will give results considerably lower than those which would be obtained months or years after construction.

In some areas, air pressure tests are being used in place of exfiltration tests (9) (10).

L. DESIGN FOR VARIOUS CONDITIONS

1. Open Cut.—Inasmuch as the load on a sewer built in open cut is a function of the bedding, trench width, backfill material, and superimposed load on the ground surface, consideration must be given to all these elements. Chapter 9, devoted to loads on pipes, presents details of this phase of design.

2. Tunnel.—A thorough knowledge of tunnel construction methods should be acquired before designing sewers for tunnel placement. This is especially necessary in order to effect economy of construction in this costly type of work. Tunneling methods are covered in detail in Chapter 9.

3. Sewers Built in Rock.—Where sewers are built in rock trenches, special attention should be given to the method of bedding to avoid damage due to contact with rock. Adequate clearances should be provided between the bottom and sides of the sewer and the adjacent rock trench. Granular bedding or a concrete cradle normally is provided. Backfill material placed 12 to 24 in. (30 to 60 cm) over the top of the sewer pipe should be free of rocks or stones to avoid damage to the pipe during the remainder of the backfill operation.

4. Exposed Sewers.—Sometimes sewers have to be built above the ground surface. In these cases, the sewer will be carried on supports or on fill material, or designed as a self-supporting span, as indicated in Chapter 7.

5. Special Foundations.—Knowledge of foundation conditions should be obtained by borings, soundings, or test pits along the route of a sewer prior to design.

Unstable foundations may be encountered in the form of silt, peat bog, quicksand, or other soft or flowing material. If the condition is known before design, a method of meeting the situation must be planned. If encountered during construction, costs usually will be higher than if anticipated beforehand.

Where the conditions are not severe, it may be possible to stabilize the trench bottom by placing a layer of crushed stone below the pipe. The stone must be fine enough, or contain fines, so that settlement will not result from the unstable bottom material flowing into the voids. Concrete or wooden cradles often will suffice to spread the load in wet or moderately soft foundations. In some cases underdrains laid under the sewer will remove water held in the soil and permit dry construction and may eliminate the need for special foundations. Joints that are tight, yet flexible, are particularly important when sewers are installed in areas with unstable soil conditions.

6. Sewers on Steep Slopes.—On steep slopes attention should be given to selecting materials to resist erosion. If drop manholes are used and are deep and the flow is heavy, special construction may be required such as baffles or steps in the manhole, a vitrified-brick manhole bottom, or a water cushion.

On steep slopes it may be necessary to provide anchorage to prevent the pipe from creeping downhill or water flowing along the pipe causing the trench to wash out. Considerable care in backfilling trenches should be exercised to minimize erosion.

7. Underwater Sewers.—See Chapter 7.

8. Force Mains.—Force mains deliver wastewater discharged from a pumping station to its destination which may be a treatment plant, a receiving stream, or a higher point in the sewerage system.

The size of the force main should be determined only after a comparative study of construction and pumping costs for several practical sizes.

Velocities normally fall within a range of from 3 to 5 fps (0.9 to 1.5 m/sec). A velocity of 2 fps (0.6 m/sec) is considered to be sufficient to prevent settling of solids, but velocities of from 2.5 to 3 fps (0.8 to 0.9 m/sec) are required to resuspend those which already have accumulated in the force main. If flushing velocities are attained once or twice per day, excessive deposits are not likely to occur.

High points in a sewage force main often are equipped with vents, usually manually controlled valves or vertical riser pipes with allowance for surges. Automatic air-relief valves on force mains carrying sewage generally do not function well. They are likely to become clogged with grease.

Other aspects of force main design including the problem of sulfide generation are examined in Chapter 7.

M. STORM CHANNEL DESIGN

This section is concerned only with the design of storm sewer channels, open-channel flow in closed conduits having been discussed previously. A primary consideration in the design of storm channels is to determine whether an improved channel is required. Many times it is more economical to utilize an existing natural watercourse and provide floodway reservations within the flood plain. These reservations may be used as parks. In other cases, particularly in areas of high property values, a fully improved and lined channel may be the most economical solution.

If a natural channel is to be improved it normally will follow the general alignment of the existing waterway. Nevertheless, every effort should be made to develop an alignment which will cause minimum interference with the existing man-made or natural obstructions and, where possible, coincide with existing property lines and easements. In most cases an improved channel will eliminate many of the local meanders of an existing waterway, with the straight portions connected by relatively long-radius curves. Existing bridges and culverts frequently are used as alignment controls; however, in cases where the existing structures are inadequate, alignment can be changed and new structures built.

The construction of improved storm channels often does not encompass the entire watershed nor proceed from the downstream end. Moreover, bridges often are built without particular regard to the channel system as a whole. When an improvement is planned for only a small reach of watershed, it is recommended strongly that its probable effect on both the existing and possible future condition in adjacent reaches be considered. Unless this is done, there is no guarantee that the new construction will be compatible with future improvements.

The most frequently used shapes for storm channels are trapezoidal, rectangular, triangular, and shapes approximating trapezoidal or rectangular but with the bottom sloping gently toward the center. The use of triangular section generally is limited to channels of small capacity. For large capacities, the trapezoidal shape is most common. Rectangular channels are confined normally to areas of limited right-of-way width because of the expense of constructing vertical walls.

The right-of-way necessary for a storm channel is dependent on the shape of the channel, the material of which the channel is constructed, and the need for access for maintenance and repair. In addition to the actual width of the channel, an access strip on one or both sides should be provided.

Whether or not a channel should be provided with a non-erodible lining depends mainly on an evaluation of the cost of right-of-way, minimum and maximum velocities, soil stability and characteristics, availability of materials, and aesthetics.

Other items which may influence the size, shape, or type of lining is the practical size of channel for efficient operation of construction equipment, the ability to maintain side slopes, and the availability of access. For example, a trapezoidal channel of considerable length should not have a bottom width less than the width of the construction machinery which will be used to build it. Channels which are grass lined are difficult to mow and maintain if the side slopes are steeper than one vertical to two horizontal.

The channel which has the best hydraulic characteristics is that which most closely approaches a half circle, i.e., a half hexagon for trapezoidal sections, and a half square for rectangular and triangular sections. However, these proportions may not be the most desirable for construction purposes and often are changed to meet local conditions.

The designer also must evaluate the hydraulic and economic aspects of freeboards, bends, transitions, and junctions. Freeboard, the vertical distance from the water surface to the top of the channel, furnishes an allowance for minor surface fluctuations and waves which will occur in a channel. In addition it will provide a factor of safety for extraordinary flows. It should not be used as a catchall for inadequate design. The water surface elevations should be computed for both uniform and non-uniform flow and an estimate should be made for the rise in water surface at channel bends. To these elevations should be added the freeboard allowance which may vary from 5 to 30 percent of the depth of flow, or about 1 ft (0.3 m) for small channels and up to 3 or 4 ft (0.9 or 1.2 m) for relatively large ones.

Bends in storm channels wherever possible should have a relatively long radius. Sharp bends cause greater surface disturbance and waves. In some cases compound transition curves or spirals may be desirable.

Flow in storm channels may at times be supercritical. When it is, a sharp change of direction produces a hydraulic jump in the form of a diagonal wave. Hydraulic jumps and vibrational forces may impose severe stresses on the conduit.

The need to dissipate energy in storm channels is sometimes necessary. This can be accomplished by stilling pools, spillways, drop structures, or impact structures.

The velocities in storm channels on relatively steep slopes often are dangerous to life. Deep channels with steep sides also may be hazards and require fencing.

The bottom of large storm channels commonly will be at an elevation lower than the water table, at least in times of wet weather, thus posing a threat of bottom rupture due to upward pressure. Weep holes should be provided for all large drains where this is a possibility.

It is undesirable to have unrestricted overland flow over the sides of the channel because of the likelihood of erosion. Wherever possible, protective swales should be constructed parallel to the channel to direct the overland flow to inlets which in turn are connected to the channel by short pipes terminating near the flow line. These pipes should be pointed in a downstream direction, and if the channel is unlined, the last section of pipe should be relatively long so that local undercutting will not dislodge it.

Unlined channels constructed in erodible material must be designed to prevent excessive velocities from developing. Permissible velocities for various materials may be established from experience or may be computed by the method of tractive force. In general, older, well-seasoned channels will stand higher velocities than new ones; deeper channels will stand higher velocities than shallower ones of similar materials. Maximum velocities may be increased considerably by the use of a grassed lining (11) (12) (13).

Chow (11) presents detailed discussions of the methods of permissible velocity and tractive force which are beyond the scope of this manual. Limiting velocities suggested by Fortier and Scoby (14) are presented in Table XXI. They were developed for well-seasoned straight channels of small slopes and for depths of flow less than 3 ft (0.9 m).

TABLE XXI.—Maximum Permissible Channel Velocities

Channel Material	Max. Velocity (after aging) (fps)		
	Clear Water	Water with Silts	Water with Sand, Gravel, or Rock Fragments
Fine sand (non-colloidal)	1.50	2.50	1.50
Sandy loam (non-colloidal)	1.75	2.50	2.00
Silt loam (non-colloidal)	2.00	3.00	2.00
Alluvial silt (non-colloidal)	2.00	3.50	2.00
Firm loam	2.50	3.50	2.25
Volcanic ash	2.50	3.50	2.00
Fine gravel	2.50	5.00	3.75
Stiff clay (very colloidal)	3.75	5.00	3.00
Graded, loam to cobble (non-colloidal)	3.75	5.00	5.00
Alluvial silt (colloidal)	3.75	5.00	3.00
Graded, silt to cobbles (colloidal)	4.00	5.50	5.00
Coarse gravel (non-colloidal)	4.00	6.00	6.50
Cobbles and shingles	5.00	5.50	6.50
Shales and hardpans	6.00	6.00	5.00

Note: Ft × 0.3048 = m.

When bridges or culverts are to be incorporated into the storm-channel design, special studies of both hydraulic and structural features will be required. The opposing problems of excessive constriction with resulting backwater problems as against excessive bridge lengths or culvert size and the effects on the channel both upstream and downstream must be examined. If the channel is erodible it will be necessary to study scour in the vicinity of the bridge or culvert, particularly if there is considerable restriction or obstruction. Attention also should be given to the possibility of debris collecting on obstructions in the channel.

Culverts frequently are designed for surcharge on the upstream side. In addition to estimating the effect of backwater in the upstream channel, the height of embankment must be checked to insure that the culvert will operate as designed hydraulically.

N. INTERCEPTING SEWERS

Intercepting sewers generally are constructed lower than all the combined-sewer mains or trunk sewers and connections are provided from the bottom of each to the interceptor with a capacity equal to the flow rate to be intercepted. Regulating devices are used to limit the rate of interception. Flows in excess of this rate are discharged directly to the receiving watercourse.

Interceptors for combined sewerage systems are expected to have sufficient capacity to carry the peak dry-weather flow of wastewater as well as groundwater infiltration. Since the peak dry-weather flow usually will range from about 3 to 4 times the average for an average flow of 1 mgd (3,785 cu m/day) to about 2 times the average for 10 mgd (37,850 cu m/day) or more, the hydraulic capacity for intercepting sewers should be at least 2 to 4 times the average flow for the design year if overflows of raw sewage into watercourses are to be prevented during dry weather.

McKee (15) has shown that a capacity of 80 to 160 times the peak dry-weather flow is required in intercepting sewers in order to prevent overflow of mixed stormwater and raw sewage during rainstorms. It is, therefore, generally uneconomical to design interceptors and treatment works to receive any appreciable part of the stormwater flow.

Stormwater runoff contains the fecal matter of animals and birds, with a corresponding content of coliform bacteria. Occasionally, engineers have provided some capacity in the interceptors and treatment works for surface runoff to control the pollution associated with the so-called first flush of street washings. Palmer (16), however, has shown from actual analyses in Detroit, Mich., that a diminution of pollution in stormwater runoff after the first flush does not occur, and successive peaks of pollution arrive later in the progress of the storm overflow.

McKee has shown for Boston, Mass., that only about 1.8 percent of the year's flow of sanitary wastewater is lost through stormwater overflows during rainstorms if the interceptors are designed with a capacity equal to 3 times the average dry-weather flow, and that this is reduced to 0.6 percent with interceptor capacity equal to 10 times the average dry-weather flow. In other words, the amount of wastewater intercepted will be increased only from 98.2 to 99.4 percent. During rainstorms, however, about 60 percent of the sanitary wastewater is lost in overflows at a rainfall intensity of 0.05 in./hr (0.12 cm/hr), and more than 90 percent when the intensity exceeds 0.3 in./hr (0.76 cm/hr). Thus, the bacterial pollution of the watercourses during rainstorms may be very great. This is particularly important if the watercourses have low velocities and do not purge themselves of pollution rapidly following rainstorms.

McKee's studies of Boston conditions show that overflows may occur 5 to 6 times per month if the interceptors are designed with capacities just equal to the peak dry-weather flow. The studies further show that an increase in interceptor capacity to 10 times the average dry-weather flow will reduce the frequency of overflows to about 3 per month. Similar studies by Palmer for Detroit show that overflows will occur about 85 times per year (7 times per month) if the interceptors are designed with a capacity of 150 percent of the average dry-weather flow and that overflows will occur about 65 times per year (5 to 6 times per month) if the interceptors have capacities of 6 times the dry-weather flow. It may be concluded that no satisfactory reduction in the pollution carried by stormwater overflow can be accomplished by any reasonable increase in interceptor capacity above that required for the peak dry-weather flow.

O. RELIEF SEWERS

An overloaded existing sewer may require relief, with the relief sewer constructed parallel to the existing line or installed to divert flows to alternate outlets. Storm and sanitary sewers alike may require relief. Relief sewers also are called supplementary sewers.

In the design of a relief sewer it must be decided whether (a) the proposed sewer is to share all rates of flow with the existing sewer, or (b) it is to take all flows in excess of some predetermined quantity, or (c) it is to divert a predetermined flow from the upper end of the system. The topography, available outlets, and available head may dictate which alternate is selected. If flows are to be divided according to some ratio, the inlet structure to the relief sewer must be designed to divide the flow. If the relief sewer is to take all flows in excess of a predetermined quantity, the excess flow may be discharged over a side-overflow weir or through a regulator to the relief sewer. If flow is to be diverted in the upper reaches of a system the entire flow at the point of diversion may be sent to the relief sewer or the flow may be divided in a diversion structure.

An examination of flow velocities in the existing and relief sewers may determine the method of relief to use. If self-cleansing velocities cannot be maintained in either or both sewers when a division of flows is used, nuisance conditions may result. If, on the other hand, the relief sewer is designed to take flows in excess of a fixed quantity, the relief sewer itself will stand idle much of the time, and deposits in it may cause nuisance. Judgment is required in deciding which method of relief to use. In some cases it might be better to make the new sewer large enough to carry the total flow and to abandon the old one.

P. ORGANIZATION OF COMPUTATIONS

The first step in the hydraulic design of a sewer system is to prepare a map showing the locations of all the required sewers and from which the area tributary to each point can be measured. Preliminary profiles of the ground surface along each line also are needed. They should show the critical elevations which will establish the sewer grades, such as the basements of low-lying houses and other buildings, existing sewers which must be intercepted, and high-water elevations in the receiving body of water or receiving trunk sewer, etc. Topographic maps are useful at this stage of the design.

Several trial designs may be required to determine which one will properly distribute the available head. Time may be saved if slopes are established tentatively by graphical means on profile paper before selecting final grades and computing the invert elevations at the ends of the sewer structures.

Sewer design computations, being repetitious, may best be done on tabular forms of which two types are in general use. The first, and probably the most common, permits both wastewater quantities and the sewer design to be placed on the same form. In the second type, computations for these two phases are shown on separate forms (17). A major objection to the second type is the number of columns of repeated material.

The form shown in Table XXII for sanitary sewer design is fairly comprehensive and can be adapted to the particular needs of the designer. A typical computation sheet for storm sewers is shown in Table XXIII. It is convenient in using these forms to record the data for the sewer stretches on alternate lines, reserving intervening lines for the data on transition losses and invert drops. An illustrative example, using a slightly modified version of Table XXIII, is found in Table XIII of Chapter 4.

In using forms of these types it is assumed that uniform flow exists in all reaches. The forms, therefore, are not recommended where a detailed analysis of the water surface profile is based on non-uniform flow.

In using Table XXII for sanitary sewer design, supplementary charts, graphs, or tables are required to calculate wastewater flows and hydraulic data. Methods for computing the quantities of sewage flow listed under Col. 8 through 15 were described in Chapter 3. If commercial or industrial wastes are to be included additional columns will be required.

Methods of calculating the hydraulic data contained in Col. 16 through 25 are set forth in Chapter 5. If the value of Col. 26 is positive, an invert drop is indicated; if negative, an invert rise is indicated, but would

TABLE XXII.—Typical Computation Form for Design of Sanitary Sewers

Line No. (1)	Location (2)	Manhole No.		Length (ft) (5)	Area			Max Flow				Min Flow				Slope of Sewer (16)	Diam (in.) (17)	Capacity Full (cfs) (18)	Velocity Full (fps) (19)	Min Velocity (fps) (20)	Max Velocity (fps) (21)	Max Depth (ft) (22)	Max Velocity Head (ft) (23)	Max Energy Head (ft) (24)	Manhole Loss: Transition + Curve + Junction (ft) (25)	Manhole Invert Drop (ft) (26)	Fall in Sewer (ft) (27)	Sewer Invert Elevation		Elevation Ground Surface	
		From (3)	To (4)		Increment (acre) (6)	Total (acre) (7)	Infiltration (mgd) (8)	Sewage (mgd) (9)		Sewage and Infiltration		Infiltration (mgd) (12)	Sewage (mgd) (13)	Sewage and Infiltration														Upper End (28)	Lower End (29)	Upper End (30)	Lower End (31)
									(mgd) (10)	(cfs) (11)			(mgd) (14)	(cfs) (15)																

TABLE XXIII.—Typical Computation Form for Design of Storm Sewers

Line No. (1)	Location (2)	Manhole No.		Length (ft) (5)	Tributary Area		Time of Flow		Impervious Area (%) (10)	Runoff (cfs/acre) (11)	Total Runoff (cfs) (12)	Slope of Sewer (13)	Diam (in.) (14)	Design Flow					Manhole Loss: Transition + Curve + Junction (ft) (21)	Manhole Invert Drop (ft) (22)	Fall in Sewer (ft) (23)	Sewer Invert Elevation		Ground Surface Elevation	
		From (3)	To (4)		Increment (acre) (6)	Total (acre) (7)	To Upper End (min) (8)	In Section (min) (9)						Capacity Full (cfs) (15)	Velocity Full (fps) (16)	Velocity (fps) (17)	Velocity Head (ft) (18)	Depth of Flow (ft) (19)	Total Energy Head (ft) (20)				Upper End (24)	Lower End (25)	Upper End (26)	Lower End (27)

Note: Ft × 0.3048 = m; acre × 0.0405 = ha; mgd × 3,785 = cu m/day; cfs × 1.7 = cu m/min; cfs/acre × 4.2 = cu m/min/ha; in. × 2.54 = cm.

not be installed because of the damming effect on low flows. Accordingly, a value of zero then is recorded in Col. 26.

In the case of combined sewers the computations should be made in the same manner as for storm sewers with the quantity of sewage flow added only when significant.

References

1. "Pollutional Effects of Stormwater and Overflows from Combined Sewer Systems." Pub. Health Service Publ. No. 1246, U. S. Govt. Printing Office, Washington, D.C. (1964).
2. "Feasibility of Curved Alignment for Residential Sanitary Sewers." Fed. Housing Admin. Rept. No. 704, U.S. Govt. Printing Office, Washington, D.C. (1959).
3. Pomeroy, R. D., "Generation and Control of Sulfide in Filled Pipes." *Sewage and Industrial Wastes,* **31,** 1082 (1959).
4. Laughlin, J. E., "Studies in Force Main Aeration." *Jour. San. Eng. Div., Proc. Amer. Soc. Civil Engr.,* **90,** SA6, 13 (1964).
5. Pomeroy, R. D., and Bowlus, F. D., "Progress Report on Sulfide Control Research." *Sew. Works Jour.,* **18,** 597 (1946).
6. Pomeroy, R. D., "Flow Velocities in Small Sewers." *Jour. Water Poll. Control Fed.,* **39,** 1525 (1967).
7. Santry, I. W., Jr., "Infiltration in Sanitary Sewers." *Jour. Water Poll. Control Fed.,* **36,** 1256 (1964).
8. "Municipal Requirements for Sewer Infiltration." *Pub. Works,* **96,** 6, 158 (1965).
9. Brown, K. W., and Caldwell, D. H., "New Techniques for the Detection of Defective Sewers." *Sewage and Industrial Wastes,* **29,** 963 (1957).
10. Ramseier, R. E., and Riek, G. C., "Low Pressure Air Test for Sanitary Sewers." *Jour. San. Eng. Div., Proc. Amer. Soc. Civil Engr.,* **90,** SA2, 1 (1964).
11. Chow, V. T., "Open-Channel Hydraulics." McGraw-Hill Book Co., Inc., New York (1959).
12. "Manual of Location and Design." Ohio State Dept. Highways, Columbus, Ohio (1956).
13. "Surface Drainage of Highways." Highway Res. Bd., Res. Rept. No. 6B, Washington, D.C. (1956).
14. Fortier, S., and Scoby, F. C., "Permissible Channel Velocities." *Tran. Amer. Soc. Civil Engr.,* **89,** 940 (1926).
15. McKee, J. E., "Loss of Sanitary Sewage through Storm Water Overflows." *Jour. Boston Soc. Civil Engr.,* **34,** 55 (1947).
16. Palmer, C. L., "The Pollutional Effects of Storm-Water Overflows from Combined Sewers," *Sewage and Industrial Wastes,* **22,** 154 (1950).
17. Babbitt, H. E., and Baumann, E. R., "Sewerage and Sewage Treatment." p. 106, 8th Ed., John Wiley & Sons, Inc., New York (1958).

CHAPTER 7. APPURTENANCES AND SPECIAL STRUCTURES

A. INTRODUCTION

Certain appurtenances are essential to the proper functioning of every system of sanitary, storm, or combined sewers. They may include manholes, terminal cleanouts, building sewers, service connections, inverted siphons, inlets, catch basins, junction chambers, diversion chambers, tide gates, and other structures or devices of special design.

Many states have established criteria through their regulatory agencies which govern to some extent the design and construction of appurtenances to sanitary sewer systems. Some public bodies exercise similar general control over the design and construction of stormwater facilities including the usual appurtenances thereto. In addition each private and public engineering office usually has its own design standards. It is to be expected, therefore, that many variations will be found in the design of even the simplest structures. The discussion to follow is limited to a general description of each of the various appurtenances, with special emphasis on the features considered essential to good design.

Many of the comments are pertinent to the design of an appurtenance whether it be for a sanitary, combined, or storm sewer. In general, refinements in design to avoid anaerobic decomposition, odor release, or corrosion are particularly applicable to sanitary sewers.

B. MANHOLES

1. Objectives

A manhole should pass at least these major tests. It should:

(*a*) Provide convenient access to the sewer for observations and maintenance operations,
(*b*) Cause a minimum of interference with the hydraulics of the sewer, and
(*c*) Be a durable structure.

2. Manhole Spacing

See Section D of Chapter 6.

3. General Shape and Dimensions

For the purposes of this section, a 3-ft diam manhole may be considered equivalent to 1.0 m, a 4-ft manhole to 1.20 or 1.25 m, and a 6-ft manhole to 1.75 m. The roughly equivalent pipe sizes are:

8 in. = 20 cm	15 in. = 38 cm	30 in. = 75 cm
10 in. = 25 cm	18 in. = 45 cm	36 in. = 90 cm
12 in. = 30 cm	24 in. = 60 cm	48 in. = 120 cm

Most manholes are circular, with the inside dimensions sufficient to perform inspecting and cleaning operations without difficulty. On small sewers a minimum inside diameter of 4 ft has been adopted widely, tapering to a cast-iron frame that provides a minimum clear opening usually specified as 24 in. (60 cm) or sometimes 22 in. It is common practice to maintain the 4-ft diam up to a conical section a short distance below the top, as shown in Figure 30a. Immediately under the frame the diameter should be at least 24 in. and it generally enlarges to 30 in. within 2 ft of the surface. However, some authorities allow a diameter of 24 in. for a distance of 24 in. below the frame. Occasionally the working space of a manhole is a 3-ft by 4-ft vault, but with the usual circular opening at the surface.

It also has become common practice in recent years to use eccentric cones, especially in precast manholes, to provide a vertical side for the steps (Figure 30b). Most often the orientation places the steps over the bench, but some designs place the steps opposite the outlet pipe.

Instead of carrying the 4-ft diam up to a cone near the surface, some engineers prefer a design that maintains this diameter to a height sufficient for a good working space, tapering then to 3 ft as shown in Figure 30c. The cast iron frame in this case has a broad base to rest on the 36-in. diam shaft. Still another design uses a removable flat reinforced concrete slab instead of a cone, as shown in Figure 30d. This is applicable whether the working space is circular or rectangular. The slab must be reinforced suitably to withstand traffic loads.

4. Frame and Cover

The manhole frame and cover normally are made of cast iron. The cover is designed with these objectives:

(a) Adequate strength to support superimposed loads. A typical traffic-weight cover for a 24-in. clear opening weighs about 160 lb (70 kg), and the frame about the same amount or somewhat more. Weights up to 440 lb (200 kg) are specified in some places as the total for cover and frame. Lighter weights may be used where there is no danger that they will be subjected to heavy loads.

(b) A good fit between cover and frame, so there will be no rattling in traffic. It usually is specified that the seat in the frame on which the cover rests and the matching face of the cover be machined to assure a good fit.

(c) Provision for opening. Most commonly this takes the form of a

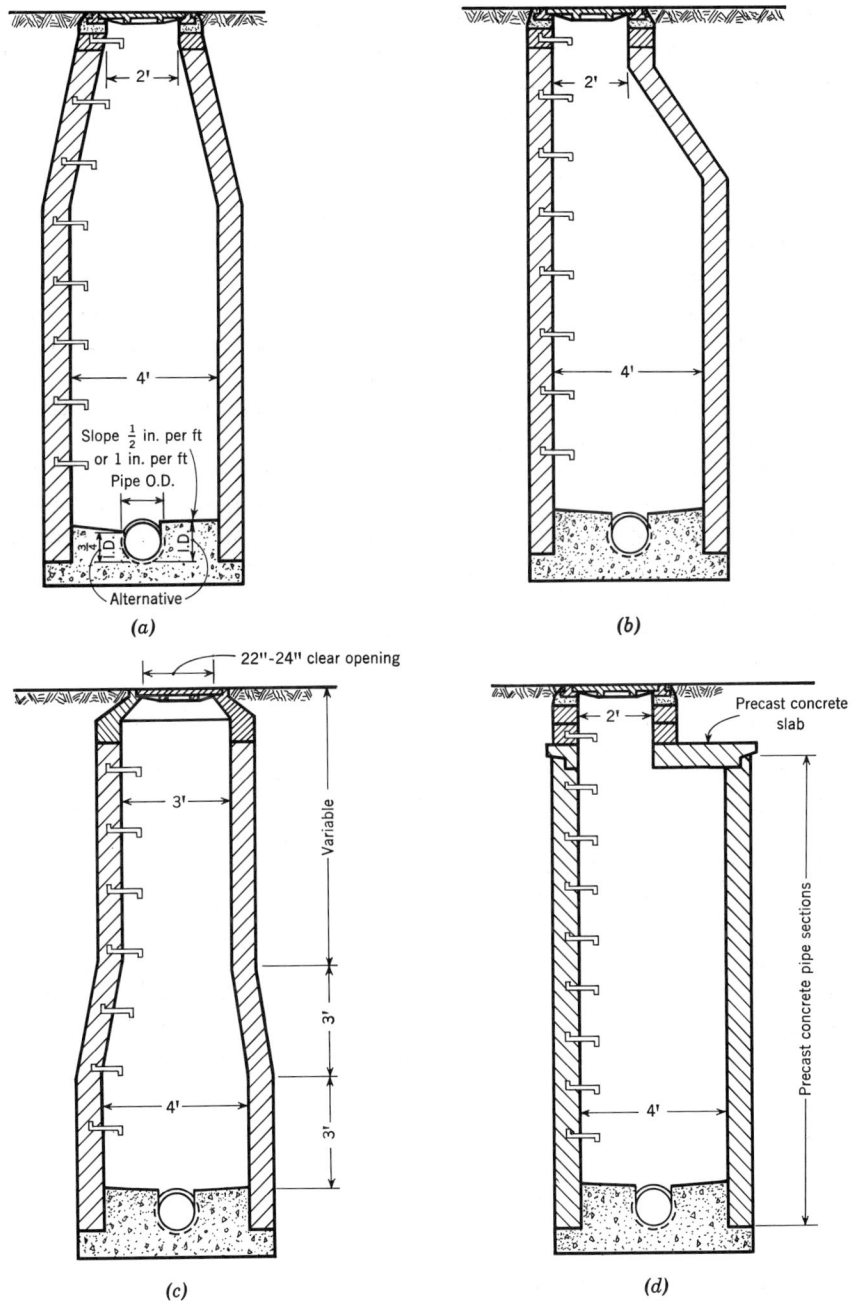

FIGURE 30.—Typical manholes for small sewers. (Ft × 0.3 = m; in. × 2.54 = cm.)

notch at the side where a pick can be used to pry the cover loose, often supplemented by a pick hole a short distance in from the edge.

(d) A reasonably tight closure. Some designs call for several holes in the cover, ostensibly to aid ventilation. At most, a ventilated cover may permit partial drying of the manhole but it cannot affect appreciably conditions in the pipe between manholes. In exceptional conditions where ventilation may be needed to relieve pressure difference or prevent oxygen depletion, special vents should be provided. On sanitary sewers, holes in the covers allow the undesirable entry of stormwater and grit, as well as the mischievous insertion of sticks; and they favor the escape of sewer air, causing odor nuisances. In some covers, the notch is designed so that it does not go all the way through, but it is undercut so that it still serves for prying. To stop earth and gravel from falling into the sewer when the manhole must be opened, and to intercept sticks or earth inserted or falling through the pick hole, a dust pan sometimes is placed under the cover. Usually this is an iron disc, slightly smaller than the manhole opening, resting on lugs at the base of the frame.

Rubber gaskets sometimes are laid on the seat under the cover to maintain tightness in low areas subject to flooding.

(e) Resistance to unauthorized entry. The principal defense against a cover being lifted by children is its weight, but more persistent and competent vandals bent on throwing debris into a manhole are not deterred easily. An emergency measure in an area particularly plagued by mischief of this sort or by illegal disposals is a covering of planks over the channel. Sometimes covers are bolted in place and occasionally a lock is provided.

5. Steps

Manhole steps should be wide enough for a man to place both feet on one step, and should be designed to prevent lateral slippage off the step. Types that have been used most commonly are made of cast iron or shaped from ¾- or 1-in. (1.9- to 2.5-cm) galvanized steel or wrought iron.

Both metals corrode in the moist atmosphere prevailing in most manholes but under normal conditions corrosion is not rapid. If the sewage contains much hydrogen sulfide and especially if there also is much turbulence, the lower steps will fail in a few years. Cast aluminum steps are being used in some places where there is little or no hydrogen sulfide; and it is reported that they are in good condition after service periods up to 15 yr. Steps formed of ⅝-in. (1.6-cm) AISI Type 304 stainless steel rods now are used in a few places. They should have a long life under most conditions.

Steps generally are spaced at intervals of 12 to 16 in. (30 to 50 cm). However, in some areas steps are omitted entirely.

6. Channel and Bench

A channel of good hydraulic properties is an important objective that frequently is not realized because of careless construction. The channel should be, insofar as possible, a smooth continuation of the pipe. In fact, the pipe sometimes is laid through the manhole and the top half removed. If this is done merely by breaking out the upper half, it may be difficult to complete a satisfactory channel; a sawed pipe is better. The completed channel should be U-shaped. Many engineers specify a channel coming up as high as the center line of the pipe on small sizes; others require that the height be three-fourths of the diameter or the full diameter. For sizes 15 in. and larger, the required channel height rarely is less than three-fourths of the diameter. It is shown in Chapter 5 that the expansion and contraction of the stream from pipe to U-shaped channel to pipe with the pipe running full should be less than three percent of the velocity head if the channel is the full depth of the pipe. It will be much greater with a half-depth channel. Close attention is required to secure good U-shaped channels.

The bench should provide good footing for a workman and a place where minor tools and equipment can be laid. It must have enough slope to drain, but not too much. A slope of 8 percent or 1 in./ft (8 cm/m) is common, but 0.5 in./ft (4 cm/m) is specified in some areas to provide a safer footing.

It was formerly ordinary practice to allow an arbitrary drop of 0.1 ft in the invert across the manhole, or a slope of 0.025 (2.5 cm/m), regardless of the slope of the adjacent pipeline. If the channel is constructed properly, this drop is unnecessary. The drop is, in fact, objectionable for it causes excessive turbulence just where it is least desirable and sacrifices head that might better be used toward the attainment of good slopes along the entire sewer.

7. Manholes on Large Sewers

With large sewers the operations and methods of maintenance are not the same as with small ones, and manhole designs are altered accordingly. Sometimes a platform is provided at one side, or sometimes over the sewer, or the manhole is tangent to the side of the sewer with no bench. Quite often the manhole is simply a vertical shaft, usually 3 ft in diameter, over the center of the sewer. A block of reinforced concrete is cast around the pipe, so designed as to form an adequate foundation to support the shaft plus transmitted traffic loads. Such manholes usually are built without steps. When entry is necessary a workman is lowered in a chair hoist or cage.

It is standard practice to use manholes of this type on sewers over 3 or 4 ft in diameter. Where a sewer is larger than 2 ft and the small-sewer type of manhole is used, the diameter of the manhole should be increased sufficiently to maintain an adequate width of bench, preferably a foot or more on each side. In sewers that a man cannot straddle, maintenance men sometimes lay planks to bridge the channel. Hence there must be adequate and well-formed benches on each side. Sometimes the entering pipe is extended two or three feet into the manhole being then mortared over to form a smooth platform. Figure 31 shows two designs that have been used.

For sewers up to 3 ft in diameter, covers sometimes are used having a clear opening of 3 ft to accommodate cleaning equipment equaling the sewer diameter. However, collapsible equipment is the rule for large

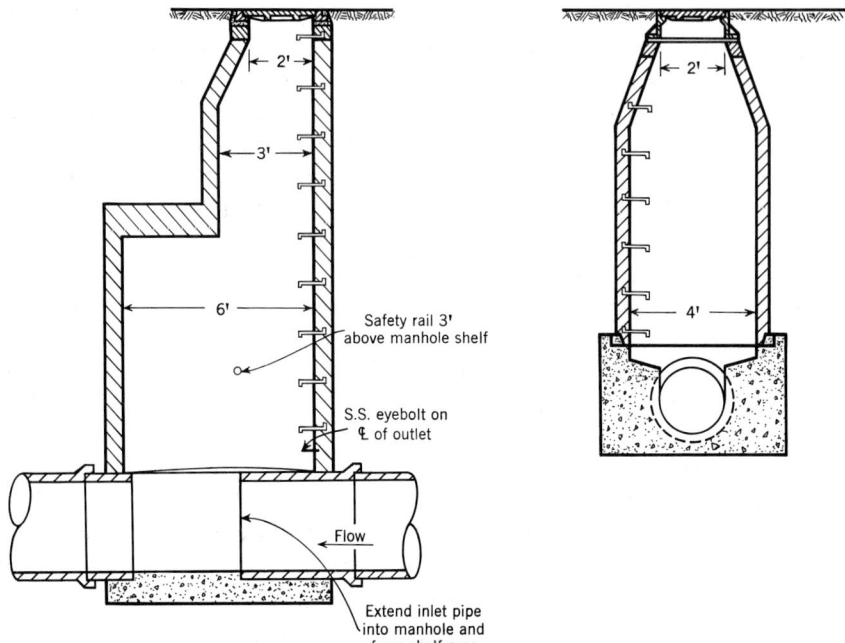

FIGURE 31.—Two manholes for intermediate-sized sewers. (Ft × 0.3 = m.)

sewers. Many authorities do not use openings larger than 2 ft; openings larger than 3 ft are used only in special cases.

8. Shallow Manholes

Irregular surface topography sometimes results in shallow manholes. If the depth to the invert is only 2 or 3 ft, all maintenance operations can be conducted from the surface. A manhole 2 or 3 ft in diameter at the bottom generally suffices in these cases; narrow slots also have been used.

If a manhole of the usual type is constructed where the height from the bench to the surface is in the range of 2 to 4 or 5 ft (0.6 to 1.2 or 1.5 m), with the shaft tapering to a 22-in. clear opening, the structure is one in which a man cannot work effectively. An extra-large cover with a 30- or 36-in. opening will help relieve this situation. A manhole of the type shown in Figure 30d is satisfactory if the head room is 4 ft.

9. Connection Between Manhole and Sewer

Differential settling of the manhole and the sewer sometimes breaks the sewer pipe. A pipe joint just outside the manhole lessens this danger. If the soil conditions are quite unstable, a second joint within 3 ft (0.9 m) of the first may be desirable. To accomplish the purpose the joints must not be rigid.

10. Construction Materials

The materials most commonly used for manhole walls include brick, precast concrete sections, and cast-in-place concrete. Concrete block construction also is used; and on sewers constructed of corrugated steel the manholes usually are made of the same material.

Precast concrete sections are available in various heights. They may be obtained with cast-in-place steps at the desired spacing. Care should be used to secure tight joints, preferably using rubber gaskets. Special transition sections are furnished to reduce the diameter of the manhole at the top as required to accommodate the frame and cover. They are usually eccentric with one side vertical. It is common practice to allow three or four courses of brick just below the rim casting to permit easier future adjustment of the top elevation. Precast bottoms also have been used in some places.

Brick or concrete block walls normally are built 8 in. (20.3 cm) thick at the shallower depths, and may increase to 12 in. (0.3 m) below 8 to 12 ft (2.4 to 3.6 m) from the surface. Joints should be filled completely with cement mortar. The outside walls of brick or block manholes should be parged with cement mortar not less than ½ in. (1.3 cm) thick. In wet areas it is advisable to apply a coating of a bituminous damp-proofing compound.

C. BENDS

Particular care must be used in the construction of curved channels to accommodate bends. Highest workmanship is necessary to produce channels that are smooth, with uniform sections, radius, and slope.

A curve of very short radius causes energy-wasting turbulence. Some authorities recommend for optimum performance that the radius of the centerline be three times the pipe diameter or channel width. Reasonably satisfactory conditions usually can be obtained if the radius is not less than 1.5 times the diameter. If the velocity is supercritical, surface turbulence and energy losses arise even with long-radius bends.

The radius of curvature of a bend within a manhole is maximized if the points of tangency of the outer curve of the channel lie at opposite ends of a manhole diameter, as illustrated for 12- and 18-in. pipes in Figure 32.

Completion of a bend within a manhole is not necessary and becomes impossible as the pipe approaches the size of the manhole. Further, when the size of the sewer is such that the manhole is only a chimney over the sewer, a manhole may be placed over the center of the curve, or on the downstream tangent, or perhaps two may be used, one upstream and one downstream. Figure 32 shows some of the possible designs for manholes on bends.

Bends of less than 90 deg can of course be accommodated more easily. For angles substantially less than 90 deg on sewers larger than 12-in. diam, the manhole may be centered over the pipe.

Bends usually are made by a series of angles in large sewers, and also in curved sewers of small diameter.

It is common to allow extra fall in the channel invert on a bend to

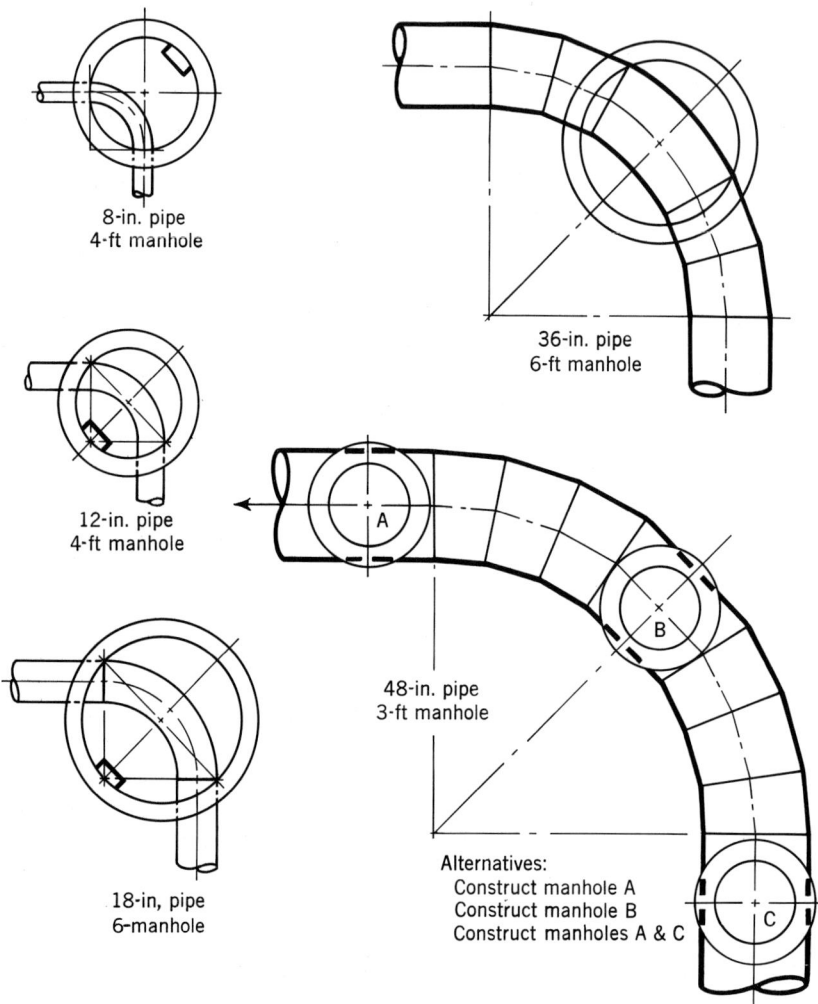

FIGURE 32.—Manhole placement for various types of bends.
(Ft × 0.3 = m; in. × 2.54 = cm.)

compensate for bend energy losses. When this is done the drop should reflect only the expected losses. Although experimental data for large conduits are scarce, it would appear that for a well-made 90-deg bend with a centerline radius of curvature not less than one pipe diameter, the loss in an open channel should not exceed 0.4 of the velocity head. Thus, for a velocity of 3 fps (0.9 m/sec), the loss would probably be not more than 0.06 ft (.02 m). A more complete discussion of energy losses in bends, including those associated with supercritical flow is found in Chapter 5.

D. JUNCTIONS

On small sewers junctions are made in ordinary manholes, the branch line being curved into the main channel. Excessive widening of the main

APPURTENANCES AND SPECIAL STRUCTURES 149

channel at the junction should be avoided. Eddying flows and accumulations of sludge and rags are the result of the poor flow patterns prevailing in many junction manholes. To minimize these objectionable conditions, the invert of the branch line should be higher than the invert of the main channel where the two join. Channels should be constructed in the manhole bottom for all lines.

Large junctions generally are constructed in cast-in-place reinforced concrete chambers entered through manhole shafts. Their hydraulic design is discussed in Chapter 5.

E. DROP MANHOLES

If an incoming sewer is considerably higher than the outgoing, a drop manhole should be used. Figure 33a shows a common type. These structures are not trouble-free. Sticks may bridge the drop pipe, starting a stoppage. Because of such a stoppage or merely because of high flow, sewage may spill out of the end of the pipe, making the manhole dangerous and objectionable as a place for a man to work. Cleaning equipment also may lodge in the drop pipe. Sometimes the drop pipe is made of a larger diameter to minimize stoppages.

Figure 33b shows a design used in some areas which largely overcomes these objections. If a stoppage should occur in the drop pipe, it is removed easily from the surface.

Drop manholes should be used sparingly and, generally, only when it is not economically feasible to steepen the incoming sewer. Some engineers eliminate drops by using vertical curves. In no case should a drop be used for a fall less than 2 ft (0.6 m).

F. TERMINAL CLEANOUT STRUCTURES

Terminal cleanouts sometimes are used at the ends of sewers. Their purpose is to provide a means for inserting cleaning tools, for flushing, or for inserting an inspection light into the sewer.

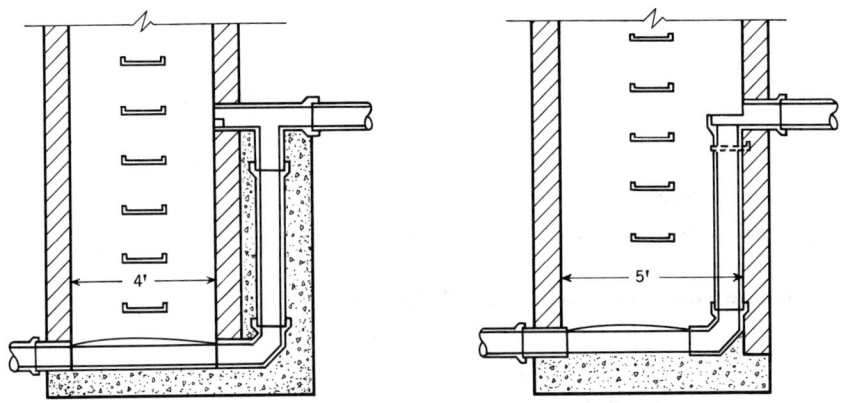

FIGURE 33.—Drop manholes. (Ft × 0.3 = m.)

150   DESIGN, CONSTRUCTION OF SANITARY, STORM SEWERS

A terminal cleanout amounts to an upturned pipe coming to the surface of the ground. The turn should be made with bends so that flexible cleaning rods can be passed through it. The diameter should be the same as that for the sewer. The cleanout is capped with a cast-iron frame and cover (Figure 34). In some districts the cleanout is brought to the surface on a continuing incline with a special cover designed to accommodate it.

In the past, tees often were used instead of pipe bends and the structures were called "lampholes." Sewer cleaning equipment cannot be passed into the sewer through a tee, so such structures are not now recommended.

Regulations in most areas allow terminal cleanouts only within 150 to 200 ft (45 to 60 m) of a manhole.

G. BUILDING SEWERS

Building sewers, also called house connections, service connections, or service laterals, are the branches between the street sewer and the property or curb line, serving individual properties, and usually are required to be 4, 5, or 6 in. in diameter (10 to 15 cm) preferably with a slope of ¼ in./ft (2 cm/m), or 2 percent. Sometimes one-percent slopes are

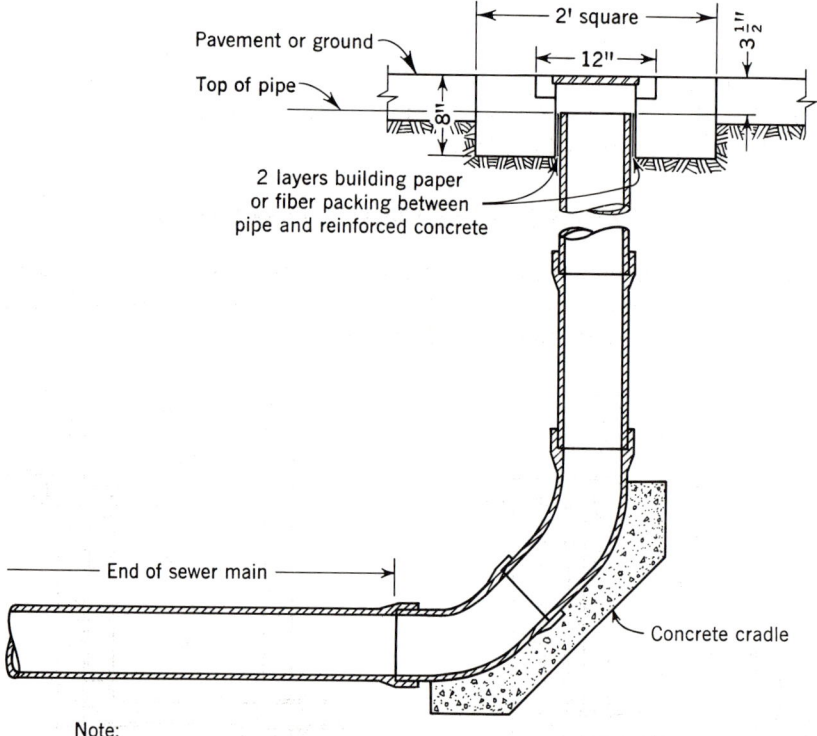

FIGURE 34.—Terminal cleanout. (Ft × 0.3 = m; in. × 2.54 = cm.)

allowed when necessary. Materials, joints, and workmanship should be equal to those of the street sewer to minimize infiltration and root penetration.

Often building sewers are constructed to the property or curb line at the time the street sewer is constructed. To meet the future needs of unsubdivided properties, wyes or tees sometimes are installed at what are presumed to be convenient intervals. Laterals or stubs not placed in use should be plugged tightly. Figure 35 shows typical connections.

If wyes or tees are not installed when the sewer is constructed, the sewer must be tapped later and deplorably poor connections often have resulted. This is especially true for those connections made by breaking into the sewer and grouting in a stub. Either a length of pipe should be removed and replaced with a wye or tee fitting, or better, a clean opening should be cut with proper equipment and a tee-saddle or tee-insert attached. Any connection other than to existing fittings must be made by experienced workmen under close supervision.

Typical connections to a deep sewer are shown in Figure 36.

In some places, test tees are required on the building sewer which permit the outlet to the street sewer to be plugged. This makes it possible to test the building sewer.

When a connection is to a concrete trunk sewer, a bell may be installed at the outside of the pipe as shown in Figure 37. Preferably this is pro-

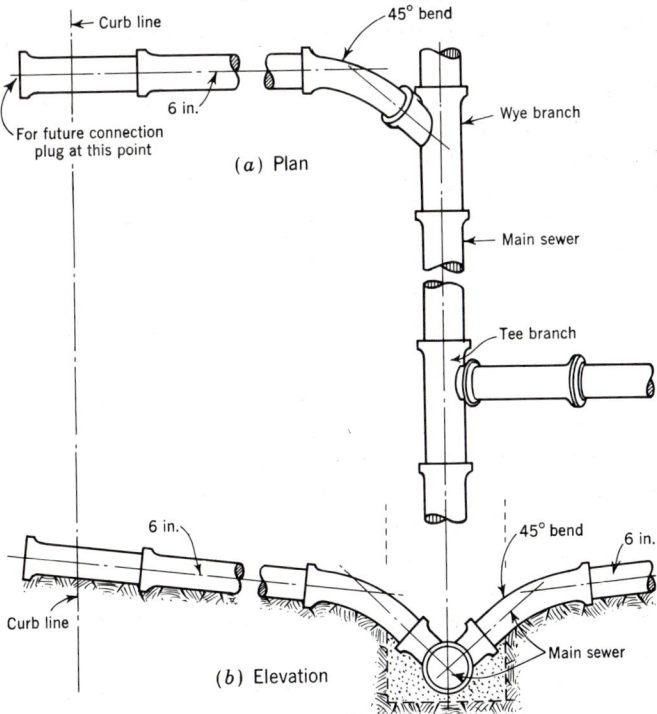

FIGURE 35.—Service connection for shallow sewer. (In. ×2.54=cm.)

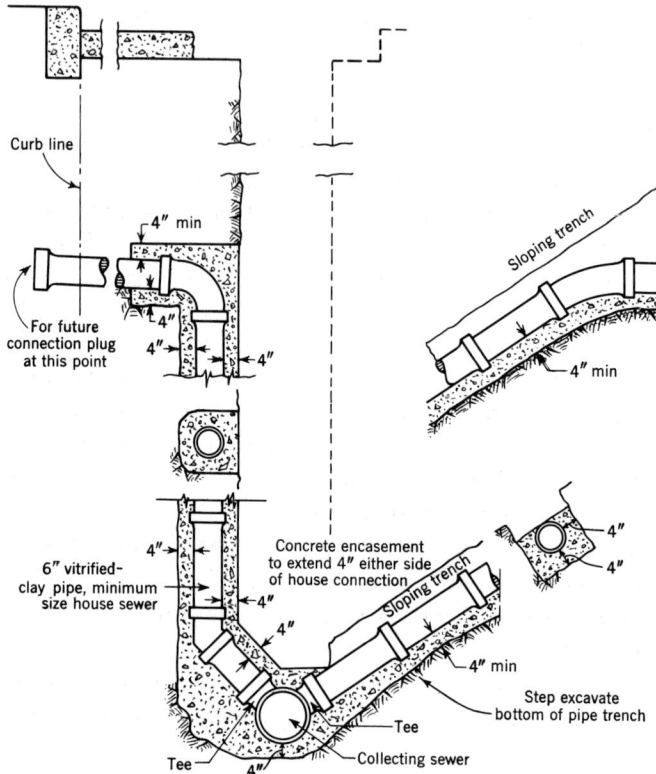

FIGURE 36.—Service connection for deep sewer. (In. × 2.54 = cm.)

vided by the manufacturer but it can be installed in the field if necessary. It must be high enough on the pipe so that the lateral will not be flooded by high flows in the sewer.

Large trunks are not used ordinarily as collecting sewers. When they are over 3 or 4 ft (0.9 to 1.2 m) in diameter they frequently are paralleled by smaller sewers that enter the trunks at manholes.

H. CHECK VALVES AND RELIEF OVERFLOWS

Where the floor of a building is at an elevation lower than the top of the next upstream manhole on the sewer system, a stoppage in the main sewer can lead to overflow of sewage in the building. Devices that sometimes are used to guard against such occurrences include backflow preventers or check valves and relief overflows.

Backflow preventers or check valves may be installed where the house plumbing discharges to the house sewer. Usually a double check valve is specified. Even so, they frequently do not remain effective over long periods of time.

Any overflow of sewage is undesirable but if a stoppage arises in a street sewer overflow must occur. It will be from a manhole in the street or into a building or else at a designated overflow point. The latter to be

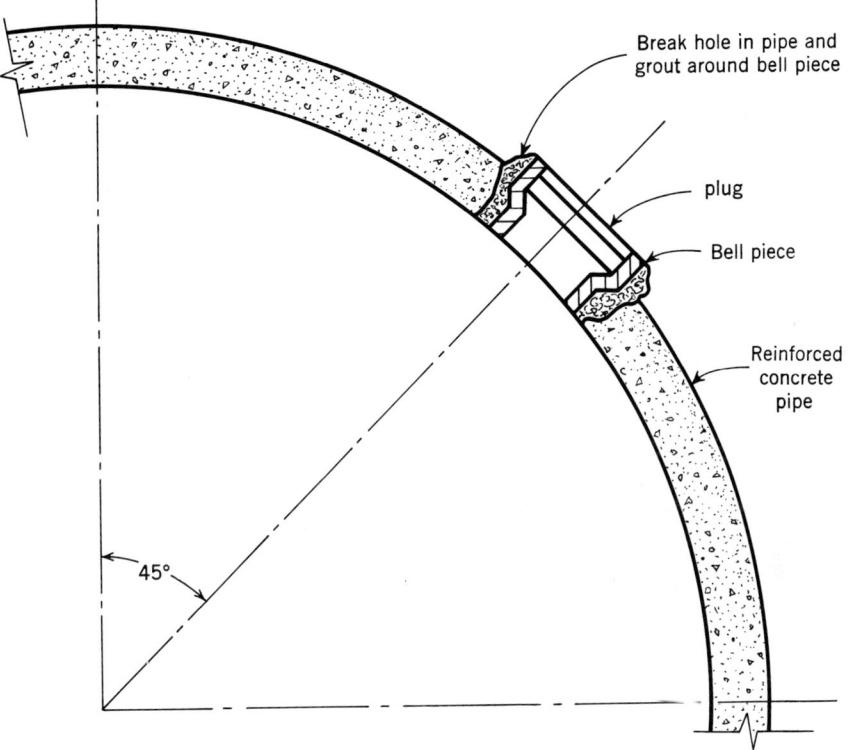

FIGURE 37.—Connection to large sewer.

effective must be at an elevation lower than the floor level being protected. At this point a relief device may be installed that encases a ball resting on a seat to close the end of a vertical riser but constructed so that the ball will rise and allow backflowing sewage to escape.

I. STORMWATER INLETS

Stormwater inlets admit surface waters to the underground conduits and comprise an important part of the storm drainage system. Their location and design should, therefore, be given careful consideration.

Inlets (Figure 38) can be classified into three major groups, each having several varieties which may be set either depressed or flush with the pavement surface.

1. Curb Inlets

Curb inlets have a vertical opening in the curb piece through which the gutter flow passes. They are called deflector inlets when equipped with diagonal notches cast into the gutter along the curb opening to form a series of ridges or deflectors. This type of inlet offers little or no obstruction to the even flow of traffic since the tops of these deflectors lie in the plane of the pavement. Figure 39 shows a curb inlet of recent design.

154   DESIGN, CONSTRUCTION OF SANITARY, STORM SEWERS

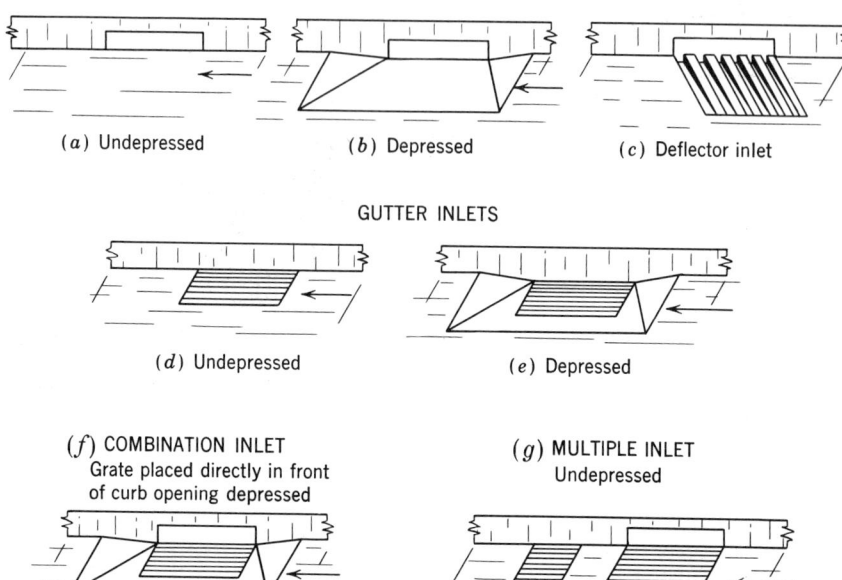

FIGURE 38.—Stormwater inlets.

FIGURE 39.—Curb inlet.

The cross slope of the gutter in front of the inlet is 1 in./ft (8.3 cm/m) out to a distance 3 ft (0.9 m) from the curb line.

2. Gutter Inlets

A horizontal opening in the gutter covered by one or more grates through which the gutter flow passes is called a gutter inlet. Tests indicate that grates with slots parallel to the curb are the least subject to plugging and provide the best inflow conditions. Width of the openings must not be greater than 1 in. (2.5 cm) so that bicycle tires will not pass through.

3. Combination Inlets

A curb and a gutter inlet acting as a unit is known as a combination inlet. Usually a gutter inlet is placed directly in front of a curb inlet, but it may be displaced in an overlapping or end-to-end position either upstream or downstream.

A multiple inlet is made up of two contiguous or closely spaced inlets acting as a unit. The actual structures for inlets can be made of brick or concrete, with cast iron covers and cast iron or steel gratings if the design employs gratings.

Engineering judgment must be used to evaluate the relative importance of clogging, nuisance to traffic, convenience, safety, and cost, for which hydraulic considerations sometimes may have to be sacrificed.

For all inlets:

(a) Use as steep a crown slope as traffic safety and comfort will permit.
(b) Design and space inlets so that 5 to 10 percent of the gutter flow reaching each inlet will pass on to the next inlet downstream, provided that this carry-over is not objectionable to pedestrian or vehicular traffic and the inlet is not in a sump.

For inlets where traffic is not expected to travel close to the curb:

(a) If clogging is not a problem, use a depressed gutter inlet or a combination inlet with curb and gutter openings contiguous and with a grate having longitudinal bars.
(b) If clogging is a problem and the design flow is small, use a depressed curb inlet. If the design flow is large, use a depressed combination inlet. The combination inlet will be the least subject to clogging and will have the greatest capacity if the curb opening is upstream from the gutter opening and the grate has longitudinal bars.

For inlets in thoroughfares where traffic moves close to the curb:

(a) If the street slope is greater than five percent use a deflector inlet if road dirt will not pack in the notches.
(b) If the street slope is less than five percent and in all streets in which clogging of the slots between the deflectors may cause difficulty,

use an undepressed gutter or combination inlet having a grate with longitudinal bars only.

For inlets in streets having little or no slope and for inlets in sumps:

(a) On flat streets pitch the gutter slightly from both directions toward the inlets. They will then behave like inlets in sumps.
(b) In sumps, use curb, or combination inlets, in order to reduce the flooding which occurs if grates clog.
(c) In sumps at the foot of sloping streets, allow ample capacity to admit water that flows past upstream inlets which are flooded or clogged.
(d) Inlets in sumps have a tendency to clog; therefore, their design capacities should be reduced accordingly. It is suggested that the design capacities of inlets in sumps should be reduced by the following amounts: 10 percent for curb inlets, 20 percent for combination inlets, and 30 percent for gutter inlets.

J. CATCH BASINS

A catch basin is a form of grit chamber intended to retain the heavy debris in stormwater which otherwise would be carried into the sewer. The periodic cleaning of catch basins is a substantial maintenance operation and in the intervals between storms the retained water may cause odor and mosquito nuisances. It now is considered generally that the disadvantages of catch basins outweigh their benefits.

In rugged areas, particularly in semi-arid climates, the stormwater discharged from canyons may carry large quantities of sand and rocks. "Debris basins" sometimes are constructed in such places by damming the water course to prevent the filling of downstream storm channels and sewers.

K. SIPHONS

Siphon in sewerage practice almost invariably refers to an inverted siphon or depressed sewer which would stand full even with no flow. Its purpose is to carry the flow under an obstruction such as a stream or depressed highway and to regain as much elevation as possible after the obstruction has been passed.

1. Single- and Multiple-Barrel Siphons

It is common practice, at least on larger sewers, to construct multiple-barrel siphons. The objective is to provide adequate self-cleansing velocities under widely varying flow conditions. The primary barrel is designed so that a velocity of 2 to 3 fps (0.6 to 0.9 m/sec) will be reached at least once each day, even during the early years of operation. Additional pipes regulated by lateral overflow weirs assist progressively in carrying flows of greater magnitude, that is, maximum dry-weather flow to maximum storm flow. The overflow weirs may be considered as submerged obstacles, causing loss in head as flow passes over them. The weir losses may be assumed equal to the head necessary to produce velocity across the crest. Weir crest elevations are dependent on the depths of flow in the upstream sewer for the design quantity increments. Sample crest length calculations have been explained by Fair and Geyer (1). The example at the end of this section illustrates the calculation of required pipe sizes.

Some engineers maintain that for sanitary sewers there is usually no need for multiple barrels. They reason that solids which settle out at low flows will flush out when higher flows obtain, except for those heavy solids that would accumulate even at high flows. Single-barrel siphons have been built with diameters ranging from 6 to 90 in. (15 to 230 cm) or more. Engineers holding to this concept in general favor a small barrel if initial flows will be much lower than in later years, with a larger barrel constructed at a later date or constructed at the outset but blocked off. In some situations a spare barrel may be desirable purely for emergency use. Rarely would it be satisfactory to use a single-barrel siphon for a combined sewer.

Figure 40 shows a plan of a multiple-barrel siphon used by the Greater Peoria Sanitary District.

2. Profile

Two considerations which govern the profile of a siphon are provision for hydraulic losses and ease of cleaning.

The friction loss through the barrel will be determined by the design velocity. For calculating this head loss, it is sound to use a conservative Hazen-Williams C of 100 (Manning n from 0.014 for small sizes to 0.018 for the largest). To the friction loss must be added losses due to side-overflow weirs in the case of multiple-barrel siphons.

Siphons may need cleaning oftener than gravity sewers. For easy cleaning by modern methods the siphon should not have any sharp bends, either vertical or horizontal; only smooth curves of adequate radius should be used. The rising leg should not be so steep as to make it difficult to remove heavy solids by cleaning tools that operate on the hydraulic principle. Some agencies limit the rising slope to 15 percent, but slopes steeper than this are considered satisfactory in most places. There should be no change of pipe diameter in the length of a barrel since this would hamper cleaning operations.

3. Air Jumpers

Positive pressure develops in the atmosphere upstream from a siphon because of the downstream movement of air induced by the sewage flow. In extreme cases this pressure may equal several inches of water. Air, therefore, tends to exhaust from the manhole at the siphon inlet, escaping in large amounts even from a pick hole. Under all except maximum flow conditions there is a drop in water surface elevation into a siphon, with

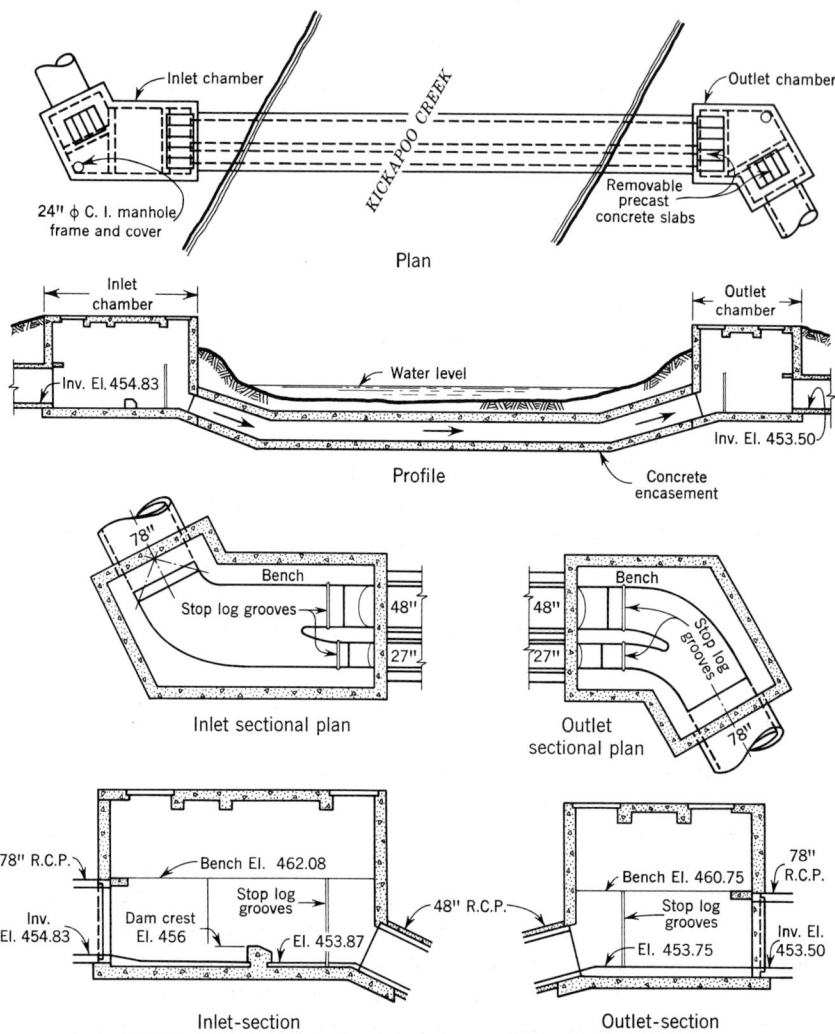

FIGURE 40.—Twin-barrel siphon. (In. × 2.54 = cm.)

consequent turbulence and release of odors. The exiting air hence can be the cause of serious odor problems. Conversely, air is drawn in at the siphon outlet.

Attempts to close the inlet structure tightly force the air out of plumbing vents or manholes farther upstream. Insofar as the attempt to close the sewer tightly is successful, oxygen depletion in the sewer atmosphere occurs, aggravating sulfide generation where this is a problem.

To overcome this difficulty a number of siphons built in recent years have used air jumpers; that is, pipes that take the air off the top of the inlet structure and return it at the end of the siphon. Usually, the jumper pipe is about half the diameter of the siphon. Sometimes the pipe can be suspended above the hydraulic grade line of the sewer but in most cases it runs more or less parallel to the siphon. In these cases provision must be made for dewatering the jumper, because it otherwise will fill

APPURTENANCES AND SPECIAL STRUCTURES 159

with condensate. Sometimes a drain can be installed to a percolation pit. One large air jumper in use comprises a 48-in. (120-cm) diam pipe paralleling a 90-in. (230-cm) siphon, 2,000 ft (610 m) long, utilizing a sump pump for dewatering.

4. Sulfide Generation

In a long siphon sulfide may be produced as in any other filled pipe. A detention time of 10 min may cause appreciable sulfide generation in a warm or high-BOD sewage, but it may require an hour if the sewage is of low temperature or low BOD. For any given flow and sewage characteristics, the sulfide concentration produced in a filled pipe is roughly proportional to the pipe diameter. For further details, see Chapter 6.

5. Provision for Draining

If it is feasible, a blowoff may be installed at the low point of the siphon to facilitate emergency maintenance operations.

The design of a siphon is illustrated below:

Example 1.—It is planned to construct an inverted siphon 400 ft (120 m) long in a 42-in. (110-cm) diam sewer. Invert elevation of the inlet is 106.24 ft (32.41 m), and outlet elevation cannot be less than 103.50 (31.60 m). Uniform flow conditions exist in the main sewer. Initial daily maximum dry-weather flow is 1.5 cfs (2.5 cu m/min), ultimate daily maximum dry-weather flow is 4.0 cfs (6.8 cu m/min), and maximum expected storm flow is 32 cfs (54 cu m/min). Losses in inlet chamber and miscellaneous losses are assumed to be 0.40 ft (0.12 m). A three-barrel siphon is to be used. What should be the pipe sizes?

For uniform flow conditions in main sewer, the depth of flow will be the same upstream and downstream from the siphon, therefore siphon losses must not exceed $106.24 - 103.50 = 2.74$ ft (0.81 m) in order to avoid an undesirable back water condition in the upstream sewer.

(a) Initial daily maximum dry-weather flow:

Diameter required for V of 3.0 fps (0.9 m/sec) $= 12\sqrt{\dfrac{4Q}{\pi V}}$

$= \sqrt{\dfrac{4 \times 1.5}{\pi \times 3.0}} = 9.57$ in. (24.3 cm).

Use 10-in. (25-cm) diam pipe.

Actual $V = \dfrac{Q}{A} = \dfrac{1.5}{0.785\,(0.833)^2} = 2.75$ fps (0.8 m/sec).

If Manning's $n = 0.013$, $h_f = (400)(0.0045) = 1.80$ ft (0.55 m); and $1.80 + 0.40 = 2.20 < 2.74$ ft.

(b) Ultimate daily maximum dry-weather flow:

Excess $Q = 4.0 - 1.5 = 2.5$ cfs (4.3 cum/min); and d required for

V of 3 fps $= 12\sqrt{\dfrac{4 \times 2.5}{\pi \times 3.0}} = 12.37$ in. (31 cm). Use 12-in. (30-cm) diam pipe.

Actual V=3.2 fps (1 m/sec); h_f=400 (0.0047) =1.88 ft (0.57 m); and 1.88+0.40=2.28<2.74 ft.

(c) Maximum storm flow:

Excess Q=32.0− (1.5+2.5) =28.0 cfs (48 cu m/min); and d (required) for V of 5 fps (1.5 m/sec) $= 12\sqrt{\dfrac{4 \times 28.0}{\pi \times 5}}$

=32.1 in. (81 cm). Use 30-in. (75-cm) diam pipe.

Actual V=5.7 fps (1.7 m/sec); h_f= (400) (0.0045) =1.80 ft (0.55 m); and 1.80+0.40=2.20<2.74 ft.

Having determined the size of each barrel as indicated, the designer can proceed to detail the inlet and outlet chambers and determine more accurately the miscellaneous hydraulic losses in transitions, over weirs, etc. If available head loss under peak flows is severely limited by some upstream or downstream control, a more careful analysis of the divided flows which occur at such times may dictate an increase in size of the larger barrels.

L. SANITARY SEWAGE DIVERSION AND STORMWATER OVERFLOW DEVICES

Many municipalities are served by systems of combined sewers. Diversion or overflow devices are used commonly to limit the flow reaching the treatment plant or other place of disposal. Excess flows are diverted to relief sewers, stormwater treatment facilities, or to the receiving waters without treatment.

There are two primary objectives which should be considered in the design of diversion and overflow structures: first, no sewage should be discharged to the receiving waters without treatment during dry weather; and second, the flow permitted to enter the wet well of a pumping station or a treatment plant during storms should be restricted to certain predetermined quantities.

An arrangement of a diversion chamber, intercepting sewers, and overflow outlet for a large combined sewer in Chicago is shown in Figure 41. The flow to the intercepting sewer is controlled by a motor-operated sluice gate with remote electrical control from the wastewater treatment works. Tide gates allow passage of stormwater to the river in storm periods and prevent backflow from the river in dry weather.

The side-overflow weir and the leaping weir, many of which are in service, are among the devices used for regulation of flow in combined sewers. Their hydraulic design is discussed in Chapter 5. But it has been found difficult to attain close control of flow with them.

Overflow weirs constructed along one or both sides of the combined sewer deliver excess flows during storm periods to the receiving waters or natural drainage courses. The crest of the weir is set at an elevation corresponding to the desired depth of flow in the sewer. The weir length must be relatively long if close regulation is to be obtained, a requirement which has led to numerous variations in weir construction.

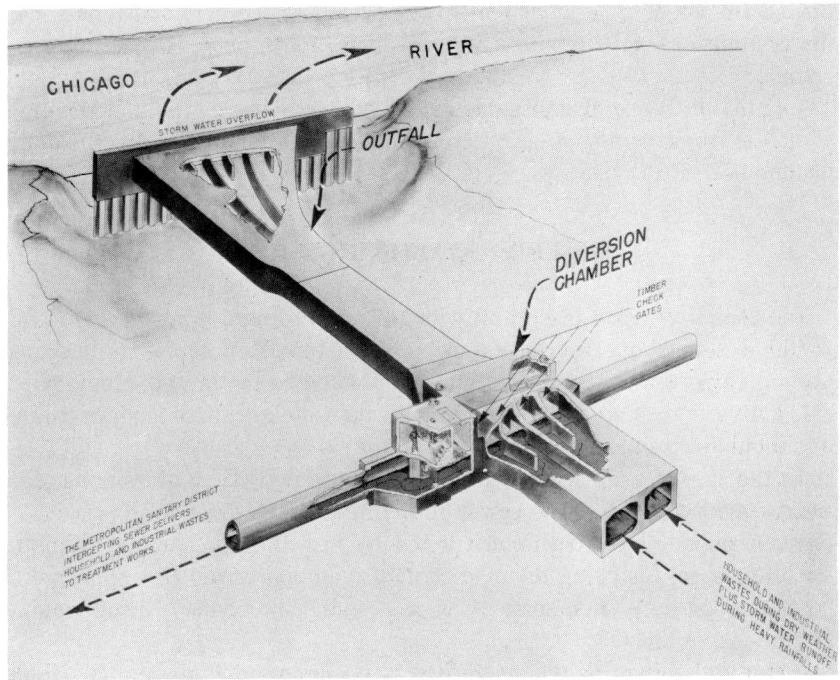

FIGURE 41.—Diversion chamber in large combined sewer.

Leaping weirs are designed to discharge the sanitary flow through an opening in the invert of the combined sewer and thence through an intercepting sewer to the point of disposal. Excess quantities of stormwater leap across the opening and are thus diverted from the intercepting sewer. Provision often is made in the design for modifying the slot opening in accordance with actual experience after installation.

Positive control of the flow in combined sewers also can be accomplished by means of proprietary automatic mechanical regulators. The moving parts are actuated by the depth of flow in the sewer. They require periodic inspection and maintenance as do all types of diversion and overflow devices.

M. FLAP OR BACKWATER GATES

Flap gates are installed at or near sewer outlets for the purpose of preventing the back-flooding of the sewer system by high tides or high stages in the receiving stream. A properly operating flap gate may make it possible to continue to pump and treat the flow discharged by a separate system of sanitary sewers or the dry-weather flow from a system of combined sewers, even though flood conditions prevail at the outlet from the plant.

Flap gates may be made of wood, cast iron, or steel. They are commercially available in sizes up to 8 ft (2.4 m) in diameter. Larger gates can be fabricated from plates and structural shapes. They should be hinged by a link-type arrangement which allows the gate shutter to seat more securely. Hinge pins, linkages, and seats should be corrosion-resistant.

The maintenance of flap gates requires regular inspection and removal of debris from the pipe and outlet chamber, lubrication of hinge pins, and cleaning of seating surfaces.

N. SEWERS ABOVE GROUND

Occasionally in rolling or hilly terrain it is desirable and economical to build sewers above the surface of the ground or across gullies and stream valleys. Such sewers often are constructed in carefully compacted fill. Culverts through the fill or other means to take care of surface drainage should be provided.

In the case of waterways, where the topography, width of crossing, and flood elevations allow, the sewer sometimes is placed over the obstacle. Sewer crossings have been constructed by installing or hanging the pipe on bridges, by fastening them to structural supports which rest on piers, by means of suspension spans with cables, and by means of a sewer pipe beam.

Structural design of overhead sewers is similar to that of comparable structures with supporting members of timber, steel, or reinforced concrete. Foundation piers or abutments should be designed to prevent overturning and settlement. The impact of flood waters and debris should be considered.

If the sewer is exposed, as on a trestle, steel pipe may be used, perhaps with mortar coating and lining for corrosion protection. Sometimes sewers of other materials are carried inside steel pipes. The steel pipe may be supported by simple piers at suitable intervals.

In recent years prestressed concrete pipe beams have been used to span waterways and other obstacles. Generally, they have been of three types:

(a) A rectangular section with a circular void extending the full length of the member, either pretensioned or post-tensioned, and similar to a hollow box highway girder. This section normally is used for smaller sewers.

(b) A pretensioned circular pipe section which may be produced in most any diameter. This type is economical for long crossings.

(c) Reinforced concrete pipe sections assembled and post-tensioned to form the required sewer pipe beam. These beams may be fabricated economically using standard pipe forms and prestressing equipment. The pipes are cast with longitudinal cable ducts in the walls. After curing, the pipes are aligned, post-tensioning cables inserted, jacked to the design tension, and anchored. Pressure grouting the ducts completes the manufacture and the sewer beam then is shipped to the job site for installation. Chapter 11 contains a picture of a prestressed concrete pipe beam constructed in this manner.

Protection against freezing and prevention of leakage are important design and construction considerations for above-ground sewers.

It has been found necessary in some designs to employ expansion jointing between above-ground and below-ground sewers. Special couplings are available for such purposes. Anchorage provision also must be made to prevent permanent creep. Expansion joints in sewers supported on bridges or buildings should match the expansion joints in the structures to which the sewer is attached.

O. UNDERWATER SEWERS AND OUTFALLS

1. Ocean Outfalls

Communities adjacent to the seacoast may decide to discharge their storm drainage and treated wastewater into the ocean. Outfalls for discharge of storm flow are usually short, with the terminal elevation in the tidal zone. No special considerations are needed in the design of these facilities over the requirements for similar units in freshwater lakes or streams except for tidal backup into the system. The effect of storm-induced tides should be evaluated as well as the maximum tides predicted by the charts.

Disinfected secondary effluents generally are discharged close to shore. Primary effluents are carried far enough to sea to avoid any undesirable effects.

For proper design it is essential to obtain detailed data on the following:

(a) Shore use and conditions to be maintained;
(b) Profiles of possible outfall routes;
(c) Nature of the bottom;
(d) Density stratification or thermoclines, by seasons; and
(e) Patterns of water movement at point of discharge and travel time to shore.

Since seawater is 2.5 percent denser than sanitary wastewater, the discharged wastewater rises rapidly, normally producing a "boil" at the surface. The rising plume aspirates and mixes with a quantity of seawater which is generally from 10 to more than 100 times the wastewater flow. Dilution increases rapidly as the "sewage field" moves away from the boil. The required length and depth of the outfall is related to the degree of treatment of the wastewater. The length must be calculated so that time and dilution will protect adequately the beneficial uses of the adjacent waters.

Rawn and Palmer (2) have published formulas for calculating the dilution, later expanded and improved by Brooks (3) and Abraham (4). Pomeroy (5) has presented a formula for predicting bacterial counts from outfalls by comparison with other outfalls under similar coastal conditions. The scope of ocean outfall problems has been reviewed by Pearson (6) and Gunnerson (7).

Where the outfall is deep and there is good density stratification (thermocline), the rising plume can pick up enough cold bottom water so that the mixture is heavier than the surface water. The rising plume stops beneath the surface or reaches the surface and then resubmerges.

To gain maximum benefit of density stratification, diffusers may be used. If, however, they merely divide the flow into many small streams in a small area (a gas-burner type of diffuser) they do little good. The flow must be dispersed widely so that huge flows of dilution water can approach at low velocity. Diffuser design has been studied by Rawn et al. (8).

These principles are well illustrated by the Los Angeles City outfall in Santa Monica Bay. The effluent is carried by a 12-ft (3.7-m) diam pipe to a point 5 miles (8 km) from shore at a depth of 190 ft (58 m), then dispersed through a Y-shaped diffuser, with the two arms totalling 8,000 ft (2,400 m) in length. Except for certain periods in winter when the thermocline is practically nonexistent, no sewage can be seen rising to the surface. The flow of effluent, essentially primary and unchlorinated, exceeds 300 mgd (114,000 cu m/day), yet the bathing waters of the highly popular beaches of the Bay remain in excellent condition. The effluent outfall is supplemented by a second outfall 22 in. in diam and 7 miles (11 km) long which is used for disposal of digested sludge at a depth of 320 ft (98 m). The topography and currents at the point chosen for this discharge are such that there is no progressive accumulation of sludge.

Outfalls into the open ocean generally are buried to a point where the water is deep enough to protect them from wave action, usually about 30 ft (10 m). Trenches in rock are formed by blasting. Beyond the buried portion, the outfall rests on the bottom, with a flanking of rock to prevent currents from undercutting it where the bottom is soft.

Ocean outfalls in the smaller sizes now are made usually of steel pipe, mortar lined and coated. Steel pipes are welded and usually can be dragged into place from the shore, as was done with the 7-mile (11.2-km) sludge line of the city of Los Angeles. Reinforced concrete pipe is used for the larger sizes. The joints for concrete outfalls usually are made with rubber gaskets similar to those used for construction on land.

2. Other Outlets

If effluent is discharged into an estuary or land-locked bay, special studies are needed to explore tidal currents, upstream flow of salt water at the bottom, available dilution, etc., in order to determine which discharge locations are compatible with various degrees of treatment.

Sewers discharging into streams with high-velocity flood flows require special thought in design to prevent undermining of the outlet structure as well as the pipe itself. Large combined sewer outlets into the Mississippi River have been built with a headwall or endwall supported on a cell or cells of steel sheet piling driven about 30 ft (9 m) below the headwall structure, with the upper 10 ft (3 m) of the cells excavated and backfilled with mass concrete.

Sewer outlets which are always or frequently above the receiving waters often are provided with masonry or concrete headwalls and wingwalls. These may prove satisfactory where erosion due to swift currents or wave action will not undercut the headwall structure. A length of metal pipe, well anchored and supported in the bank and cantilevered out to drop its flow into a cushion of dry-weather flow in the stream, may be the best solution to the outlet problem for small pipe outlets well above the more frequent flood flows in the receiving stream.

Combined sewer outlets, normally or frequently above the surface of the receiving waters, sometimes incorporate a dry-weather flow interceptor and a dry-weather outlet built beneath the combined sewer to discharge dry-weather flow beneath the surface of the receiving stream or lake.

P. MEASURING WASTEWATER FLOWS

1. General Methods

Depending on the accuracy desired any one of several methods may be used to measure the flow in a sewer. A crude approximation may be obtained from the size, slope, and observed depth of flow, using an appropriate hydraulic formula. The accuracy can be improved by using a dye to determine the velocity. But because the depth of flow measured at manholes does not necessarily show the average depth in the intervening sewer, results are uncertain. Other methods include addition of salt or some other tracer at a constant rate and determining the downstream concentration of the added constituent.

Of more general utility are weirs and flumes. In both types the flow is caused to pass through a critical control section. The total energy head upstream from the control section determines the velocity and hence the discharge. When possible, the upstream condition is measured at a point where the velocity is low so that the surface elevation can be taken as a measure of total energy head. Where this cannot be done, a correction is made for the velocity of approach. For the case of a rectangular section under the ideal conditions of no friction loss and uniform velocity across the section, the discharge is given by the equation

$$Q = 2/3 B \sqrt{2/3 g H^3} \dots\dots\dots\dots\dots\dots\dots 1$$

where B is the width and H is the total energy head at the control section. The equation is the same in both metric and English units provided consistent units are used.

For a sharp-crested weir the effective control section is somewhat upstream from the actual crest and the discharge is somewhat higher than shown by the theoretical equation. Omitting refinements, the formula for a rectangular weir is approximately $Q = 3.33 B H^{3/2}$ ft (English units); thus, the discharge is 8 percent greater than for the ideal case. Head-discharge relationships for weirs of rectangular, triangular, and trapezoidal sections may be found in hydraulics handbooks.

Palmer and Bowlus (9) showed that if a flume is constructed to make the flow converge into a throat having a level floor for a length roughly equal to the pipe diameter (sometimes to the average depth of flow), and with flow conditions that allow the stream to leave the throat at supercritical velocity, a control section will be established in the throat, with the discharge conforming closely to the theoretical equations. Palmer-Bowlus flumes have been constructed in rectangular, trapezoidal, and polygonal shapes. They also are made merely by laying a slab in the bottom of a channel; and they have been formed by narrowing the sides of a rectangular channel with no modification of the invert. Figure 42 shows some of these designs.

Ludwig and Ludwig (10) showed how to calculate the head-discharge relations for the trapezoidal type in sewers. Wells and Gotaas (11) experimented with both trapezoidal and slab types. They found that these flumes installed in sewers can meter flows up to 90 percent of the sewer capacity; and under proper conditions of use, the flows determined from the calculated rating curve are within three percent of actual.

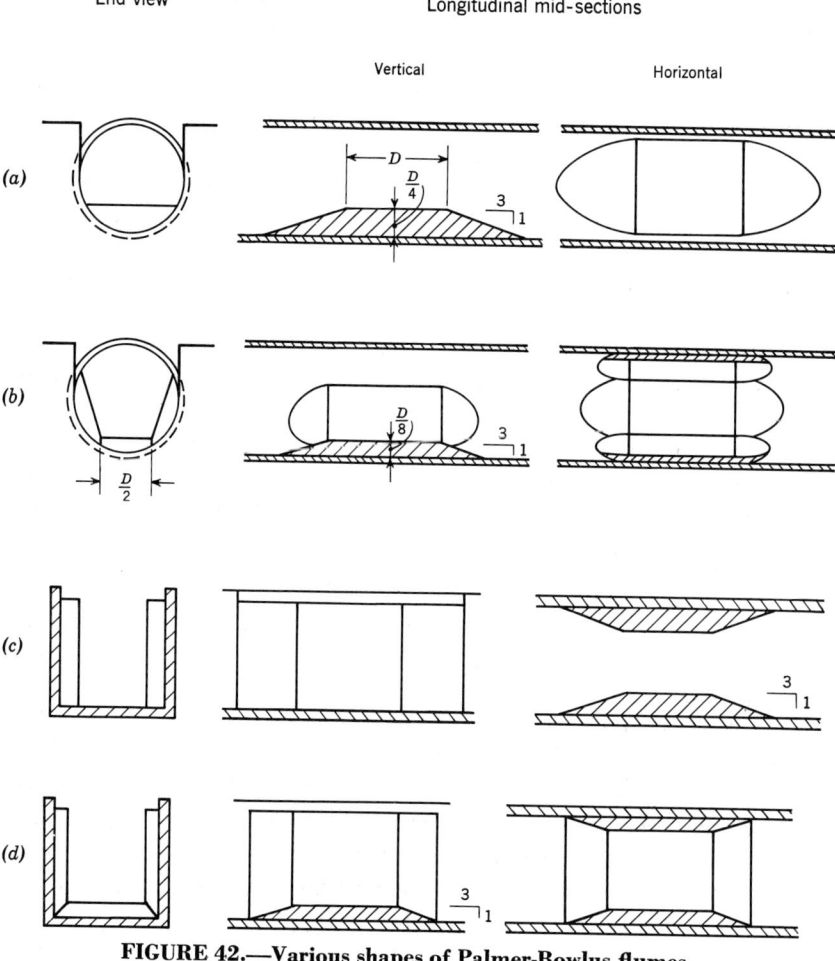

FIGURE 42.—Various shapes of Palmer-Bowlus flumes.

Another type of measuring device widely used in treatment works is the Parshall flume, which uses a converging throat with the floor sloping downward. It is rectangular in section, with all dimensions in rigidly specified proportions. Empirical head-discharge tables are found in hydraulics handbooks. The discharge is about 10 percent greater than would be calculated for a control section in the throat showing that the section is in effect somewhat upstream.

2. Temporary Metering Stations

Palmer-Bowlus flumes are particularly suitable for installation in existing sewers. Cast iron, stainless steel, and fiberglas inserts are manufactured which when installed form trapezoidal flumes. Concrete slab inserts are constructed readily and are useful for large flows. The floor of the slab is most commonly one-fourth the pipe diameter above the invert. A thinner slab may be used for relatively low flows. Table XXIV provides data from which rating curves for the $D/4$ slab type can be plotted for any desired pipe diameter. The tabulated data apply to the case where the depth measurement is made in the U-shaped channel of the manhole upstream from the control section located in the pipe. If the depth is measured in a pipe, the curve would be slightly different at high flows. If the manhole bench comes up only half way, there will be substantial errors when the bench is flooded.

It is necessary that the flow leave the throat at supercritical velocity. Existence of this condition is proved if a hydraulic jump is seen downstream. Nevertheless, absence of a jump is sometimes due to continuing supercritical velocities.

Weirs, too, can be installed in sewers. For a small sewer, a weir has the

TABLE XXIV.—Generalized Rating Data for Slab-Type * Palmer-Bowlus Flume

*Slab depth $t = D/4$. Calculated for slab in a pipe, with water depth, h, measurement in upstream U-shaped channel.

Head above Control Elevation, as Proportion of Pipe diameter $(h - t)/D$	Flow Factor, $Q/D^{5/2}$ (cfs/ft$^{5/2}$)
0.05	0.031
0.10	0.091
0.15	0.173
0.20	0.276
0.25	0.398
0.30	0.538
0.35	0.697
0.40	0.866
0.45	1.051
0.50	1.247
0.55	1.452
0.60	1.667
0.65	1.887

Note: Cfs/ft$^{5/2} \times 33.1 =$ cu m/min/m$^{5/2}$.

advantage that it can be fabricated easily with local materials and securely fastened in place, the periphery being caulked with rags or jute. Prefabricated weirs with clamps and gaskets for rapid installation are available. For relatively small flows in a sewer, a weir is usually of the 90 deg V-notch type, but a rectangular weir may serve better for larger flows. Velocity of approach must be considered in calculating the head-discharge relationship.

For accurate measurements weirs require considerably greater head loss than Palmer-Bowlus flumes and are more limited in range. A major disadvantage of a weir is that rags, hair, and paper tend to hang on the crest, especially at low flows. This difficulty may be minimized if the crest is smoothed carefully. Another disadvantage is that the upstream channel accumulates sludge, progressively modifying velocity of approach. Negligible solids deposition occurs with a suitably designed Palmer-Bowlus flume.

To set up a temporary recording station using either a weir or a flume, the insert is placed in the outlet of a manhole. A clock-operated stage recorder, connected to a float mounted on a hinged arm, is placed in or over the manhole. The recorder should be set to read zero when the float level is the same as the control elevation, that is, the elevation of the weir crest or the floor of the throat. In setting the recorder, measurements can be made from the invert of the sewer or other convenient reference point after first determining the elevation difference between the control elevation and reference point with the aid of a carpenter's level.

3. Permanent Metering Stations

Permanent metering stations often are required on trunk sewers, especially where payments from contributing parties must be based on flow. The measuring device, usually housed in a concrete vault, may be a weir, a Palmer-Bowlus flume, or a Parshall flume. A float well should be provided outside the channel. Often it is placed outside the vault that houses the channel. The recorder should be outside the vault to escape the humid and corrosive atmosphere. Sometimes the float well is at the side of the road with the recorder housed in a small building over it.

The float well is connected to the channel with a pipe generally in the size range of 0.5 to 1 in. (1.3 to 2.5 cm) in diam, terminating in a smooth opening flush with the wall of the channel. Fresh water must be provided for periodic or continuous flushing of the well and connecting pipe. Safety of the fresh water supply must be protected by an air gap located higher than the highest water level under flooding conditions which generally means higher than the ground surface. A leveling drain also should be provided, so arranged that when the connection to the channel is shut off and the leveling valve opened, the water will discharge to a sump or other low point until the level in the well exactly equals the floor of the flume or crest of the weir.

Where the wastewater flows in a completely filled pipe, as in a force main, inverted siphon, or a section intentionally depressed for the purpose, other methods of metering are applicable. For raw wastewater, impulse-type and orifice meters are unsuitable because of fouling problems. Until recent years, Venturi meters were used most commonly. In this meter, both the signal and the head loss increase as the square of the flow. Thus, it is difficult to measure accurately a wide range of flows without excessive head loss at the higher flows. In recent years the magnetic flow meter has gained in popularity. Suitable accuracy is attainable over a wide range and the directly obtained electrical signal facilitates amplification, transmission, and recording. The magnetic meter imposes no obstacle to the flow. Nonetheless, it must be kept clean because accumulations of slime or grease can affect the readings.

References

1. Fair, G. M., and Geyer, J. C., "Water Supply and Waste-Water Disposal." 1st Ed., John Wiley & Sons, Inc. (1954).
2. Rawn, A M, and Palmer, H. K., "Predetermining the Extent of a Sewage Field in Sea Water." *Trans. Amer. Soc. Civil Engr.*, **94,** 1036 (1930).
3. Brooks, N. H., "Diffusion of Sewage Effluent in an Ocean Current." Proc. 1st Conf. on Waste Disposal in Marine Environment, Univ. Calif., Berkeley, Pergamon Press Ltd., London, England (1960).
4. Abraham, G., "Jet Diffusion in Stagnant Ambient Fluid." Delft Inst. Hydr. Lab. Publ. No. 29, Delft, The Netherlands (1963).
5. Pomeroy, R. D., "The Empirical Approach for Determining the Required Length of an Ocean Outfall." Proc. 1st Conf. on Waste Disposal in Marine Environment, Univ. Calif., Berkely, Pergamon Press Ltd., London, England (1960).
6. Pearson, E. A., "An Investigation of the Efficacy of Submarine Outfall Disposal of Sewage and Sludge." Calif. Water Poll. Contr. Bd. Publ. No. 14, Sacramento, Calif. (1956).
7. Gunnerson, C. G., "Marine Disposal of Wastes." *Jour. San. Eng. Div., Proc. Amer. Soc. Civil Engr.* **87,** SA1, 23 (1961).
8. Rawn, A M, Bowerman, F. R., and Brooks, N. H., "Diffusers for Disposal of Sewage in Sea Water." *Jour. San. Eng. Div., Proc. Amer. Soc. Civil Engr.,* **86,** SA2, 65 (1960).
9. Palmer, H. K., and Bowlus, F. D., "Adaption of Venturi Flumes to Flow Measurements in Conduits." *Trans. Amer. Soc. Civil Engr.*, **101,** 1195 (1936).
10. Ludwig, J. H., and Ludwig, R. G., "Design of Palmer-Bowlus Flumes." *Sewage and Industrial Wastes,* **23,** 1096 (1951).
11. Wells, E. A., Jr., and Gotaas, H., "Design of Venturi Flumes in Circular Conduits." *Trans. Amer. Soc. Civil Engr.*, **123,** 749 (1958).

CHAPTER 8. MATERIALS FOR SEWER CONSTRUCTION

A. INTRODUCTION

Factors which should be considered in the selection of materials for sewer construction are:

1. Flow characteristics—friction coefficient,
2. Life expectancy and use experience,
3. Resistance to scour,
4. Resistance to acids, alkalis, gases, solvents, etc.,
5. Ease of handling and installation,
6. Physical strength,
7. Type of joint—watertightness and ease of assembly,
8. Availability and ease of installation of fittings and connections,
9. Availability in sizes required, and
10. Cost of materials, handling, and installation.

These factors, for example, become inherent when consideration is given to the character of wastewater or stormwater to be transported and the construction conditions to be met or methods used.

No single material will meet all conditions that may be encountered in sewer design. Selections should be made for the particular application and different materials may be selected for parts of a single project. Cost of pipe is usually a minor factor for smaller sewers, the differences in cost among the various pipe materials being a small part of the total project cost.

New materials are continually being offered for use in sewer construction. But not all of them can be evaluated at this time. Accordingly, discussions in this chapter are generally confined to some of the more commonly used materials, listed below alphabetically:

1. Asbestos-cement pipe

   (a) Gravity
   (b) Pressure

2. Brick masonry

3. Clay (vitrified)

4. Concrete

   (a) Gravity
      (1) Plain
      (2) Reinforced
   (b) Pressure
   (c) Cast-in-place

5. Iron and steel
   - (*a*) Cast iron
   - (*b*) Ductile iron
   - (*c*) Fabricated steel
     - (1) Corrugated
     - (2) Plain

6. Organic materials
   - (*a*) Solid wall plastic
   - (*b*) Truss pipe

B. ASBESTOS-CEMENT PIPE

Pipe of asbestos fiber and cement, in diameters from 4 through 36 in. (10 to 91 cm) and in some areas up to 42 in. (107 cm), is used for both gravity and pressure sewers. Asbestos-cement or cast iron fittings are used for asbestos-cement gravity sewers; while cast iron fittings are used for asbestos-cement force mains. Asbestos-cement pipe jointing is accomplished by compressing rubber rings between factory-machined pipe ends and sleeves or bell ends of the same material as the pipe.

For gravity sewer systems, asbestos-cement pipe is manufactured in five strength classifications—Class 1,500, 2,400, 3,300, 4,000, and 5,000. The class designation represents the minimum three-edge bearing crushing strength in lb/lin ft of pipe, regardless of size.

For pressure systems, asbestos-cement pressure pipe is manufactured in three classes of operating pressure—Class 100, 150, and 200. The class designation for pressure pipe represents operating pressures in psi.

The advantages of asbestos-cement pipe are light weight and ease of handling, long laying lengths in some situations, tight joints, and a wide range of available strength classifications. Because of its regularity of dimensions, the use of asbestos-cement pipe with rubber-ring type joints results in minimum infiltration when properly installed.

The disadvantages of asbestos-cement pipe are that it may be subject to corrosion where acids or hydrogen sulfide are present. Protective linings of proven performance should be used where excessive corrosion is likely to occur.

Asbestos-cement pipe is specified by pipe diameter, class or strength, and type of joint. The pipe should conform, insofar as appropriate, to the following standard specifications:

1. "Pipe, Asbestos-Cement," Federal Specification SS–P–351a (Pressure); and "Pipe, Asbestos-Cement, Sewer, Nonpressure," SS–P–331c.
2. "Tentative Standard Specification for Asbestos-Cement Water Pipe," AWWA * C400T.

---

* American Water Works Association, New York, N.Y.

3. "Asbestos-Cement Pressure Pipe," ASTM * C296.
4. "Asbestos-Cement Nonpressure Sewer Pipe," ASTM C428.
5. "Testing Asbestos-Cement Pipe," ASTM C500.

Additional information may be obtained from the publications of the industry and various manufacturers' installation guides (1).

C. BRICK MASONRY

Before concrete came into common use, brick masonry was used for sewers of large diameter. Many old brick sewers are still in use but many have failed through disintegration of the mortar joints. Because of high material and labor cost and other factors, brick is now used for sewer construction in the United States only in special applications.

In areas where brick masonry is subjected to alternate freezing and thawing such as in the upper portion of manholes, hard brick with low absorption qualities should be specified.

The standard specification for sewer brick (made from clay or shale) is ASTM C32.

D. CLAY PIPE

Practically all clay sewer pipe is vitrified as part of the manufacturing process and discussion throughout this manual refers to vitrified clay pipe. Clay pipe is manufactured in diameters from 4 through 36 in. (10 to 91 cm) and in some areas up to 42 in. (107 cm). It is manufactured in standard and extra-strength classifications, although in some places the manufacture of standard-strength pipe has been discontinued in the smaller sizes.

The strength of clay pipe varies with the diameter and strength classification. The standard specifications or manufacturer should be consulted in this regard.

The laying length of clay pipe varies from 3 to 6 ft according to locale, longer lengths being preferred. Clay fittings are available to meet most requirements, with special fittings manufactured on request.

Pipe dimensions and tolerances vary somewhat in different areas and exact dimensions should be obtained from local manufacturers.

In the past most clay pipe was salt glazed. There is a current trend to use unglazed pipe because it is resistant to corrosion and is non-absorptive.

The resistance of clay pipe to corrosion from acids and alkalies gives it a distinct advantage over other materials in handling waste which has a high acid concentration or when hydrogen sulfide generation may result in the formation of sulfuric acid. It also has excellent resistance to erosion and scour.

---

* American Society for Testing and Materials, Philadelphia, Pa.

Disadvantages of clay pipe are its limited range of sizes and strengths and the fact that it is more brittle than other sewer pipe materials.

Clay pipe is specified by the pipe diameter, either full or nominal, class or strength, and type of joint. It should conform, insofar as appropriate, to the following standard specifications:

1. "Recommended Practice for Installing Vitrified Clay Sewer Pipe," ASTM C12.
2. "Standard Strength Clay Sewer Pipe," ASTM C13.
3. "Extra Strength Clay Pipe," ASTM C200.
4. "Standard and Extra Strength Perforated Clay Pipe," ASTM C211.
5. "Methods of Testing Clay Pipe," ASTM C301.
6. "Compression Joints for Vitrified Clay Bell and Spigot Pipe," ASTM C425.
7. "Compression Couplings for Vitrified Clay Plain End Pipe," ASTM C594.
8. "Pipe, Clay, Sewer," Federal Specification SS-P-361b.

Additional information may be obtained from the publications of the industry and various manufacturers such as the "Clay Pipe Engineering Manual" (2).

E. CONCRETE PIPE

Non-reinforced concrete pipe from 4 to 24 in. (10 to 61 cm) and reinforced pipe from 12 to 144 in. (30 to 366 cm) in diameter is available generally in circular cross sections for gravity sewers. Concrete pipe is manufactured by a number of methods including centrifugal spun, vertical or horizontally cast, and combinations of casting, vibrating, packing, and spinning techniques. The type of manufacturing process used depends on the locale, size of pipe, and the pertinent specifications.

Reinforced concrete pressure pipe and prestressed concrete pressure pipe are used for force mains, submerged outfalls, inverted siphons, and other applications where internal pressures are encountered and for gravity sewers where there are exceptional tightness requirements.

Reinforced concrete elliptical and arch pipe also is available generally in sizes equivalent in capacity to the circular pipe.

Pressure and non-pressure pipe may be manufactured to any reasonable strength requirement by varying the wall thickness, concrete strength, or the percentage and shape of the reinforcing steel or the prestressed elements. Concrete fittings and appurtenances such as wyes, tees, and manhole sections are available generally. A number of jointing methods are available depending on the tightness required and the operating pressure.

The advantages of concrete pipe are the relative ease with which the required strength may be provided and the wide range of sizes and laying lengths available, the latter varying from 4 to 24 ft (1.2 to 7.4 m), depending on the type of pipe and the manufacturer. Specially designed concrete pipe can be adapted for jacking purposes.

A disadvantage is that it may be subject to corrosion where acids or hydrogen sulfide are present. Protective linings of proven performance should be used where excessive corrosion is likely to occur. Only dense high-quality concrete should be used in concrete exposed to wastewater.

Concrete pipe is specified by pipe diameter, class or strength, the method of jointing, and any special lining or concrete requirement such as method of manufacture when important. The pipe should conform, insofar as appropriate, to the following standard specifications:

1. "Pipe, Concrete, (Nonreinforced, Sewer)," Federal Specification SS–P–371c; "Pipe, Concrete, (Reinforced, Sewer)," SS–P–375b; and "Pipe, Pressure, Reinforced Concrete, Pretensioned Reinforcement (Steel Cylinder Type)," SS–P–381.
2. "Standard for Reinforced Concrete Water Pipe—Steel Cylinder Type, Not Prestressed," AWWA C300; "Standard for Reinforced Concrete Water Pipe—Steel Cylinder Type, Prestressed," AWWA C301; and "Standard for Reinforced Concrete Water Pipe—Noncylinder Type, Not Prestressed," AWWA C302.
3. "Concrete Sewer, Storm Drain, and Culvert Pipe," ASTM C14; "Reinforced Concrete Culvert, Storm Drain, and Sewer Pipe," ASTM C76; "Reinforced Concrete Low-Head Pressure Pipe," ASTM C361; "Joints for Circular Concrete Sewer and Culvert Pipe, Using Flexible, Watertight, Rubber Gaskets," ASTM C443; "Determining Physical Properties of Concrete Pipe or Tile," ASTM C497; "Reinforced Concrete Arch Culvert, Storm Drain, and Sewer Pipe," ASTM C506; "Reinforced Concrete Elliptical Culvert, Storm Drain, and Sewer Pipe," ASTM C507; "Perforated Concrete Pipe," ASTM C444; and "Precast Reinforced Concrete Manhole Sections," ASTM C478.

For general information the engineer is referred to "Concrete Pipe Handbook" (3) and "Concrete Sewers" (4).

Cast-in-place reinforced concrete sections are employed generally only when standard concrete pipe is not available or when special conditions preclude its use. Cast-in-place sections can be formed to almost any shape. Wide, flat culvert bottoms, however, should be avoided by providing a V-shaped invert of suitable dimensions for better low-flow characteristics. All forming should be smooth, unyielding, and tight.

Specifications for cast-in-place concrete generally set forth a 28-day strength, a minimum cover of 2 to 3 in. (5 to 8 cm) for reinforcing steel, a minimum slump consistent with desired workability, placement procedures to avoid separation, and a requirement that the concrete be vibrated in place with approved mechanical vibrators to obtain a dense concrete structure free of voids. Air-entraining cement or air-entrained admixtures may be used to insure a more dense concrete which will flow into place more easily. In long tunnels or in open cut with limited working space, concrete may be placed by pumping.

Methods of corrosion protection for cast-in-place concrete are the same as those described for concrete pipe in Section H.

The engineer is referred to standard ASTM specifications for reinforcing steel, concrete aggregates, and Portland cement.

F. IRON AND STEEL

1. Cast Iron Pipe

Cast iron pipe in diameters from 2 to 48 in. (5 to 122 cm) with a variety of jointing methods is used for pressure sewers, for sewers above ground, for submerged outfalls, for wastewater treatment plant piping, and for gravity sewers where very tight joints are essential. It is available in a number of thicknesses, classes, and strengths. The advantages of cast iron pipe include long laying lengths in some situations, tight joints, ability when properly designed to withstand relatively high internal design pressures and external loads, and corrosion resistance in most natural soils.

The disadvantages are that it may be subject to corrosion by acid or highly septic wastewater and by corrosive soils. A cement lining with a bituminous seal coating usually is specified on the interior of the pipe.

Cast iron pipe and fittings are specified by pipe diameter; thickness, class, or strength; the method of jointing (whether bell and spigot, flanged, mechanical or victaulic coupled, rubber or neoprene ring-type push-on joint, or ball and socket); the type of lining (whether cement, bituminous enamel lining, or both); and the types of exterior coating (whether asphalt varnish or otherwise).

The pipe should conform, insofar as appropriate, to the following standard specifications:

1. "Pipe, Cast-Iron, Pressure, (For Water and Other Liquids)," Federal Specification WW–P–421b.
2. "Cast Iron Pit Cast Pipe for Water and Other Liquids," USASI* A21.2; "Cast Iron Pipe Centrifugally Cast in Metal Molds, for Water or Other Liquids," USASI A21.6; "Cast-Iron Pipe Centrifugally Cast in Sand-Lined Molds, for Water or Other Liquids," USASI A21.8; "Cement-Mortar Lining for Cast-Iron Pipe and Fittings," USASI A21.4; "Cast-Iron Fittings, 2 Inch through 48 Inch for Water and Other Liquids," USASI A21.10; and "Rubber Gasket Joints for Cast-Iron Pressure Pipe and Fittings," USASI A21.11.
3. "Cast-Iron Pressure Fittings," AWWA C100; and "Cast Iron Water Pipes and Special Castings," AWWA 7C.1–1908.

Additional information relative to the selection of the proper pipe, joint, and coating may be found in the "Handbook of Cast Iron Pipe" (5).

---

*United States of America Standards Institute, New York, N.Y.

2. Ductile Iron Pipe

Ductile iron pipe is manufactured by adding cerium or magnesium to cast iron. The pipe is available generally in diameters from 2 to 48 in. (5 to 122 cm) and in lengths to 20 ft (6.1 m). Cast iron fittings are used with ductile iron pipe and jointing methods of pipe and fittings are similar to cast iron.

The advantages of ductile iron pipe include long laying lengths in some instances, tight joints, corrosion resistance comparable to that of cast iron, ability when designed properly to withstand relatively high internal design pressures and external loads, ductility, machinability, and toughness.

The disadvantages are that it may be subject to corrosion by acid or highly septic wastewater and by corrosive soils. A cement lining with a bituminous seal coating usually is specified on the interior of the pipe and an exterior coating similar to that used on cast iron pipe. Its cost is somewhat more than cast iron in the smaller sizes.

Ductile iron pipe should conform, insofar as appropriate, to the following standard specifications:

1. "Ductile-Iron Pipe, Centrifugally Cast in Metal Molds or Sand-Lined Mold, for Water or Other Liquids," USASI A21.51 (or AWWA C151).
2. "Thickness Design of Ductile-Iron Pipe," USASI A21.50 (or AWWA H3).
3. "Ductile Iron Castings," ASTM A536.

3. Fabricated Steel Pipe

(a) Corrugated Steel Pipe, Arches, and Pipe Arches.—Galvanized corrugated steel and iron are fabricated in a variety of conduit shapes with a choice of additional protective coatings when deemed necessary. Available sizes and shapes include: circular, in diameters from 8 to 20 in. (21 to 305 cm); pipe arches manufactured from circular pipe from 15 to 108 in. (38 to 274 cm) in diameter; structural plate structures of 60 to 252 in. (152 to 640 cm) diameter; structural plate arches from 5 to 25 ft (1.5 to 7.6 m) in span with concrete base; and tunnel liner plates, circular, arch, or horseshoe in shape from 4 to 15 ft (1.2 to 4.6 m) in span.

Strengths to meet a variety of design loads may be obtained by specifying from a range of gages (metal thicknesses), types of joint, and methods of ellipsing or strutting the pipe before installation.

Pipe sections generally are furnished up to 20 ft (6.1 m) in length in multiples of 2 ft (0.6 m). The sections are joined by coupling bands which may be single piece, two piece, or of an internal expanding type used in lining work. The band used for sanitary sewer construction with corrugated metal pipe is designed to withstand exterior water pressure much better than interior, and therefore may give misleading results if subjected to an exfiltration test.

Large sizes (structural plate conduits) are field-bolted. To increase durability and resist corrosion, galvanized pipe and plates can be coated. The interior can be lined with bituminous material to cover the crests of the corrugations, forming a smooth surface to improve flow characteristics, and, in addition, the invert may be paved with bituminous material.

Appurtenances include tees, wyes, elbows, and manholes fabricated from the same corrugated material. Corrugated pipe may be designed specially for jacking purposes.

The advantages of corrugated metal pipe, arches, and pipe arches include light weight, long laying lengths in some situations, strength, flexibility, and usefulness as a lining for the repair of existing structures.

The principal disadvantages are the difficulty in making house connection taps, relatively poor hydraulic coefficient, and the vulnerability of thin metal sheets to corrosion. Bituminous linings sometimes are used to improve the hydraulic coefficient and to impart corrosion resistance. External corrosion protection may be necessary depending on soil conditions. Bituminous coatings are flammable and may be damaged or destroyed by petroleum wastes or solvents. Continuous adequate lateral support is essential for the structural stability of the pipe.

Corrugated metal pipe, arches, or pipe arches are specified by size (nominal diameter, span and rise, or arc length), shape (full circle, pipe arch, or segmental plate arch), gage of metal (depending on strength requirements), assembly of sections (band or bolts), coatings or linings, and couplings (width of single piece and two piece).

The pipe should conform, insofar as appropriate, to the following standard specifications:

1. "Pipe, Corrugated (Iron or Steel, Zinc Coated)," Federal Specification WW–P–00405.
2. "Corrugated Metal Culvert Pipe," AASHO * M36.
3. "General Requirements for Delivery of Zinc Coated (Galvanized) Iron or Steel Sheets, Coils and Cut Lengths Coated by the Hot-Dip Method," ASTM A525.
4. "Zinc-Coated Steel Wire Strand," ASTM A475.

Coatings of zinc and asphalt, where required, should be in accordance with ASTM and American Railway Engineering Association (AREA) specifications. Additional information relating to corrugated metal pipe can be obtained from the "AISI Steel Highway Construction and Drainage Products Handbook," (6) and "Handbook of Drainage & Construction Projects" (7).

(b) Welded Steel Pipe.—Steel pipe has been designed in many ways and according to many different standards. Site locations and installation and operating conditions generally determine the practices which are followed. The pipe usually is jointed by welding or by flexible connections with Dresser or victaulic couplings.

Steel pipe is available in a variety of sizes, thicknesses, strengths, and

---

* American Association of State Highway Officials, Washington, D.C.

lengths. Its advantages include all of the above and its ease of fabrication, including fittings, to meet almost any design and shape requirement.

Its principle disadvantage is the necessity of providing adequate protection by coatings and linings of proven performance to withstand internal and external corrosion.

Steel pipe generally is not used for gravity sewers, but is used for pressure mains, river crossings, bridge crossings, and pumping station and treatment plant piping.

Steel pipe is specified by pipe diameter, thickness, strength, type of joint, coating or wrapping, lining, and length. It should conform, insofar as appropriate, to the following standard specifications:

1. "Fabricated Electrically Welded Steel Water Pipe," AWWA C201.
2. "Mill-Type Steel Water Pipe," AWWA C202.
3. "Coal-Tar Enamel Protective Coatings for Steel Water Pipe," AWWA C203.
4. "Cement-Mortar Protective Lining and Coating for Steel Water Pipe," AWWA C205.
5. "Welded and Seamless Steel Pipe," ASTM A53.

The engineer also is referred to "Design Standards for Steel Water Pipe" (8).

G. ORGANIC MATERIALS

In recent years organic materials, particularly plastics, have appeared on the market. Two products are described because their use has increased in some parts of the country; unfortunately, experience is limited and long-term results are not known.

(a) Solid Wall Plastic Pipe.—This pipe is available in sizes up to 12-in. (31-cm) diameter. It is light and usually comes in 10-ft (3.2-m) lengths. The jointing is generally of the sleeve type, chemically welded. Watertight seals are provided to prevent leakage at plastic-to-concrete manhole connections.

The advantages of solid wall plastic pipe include its light weight, tight joints, long laying lengths in some situations, and corrosion-resistant nature.

Disadvantages include thin walls, susceptibility to sunlight-induced changes (ultra-violet rays) which may affect shape and impact strength, limited sizes, and limited experience in its performance. Continuous adequate lateral support is essential for the structural stability of the pipe.

Solid wall plastic pipe is specified by pipe diameter, wall thickness if necessary, and type of plastic. The engineer is referred to ASTM specifications and manufacturers' publications for additional information and the plastics available.

(b) Truss Pipe.—This pipe is available in diameters from 8 to 15 in. (20 to 38 cm). It currently is manufactured by extruding ABS (Acrylonitrile-Butadiene-Styrene) thermoplastic into a truss with inner and outer walls connected by webs. The voids are filled with a lightweight filler. Jointing is accomplished by a chemical weld sleeve or compression gasket. Fittings are available. Watertight seals are provided to prevent leakage at plastic-to-concrete manhole connections.

Truss pipe has the advantage and disadvantages of solid wall plastic pipe.

Truss pipe is specified by pipe diameter. It currently is available in a single wall thickness, depending on diameter. The engineer is referred to manufacturers' publications and the following specification for additional information: "Acrylonitrile-Butadiene-Styrene (ABS) Composite Pipe," ASTM O2680.

H. CORROSION PROTECTION

1. General

Insofar as possible, systems should be designed and operated to be sulfide-free, a problem discussed in detail in Chapter 6. When corrosion cannot be prevented by design, maintenance, or control of wastes entering the sewer, consideration must be given to corrosion-resistant materials, vitrified clay, or to protective linings of proven performance. Plastic pipes also may be used if acceptable in all other respects.

2. Concrete and Asbestos-Cement Pipe Protection

Where corrosion is anticipated, efforts to protect or extend the life of pipe may go in one of two directions: application of an acid-resistant barrier or alterations in the composition of the pipe materials.

(*a*) **Protective Barriers.**—Protecting concrete and asbestos-cement pipe against acid attack by means of a barrier is difficult. Coatings and linings of bituminous and coal-tar products, vinyl and epoxy resins, and paints have been used with varied success for the protection of pipes and structures. If any acid seeps or diffuses through the lining at any point, even through a pinhole, reaction with the cement ensues and the effectiveness of the lining is destroyed. To be effective, the lining, including joints, must be sealed completely to protect the sewer system throughout its expected life, usually 50 yr or longer.

A lining used on large-diameter concrete pipe that has proved reasonably satisfactory is a plasticized polyvinyl chloride sheet, having T-shaped protections on the back which key into the pipe wall at the time of manufacture. All lining seams and joints between pipes must be sealed completely. It is reasonable to expect this lining to have a life of at least 50 yr if properly installed.

(*b*) **Composition of Materials.**—Type II Portland cement sometimes is specified for the manufacture of pipe because it is somewhat superior to Type I in resisting sulfate attack. Sulfate attack, however, is not common in concrete sewers, in all areas. The type of cement does not affect resistance to acid attack.

On concrete pipe, extra wall thickness (sacrificial concrete) sometimes is specified to increase pipe life in the event corrosive conditions develop. On reinforced concrete this takes the form of added cover over the inner reinforcing steel.

Another method of modifying the composition of concrete is by the use of limestone or dolomite aggregate in the manufacture of the pipe (9) (10). The use of such aggregates increases the amount of acid-soluble material in the concrete which prolongs the life of the pipe in corrosive environments. The rate of acid attack of limestone or dolomite aggregate pipe may be only about one-fifth as great as when granitic aggregate is used. Unfortunately, not all limestone and dolomite aggregrates exhibit the same resistance to this form of corosion (11). Accordingly, tests should be made before limestone or dolomitic aggregate is used. Much limestone aggregate pipe has been installed in the United States, especially in California and Arizona.

The practicability of using limestone or dolomite aggregate in pipe also varies throughout the country with its availability.

3. Iron and Steel Pipe Protection

The interior of cast iron and ductile iron pipe usually is lined with cement mortar as described in Section F. Steel pipe sometimes is lined similarly. Smooth-walled steel pipe also may be protected by cementing plasticized polyvinyl chloride sheets to the pipe and sealing the joints.

Corrugated metal pipe may be coated inside and out with bituminous material. For added protection asbestos fibers may be embedded in the molten zinc before it is bituminous coated (asbestos bonded). Such coatings should be of impermeable material of sufficient thickness and free of flaws such as holidays and pin holes.

4. Other Linings

Other protective linings have been used or are being offered for use on various pipe materials. Before using them, the engineer should require that information be submitted on the linings' life expectancy and other relevant data, such as the results of tests in which pipes and joints have been held upright and kept filled with five-percent sulfuric acid for at least a year.

I. JOINTS

A good joint must be water-tight, root-resistant, flexible, and durable. In the past, joints for bell-and-spigot pipe in sanitary sewers generally were made with cement mortar. With careful workmanship, such joints were initially water-tight and root-resistant, but often the pipe and joints later failed because of the rigidity of the joints. Where sulfide conditions prevailed, breakage of bells or splitting of the pipe occurred because sulfuric acid diffused to the mortar and caused expansion, even though the pipe material itself was immune to acid attack. On the other hand, for storm drains, these objections may not be as important and mortar joints still are used extensively.

Eventually bituminous materials were developed and used extensively to overcome the objections to mortar joints. Hot-poured formulations were preferred over cold types. Using the proper material, poured against primed clay or concrete surfaces in the absence of moisture, satisfactory joints were obtained. Under average construction conditions, however, water tightness could not be assured. Nevertheless, both hot-poured and cold-packed bituminous joints are used in various areas of the country and represent a considerable improvement over mortar joints.

In the 1950's molded joints on clay pipe became available, first as a plasticized polyvinyl chloride (PVC) cast on the bell and spigot so that a tight joint was made when the pipes were pushed together. Because of dimensional changes of the plasticized material, the joints did not always remain tight, particularly when subjected to shear stresses.

Polyurethane has been substituted since for polyvinyl chloride, producing a joint that appears to be fully satisfactory because it has the ability to maintain compression without yielding. Clay pipe manufacturers now produce pipe having a polyester compound cast on to the spigot and into the bell, with a rubber ring making a seal between these surfaces. The polyester serves to provide dimensional uniformity in the joint space.

Collar joints are sometimes used on clay pipe. A rubber sleeve joint is frequently used to join pipes of different materials.

Asbestos-cement manufacturers adopted a collar joint for their pipe, equipped with two rubber rings to make seals around the two inserted spigots. The tightness of the rubber ring joint is attested by its extensive use in water distribution systems.

Concrete pipe manufacturers employ the rubber-ring principle and compression rubber gasket joints on bell and spigot and tongue and groove concrete pipe are being used successfully. Steel and cast iron pipes, too, are joined commonly by gaskets of this type.

The round-rubber gasket joints often have been referred to as O-ring joints, alluding to the circular cross-section of the gasket. The circular section, however, is not essential and many gaskets are of other shapes. The important requirement is that the gasket be under continuous, substantial compression and that it be restrained against moving either way in the assembled joint. The joint must be able to pass a test for tightness not only when the pipes are in perfect alignment, but also when subjected to maximum shear.

The term rubber as used here, refers to any rubber-like elastomer. Neoprene is used when it is believed that there is danger of attack by hydrocarbons. Moreover, some engineers specify that neoprene or oil-resistant gaskets be used in ordinary sanitary sewers.

Tight joints are necessary if a low rate of infiltration is to be maintained. The data contained in Table XXV were taken from a study by Santry (12) on small-sized clay and concrete pipe joints and give some idea of the magnitude of the problem.

TABLE XXV.—Infiltration at Joints

Joint Material	6-In. Diam Clay Pipe		Joint Material	6-In. Diam Concrete Pipe	
	Head Above Flow Line (in.)	Avg Infiltration Rate (gpd/in. diam/mile)		Head Above Flow Line (in.)	Avg Infiltration Rate (gpd/in. diam/mile)
Jute only	3	8,270	Jute only	3	6,710
	9	71,050		9	52,800
	15	155,250		15	118,000
	21	258,000		21	205,500
	27	356,000		27	278,000
Cement	3	3,360	Cement	3	680
	9	15,000		9	4,950
	15	28,700		15	10,450
	21	41,200		21	16,500
	27	53,200		27	22,000
Hot Pour	3	1,330			
	9	1,660	Cold Mastic	3	990
	15	3,410		9	1,450
	21	4,720		15	3,210
	27	5,520		21	5,130
				27	7,810
PVC Compression Joint	3	0	Hot Pour	3	0
	9	645		9	107
	15	1,450		15	235
	21	1,850		21	419
	27	2,400		27	513
Compression Joint B	0 to 27	Negl.	Rubber Gasket	0 to 27	Negl.

Note: In. $\times$ 2.54 = cm; gpd/in. diam/mile $\times$ 0.000925 = cu m/day/cm diam/km.

These results were obtained under laboratory conditions, but the superior performance of the compression-type joints is apparent.

The joint designated as PVC compression joint was the initial design used by one manufacturer prior to the establishment in 1958 of tentative ASTM C425 specification. The use of both a lubricant and a sealer was required in the installation of this joint. Compression Joint B was a joint conforming to ASTM C425 requirements.

References

1. "Installation Guide—Transite Ring-Tite Pressure Pipe," Johns-Manville Co., New York, N.Y. (1960).
2. "Clay Pipe Engineering Manual," National Clay Pipe Institute, Crystal Lake, Ill.
3. "Concrete Pipe Handbook," American Concrete Pipe Assn., Arlington, Va.
4. "Concrete Sewers," Portland Cement Assn., Chicago, Ill.
5. "Handbook of Cast Iron Pipe," Cast Iron Pipe Research Assn., Chicago, Ill.
6. "AISI Steel Highway Construction and Drainage Products Handbook."
7. "Handbook of Drainage & Construction Products," Armco Drainage and Metal Products, Inc., Middletown, Ohio (1958).
8. Barnard, R. E., "Design Standards for Steel Water Pipe." *Jour. Amer. Water Works Assn.*, **40**, 24 (1948).

9. Stutterheim, N., and Van Aardt, J. H. P., "Corrosion of Concrete Sewers and Some Remedies." *S. African Ind. Chem.* No. 10 (1953).
10. Pomeroy, R. D., "Protection of Concrete Sewers in the Presence of Hydrogen Sulfide." *Water and Sew. Works* **107,** 400-3 (1960).
11. Swab, B. H., "Effects of Hydrogen Sulfide on Concrete Structures." *Jour. San. Eng. Div., Proc. Amer. Soc. Civil Engr.,* **87,** SA5, 1 (1961).
12. Santry, I. W., Jr., "Infiltration in Sanitary Sewers." *Jour. Water. Poll. Control Fed.,* **36,** 1256 (1964).

CHAPTER 9. STRUCTURAL REQUIREMENTS

A. INTRODUCTION

The structural design of a sewer requires that the supporting strength of the conduit as installed, divided by a suitable factor of safety, must equal or exceed the loads imposed on it by the weight of earth and any superimposed loads.

This chapter presents generally accepted criteria and methods for determining loads and supporting strength, as well as procedures for combining these elements with the application of a factor of safety to produce a safe and economical design.

Methods are presented for estimating probable maximum loads due to gravity earth forces and for both static and moving superimposed loads. Where so noted, the methods apply to rigid and flexible conduits in the three most common conditions of installation: in a trench in natural ground; in an embankment; and in a tunnel.

The supporting strength of buried conduits is a function of installation conditions as well as the strength of the pipe itself. This chapter presents procedures for determining the field or installed supporting strength of rigid sewer pipe based on its established relationship to the laboratory test strength. It also presents a brief discussion of the method of determining the safe supporting strength of flexible pipe based on a semi-empirical equation for deflection.

Since installation conditions have such an important effect on both load and supporting strength, a satisfactory sewer construction project requires attainment of design conditions in the field. Therefore, this chapter also includes a section on recommendations for construction and field observations to achieve this goal.

This chapter does not include information on reinforced concrete design or design of the conduit section. Reference should be made to standard textbooks and to ASTM specifications or industry handbooks for such design data.

B. LOADS ON SEWERS DUE TO GRAVITY EARTH FORCES

1. General Method

Marston (1) (2) developed methods for determining the vertical load on buried conduits due to gravity earth forces in all of the most commonly encountered construction conditions. His methods are based on both theory and experiment and have achieved acceptance as being the most useful and reliable. In general, the theory states that the load on a buried conduit is equal to the weight of the prism of earth directly over it, called the interior prism, plus or minus the frictional shearing forces transferred to that prism by the adjacent prisms of earth—the magnitude and direction of these frictional forces being a function of the relative settlement between the interior and adjacent earth prisms. The theory makes the following assumptions:

(a) The calculated load is the load which will develop when ultimate settlement has taken place.
(b) The magnitude of the lateral pressures which induce the shearing forces between the interior and adjacent earth prisms is computed in accordance with Rankine's theory.
(c) Cohesion is negligible except for tunnel conditions.

The general form of Marston's equation is

$$W = CwB^2 \quad \dots \dots \dots \dots \dots \dots \dots \dots \dots \dots 1$$

in which W is the vertical load per unit length acting on the conduit due to gravity earth loads; w is the unit weight of earth per unit volume; B is the trench width or conduit width, depending on installation conditions; and C is a dimensionless coefficient that measures the effect of:

(a) Ratio of the height of fill to width of trench or conduit,
(b) Shearing forces between interior and adjacent earth prisms, and
(c) Direction and amount of relative settlement between interior and adjacent earth prisms for embankment conditions.

2. Types of Loading Conditions

Although the general form of Marston's equation includes all the factors necessary to analyze all types of installation conditions, it is convenient to classify these conditions, write a specialized form of the equation, and prepare separate graphs and tables of coefficients for each.

The accepted system of classification is shown diagrammatically in Figure 43 and is described briefly below:

Trench conditions are defined as those in which the conduit is installed in a relatively narrow trench cut in undisturbed ground and covered with earth backfill to the original ground surface.

Embankment conditions are defined as those in which the conduit is covered with fill above the original ground surface or when a trench in undisturbed ground is so wide that trench wall friction does not affect the load on the pipe. The embankment classification is further subdivided into two major subclassifications—positive projecting and negative projecting. Conduits are defined as positive projecting when the top of the conduit is above the adjacent original ground surface. Negative projecting conduits are those installed with the top of conduit below the adjacent original ground surface in a trench which is narrow with respect to the size of pipe and depth of cover (Figure 43) and when the native material is of sufficient strength that the trench shape can be maintained dependably during the placing of the embankment.

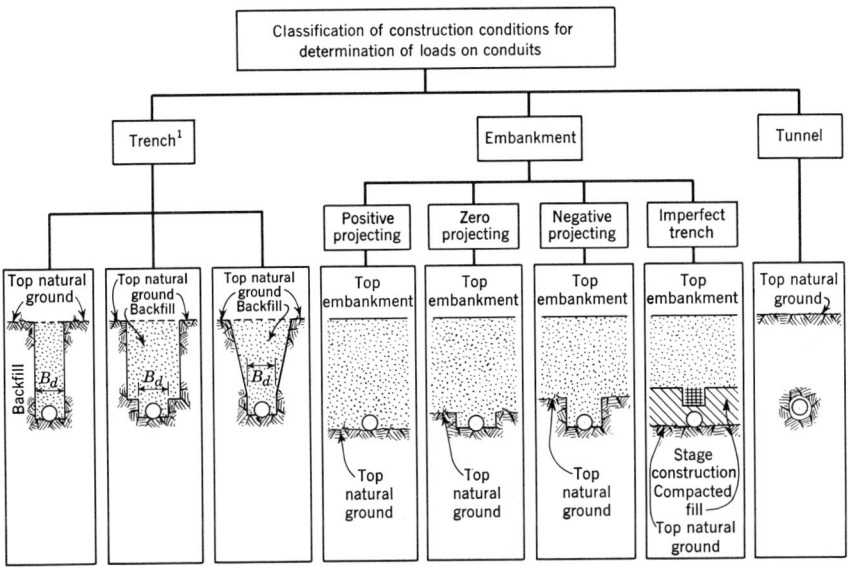

¹ B_d within required limit, otherwise conduit is positive projecting

FIGURE 43.—Classification of construction conditions.

A special case, called the imperfect trench condition, may be employed to minimize the load on a conduit under embankments of unusual height.

3. Loads for Trench Conditions

Sewers usually are constructed in ditches or trenches which are excavated in natural or undisturbed soil, and then covered by refilling the trench to the original ground line. This construction procedure often is referred to as "cut and cover," or "cut and fill."

(a) **Load-Producing Forces.**—The vertical load to which a sewer pipe is subjected, when so constructed, is the resultant of two major forces: the first is the weight of the prism of soil within the trench and above the top of the pipe; and the second is the friction or shearing forces generated between the prism of soil in the trench and the sides of the trench.

The backfill soil has a tendency to settle in relation to the undisturbed soil in which the trench is excavated. This downward movement or tendency for movement induces upward shearing forces which support a part of the weight of the backfill. Thus, the resultant load on the horizontal plane at the top of the pipe within the trench is equal to the weight of the backfill minus these upward shearing forces, as indicated in Figure 44.

(b) **Marston's Formula.**—Marston's formula for loads on rigid conduits in trench conditions is

$$W_c = C_d w B_d^2 \qquad \qquad \qquad \qquad \ldots 2$$

in which W_c is the load on the pipe in lb/ft (kg/m); w is the unit weight of backfill soil in lb/cu ft (kg/cu m); B_d is the width of trench at the top of the pipe in ft (m); C_d is a dimensionless load coefficient which is a

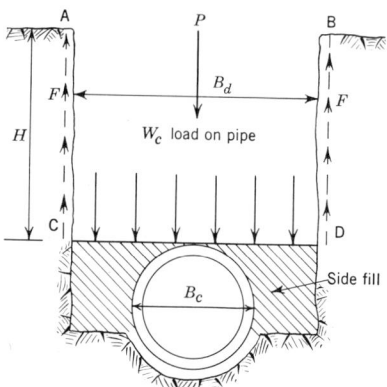

FIGURE 44.—Load-producing forces: P = weight of backfill $ABCD$; F = upward shearing forces on AC and BD; and $W_c = P - 2F$.

function of the ratio of height of fill to width of trench and of the friction coefficient between the backfill and the sides of the trench. The load coefficient, C_d, is computed as follows:

$$C_d = \frac{1 - e^{-2k\mu'H/B_d}}{2k\mu'} \quad \quad \quad \quad \quad \quad \quad \quad \quad 3$$

in which e is the base of natural logarithms; k = Rankine's ratio of lateral pressure to vertical pressure

$$= \frac{\sqrt{\mu^2+1}-\mu}{\sqrt{\mu^2+1}+\mu} = \frac{1-\sin\phi}{1+\sin\phi} \quad \quad \quad 4$$

in which $\mu = \tan\phi$ = the coefficient of internal friction of backfill material; $\mu' = \tan\phi'$ = the coefficient of friction between backfill material and sides of trench (μ' may be equal to or less than μ, but never greater than μ); and H is the height of fill above top of pipe in ft (m). The value of C_d for various ratios of H/B_d and various types of soil backfill may be obtained from Figure 45.

The trench load formula, Equation 2, gives the total vertical load on a horizontal plane at the top of the pipe. If the pipe is rigid it will carry practically all this load. If the pipe is flexible and the soil at the sides is compacted to the extent that it will deform under vertical load the same amount as the pipe itself, the side fills may be expected to carry their proportional share of the total load. Under these circumstances the trench load formula may be modified to

$$W_c = C_d w B_c B_d \quad \quad \quad \quad \quad \quad \quad \quad 5$$

in which B_c is the outside width of pipe in ft (m).

The term "side fill" refers to the soil backfill which is placed between the sides of a pipe and the sides of the trench. The character of this material and the manner of its placement have two important influences on the structural behavior of a pipe.

STRUCTURAL REQUIREMENTS 189

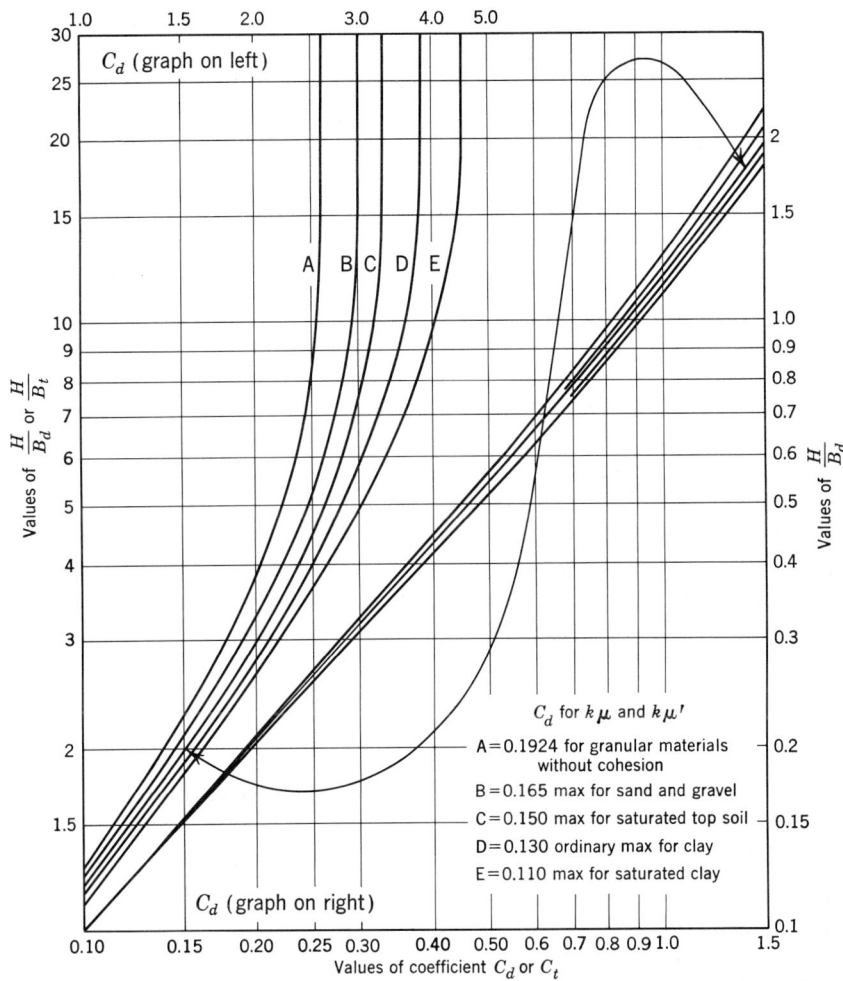

FIGURE 45.—Computation diagram for earth loads on trench conduits (completely buried in trenches).

First, the side fill may carry a part of the total vertical load on the horizontal plane at the elevation of the top of the pipe. If the side fill is relatively yielding as compared with the stiffness of the pipe, practically all the load will be carried by the pipe. If the side fill is of about the same stiffness as the pipe to the extent that it will deform under vertical load the same amount as the pipe itself (a condition which may exist when the pipe is flexible and the side fill is tamped), it may carry its proportional share of the total load on the plane.

Second, the side fill plays an important role in helping the pipe carry vertical load. Every pound of pressure which can be brought to bear against the sides of an elastic ring increases the ability of the ring to carry vertical load by nearly the same amount. This fact points up the desirability of tamping the side fills in the case of rigid pipe, and the absolute necessity in the case of flexible pipe. However, caution should be exercised when tamping the soil at the sides of non-reinforced rigid pipe to prevent damage to the pipe.

(c) Influence of Width of Trench.—Examination of Equation 2 indicates the important influence width of the trench exerts on the load as long as the trench condition formula applies. This influence has been verified by extensive experimental evidence. These experiments also have indicated that the width of trench at the top of the pipe is the controlling factor.

Depending on height of fill above the pipe, the width of trench below the top of the pipe must not be permitted to exceed the safe limit for the strength of pipe and class of bedding used. The minimum width must be consistent with the provision of sufficient working space at the sides of the pipe to assemble joints properly, to insert and strip forms, and to compact backfill. The designer must allow reasonable tolerance in width for variations in field conditions and accepted construction practice.

The position of the lower wale usually will determine the proper width of trench from face-to-face of sheeting, where sheeting and bracing are required. A working-room allowance of 12 in. (30 cm) from each side of the pipe, pipe cradle, or monolithic conduit to the face of the sheeting is a workable minimum for small- and medium-sized pipe for trenches up to about 14 ft (4 m) deep.

At any given depth and for any given conduit size there is a certain limiting value to the width of trench beyond which no additional load is transmitted to the conduit. This limiting value is called the "transition width" (3). There are sufficient experimental data at hand to show that it is safe to calculate the imposed load by means of the trench-conduit formula (Equation 2) for all widths of trench less than that which gives a load equal to the load calculated by the projecting-conduit formula (Equation 6). In other words, as the width of the trench increases, other factors remaining constant, the load on a rigid conduit increases in accordance with the theory for a trench conduit until it equals the load determined by the theory for a projecting conduit. The width of trench at which this transition occurs may be determined from the diagram in Figure 46. [The term, $r_{sd}p$, is defined in Section B4b(2).]

It is advisable in the structural design of conduits to evaluate the effect of the transition width on both the design criteria and the construction latitude. A contractor, for instance, may wish to place well points for drainage in the trench. If this requires a wider trench than usual, a stronger pipe or higher class of bedding may be necessary.

It may be economical and proper to excavate the trench with sloping sides in undeveloped areas where no inconvenience to the public or danger to property, buildings, subsurface structures, pavements, etc., will result. A subditch (Figure 47) may be employed in such cases to minimize the load on the pipe.

When sheeting of the trench at the pipe is necessary, it should extend about 1.5 ft (45 cm) above the top of the conduit. It is recommended that this sheeting and bracing be left in place. The load on the conduit should

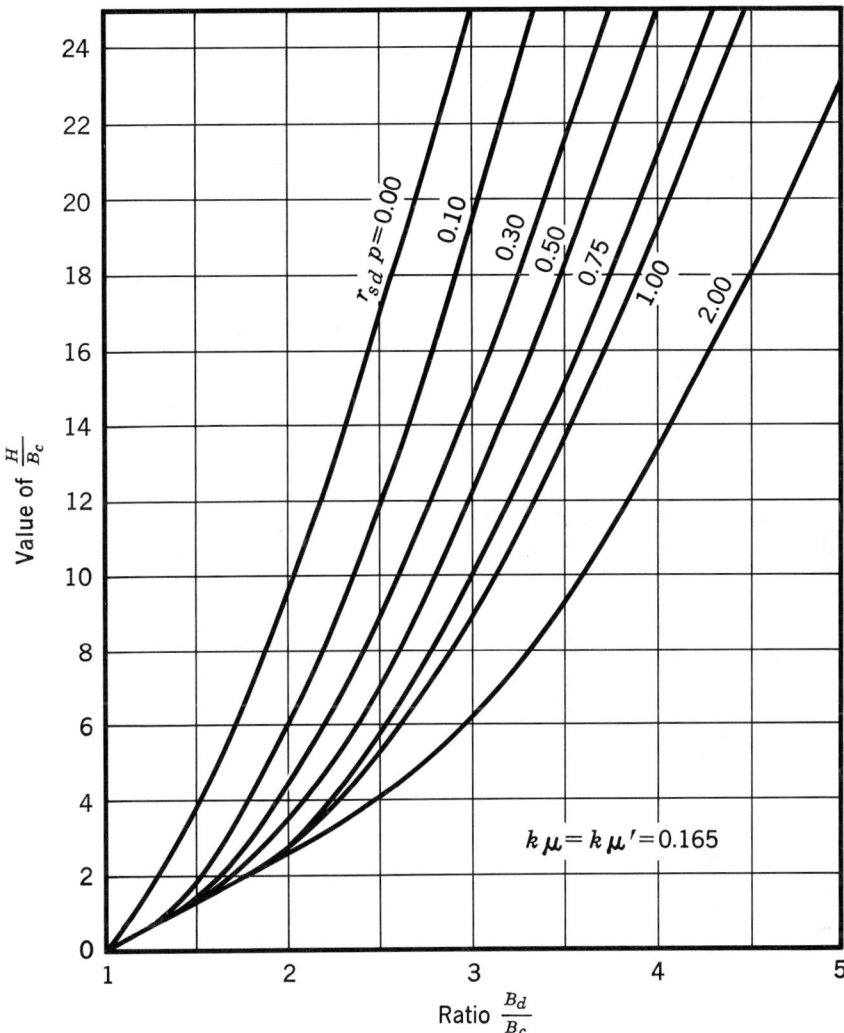

FIGURE 46.—Values of B_d/B_c at which the trench conduit and projecting conduit load formulas give equal loads (3).

be computed for a width of trench B_d equal to the distance to the outside of the sheeting if it is to be removed, or to the inside if left in place.

If a shield is employed in pipe-laying operations, the width of the shield controls the width of the trench at the top of the pipe. This width should be the width factor used in computing loads on the pipe.

Conduits which are to be constructed in sloping sided trenches with the slopes extending to the invert, or to any plane above the invert but below the top of the structure, should be designed for loads computed by using the actual width of the trench at the top of the pipe, or by the projecting-conduit formula (Equation 6), whichever gives the least load on the pipe.

If for any reason the trench becomes wider than that specified and for

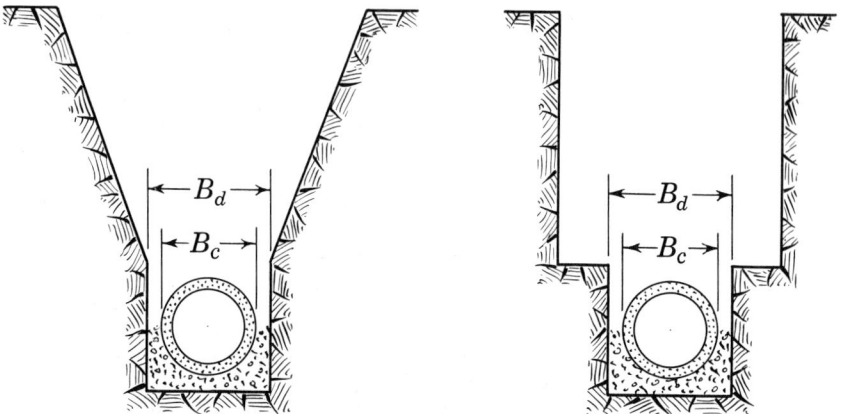

FIGURE 47.—Examples of subditch.

which the pipe was designed, the load on the pipe should be checked and a stronger pipe furnished or higher class bedding used, if necessary.

Example 1. Determine the load on a 24-in. diam rigid pipe under 14 ft of cover in trench conditions.

Assume that the pipe wall thickness is 2 in.; $B_c = 24+4 = 28$ in. $= 2.33$ ft; $B_d = 2.33+2.00 = 4.33$ ft; and $w = 120$ lb/cu ft for saturated top soil. Then $H/B_d = 14/4.33 = 3.24$; C_d (from Figure 45) $= 2.1$; and $W_c = 2.1 \times 120 \times 4.33^2 = 4{,}720$ lb/ft (7,030 kg/m).

Example 2. Determine the load on the same size conduit laid on a concrete cradle and with trench sheeting to be removed.

Assume that the wall thickness is 2 in.; the cradle projection of the outside pipe is 8 in. (4 in. on each side); and the maximum clearance between cradle and outside of sheeting is 14 in. Then $B_d = 24 + (2 \times 2$ in.$) + 8 + (2 \times 14) = 64$ in. $= 5.33$ ft.

As this seems to be an extremely wide trench, a check should be made on the transition width of the trench; $B_c = 2.33$ ft; $H = 14$ ft; $r_{sd}p = 0.5$; and $H/B_c = 14/2.33 = 6.0$.

From Figure 46, $B_d/B_c = 2.39$ (the ratio of the width of the trench to the width of the conduit at which loads are equal by both ditch-conduit theory and projecting-conduit theory); $B_d = 2.33 \times 2.39 = 5.80 > 5.33$; $H/B_d = 14/5.33 = 2.61$; C_d (from Figure 45) $= 1.85$; and $W_c = 1.85 \times 120 \times 5.33^2 = 6{,}300$ lb/ft (9,350 kg/m).

Example 3. Determine the load on the same conduit if (rough) sheeting is left in place.

B_d becomes 4 in. less $= 5.00$ ft; $H/B_d = 14/5.00 = 2.8$; C_d (from Figure 45) $= 1.92$; and $W_c = 1.92 \times 120 \times 5.00^2 = 5{,}750$ lb/ft (8,570 kg/m).

Example 4. Determine the load on a 30-in. diam flexible conduit installed in a trench 4 ft 6 in. wide at a depth of 12 ft.

Assume the soil is clay weighing 120 lb/cu ft and that it will be well compacted at the sides of the pipe. Then $H=12$ ft; $B_d=4.5$ ft; $H/B_d=2.67$; $C_d=1.9$; and $W_c=1.9\times 120\times 4.5\times 2.58=2,650$ lb/ft (3,950 kg/m).

(*d*) **Soil Characteristics—Trench Conditions.**—The load on a sewer pipe is influenced directly by the unit weight of the soil backfill. This value varies widely for different soils, from a minimum of about 100 lb/cu ft (1,600 kg/cu m) to a maximum of about 135 lb/cu ft (2,200 kg/cu m). The average maximum unit weight of the soil which will constitute the backfill over the pipe may be determined by density measurements in advance of the structural design of the pipe. A design value of not less than 120 or 125 lb/cu ft (1,900 or 2,000 kg/cu m) is recommended if such measurements are not made.

The load also is influenced by the coefficient of friction between the backfill and the sides of the trench and by the coefficient of internal friction of the backfill soil. Ordinarily these two values will be nearly the same and may be so considered for design purposes, as in Figure 45, but in special cases this may not be true. For example if the backfill is sharp sand and the sides of the trench are sheeted with finished lumber, μ may be substantially greater than μ'. Unless specific information to the contrary is available, values of the products $k\mu$ and $k\mu'$ may be assumed to be the same and equal to 0.130 (ordinary maximum for clay, Figure 45). If the backfill soil is a "slippery" clay and there is a possibility that it will become very wet shortly after being placed, $k\mu$ and $k\mu'$ equal to 0.110 (maximum for saturated clay, Figure 45) should be used.

4. Loads for Embankment Conditions

(*a*) **General.**—A sewer is described as a projecting conduit when it is installed in a wide trench or in such a manner that the top of the conduit is at or near the natural ground surface or the surface of thoroughly compacted soil and subsequently is covered with an embankment. If the top of the conduit projects some distance above the natural ground surface or if it is installed in a wide trench, it is a positive projecting conduit. There are, however, other methods of installing conduits under embankments which have the favorable effect of minimizing the load on the conduit. In these cases, the installation is classified as a negative projecting conduit or an imperfect trench conduit (Figure 43).

These variations of embankment conditions will be treated separately for convenience in computation.

(*b*) **Positive Projecting Conduits.**—The load on a positive projecting conduit is equal to the weight of the prism of soil directly above the structure, plus (or minus) vertical shearing forces which act on vertical planes extending upward into the embankment from the sides of the conduit. These vertical shearing forces ordinarily do not extend to the top

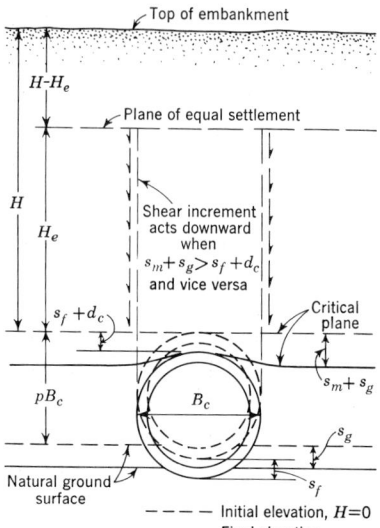

FIGURE 48.—Settlements that influence loads on positive projecting conduits: s_g = settlement of natural ground adjacent to conduit, s_m = compression of columns of soil of height pB_c, d_c = deflection of the conduit, and s_f = settlement of bottom of conduit.

of the embankment, but terminate in a horizontal plane at some elevation above the top of the conduit known as the "plane of equal settlement" as shown in Figure 48. The shear increment acts downward when $s_m + s_g > s_f + d_c$ and vice versa. In this expression s_m is the compression of the columns of soil of height pB_c; s_g is the settlement of the natural ground adjacent to the conduit; s_f is the settlement of the bottom of the conduit; and d_c is the deflection of the conduit.

1. Marston's Formula.—Marston's formula for loads on rigid and flexible positive projecting conduits is written

$$W_c = C_c w B^2_c \dots\dots\dots\dots\dots\dots\dots\dots\dots\dots 6$$

in which W_c is the load on the conduit in lb/ft (kg/m); w is the unit weight of the soil in lb/cu ft (kg/cu m); B_c is the outside width of the conduit in ft (m); and C_c is the load coefficient. Values of C_c may be obtained from Figure 49. In this diagram, H is the height of fill above the top of the conduit in ft (m); B_c is the outside width of conduit in ft (m); p is the projection ratio; and r_{sd} is the settlement ratio (the latter two terms are defined immediately below).

2. Influence of Environmental Factors.—The shear component of the total load on a sewer under an embankment depends on two factors associated with the conditions under which the conduit is installed. These are the projection ratio, p, and the settlement ratio, r_{sd}.

The projection ratio, p, is defined as the ratio of the distance that the

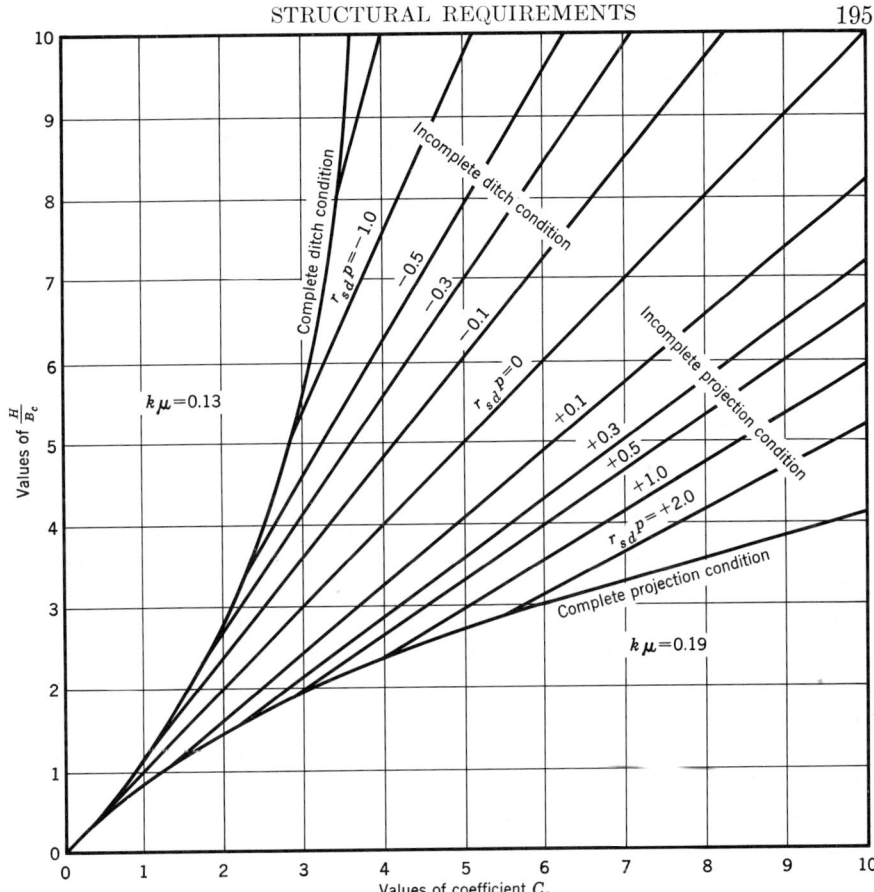

FIGURE 49.—Diagram for coefficient C_c for positive projecting conduits.

top of the conduit projects above the adjacent natural ground surface, or the top of thoroughly compacted fill, or the bottom of a wide trench, to the vertical outside height of the conduit. It is a physical factor that can be determined in advanced stages of planning when the size of the conduit and its elevation have been established.

The settlement ratio, r_{sd}, indicates the direction and magnitude of the relative settlements of the prism of soil directly above the conduit and of the prisms of soil adjacent thereto. These relative settlements generate the shearing forces which combine algebraically with the weight of the central prism of soil to produce the resultant load on the conduit. The settlement ratio is the quotient obtained by taking the difference between the settlement of the horizontal plane in the adjacent soil which was originally level with the top of the conduit (the critical plane) and the settlement of the top of the conduit and dividing the difference by the compression of the columns of soil between the natural ground surface and the level of the top of the conduit. The formula for the settlement ratio is

$$r_{sd} = \frac{(s_m + s_g) - (s_f + d_c)}{s_m} \quad \dots \dots \dots \dots \dots \dots .7$$

in which r_{sd} is the settlement ratio (positive projecting conduit); s_g is the settlement of the natural ground adjacent to the conduit; s_m is the compression of the columns of soil of height pB_c; (s_m+s_g) is the settlement of the critical plane; d_c is the deflection of the conduit, that is, the shortening of its vertical dimension; s_f is the settlement of the bottom of the conduit; and (s_f+d_c) is the settlement of the top of the conduit.

The elements of the settlement ratio are shown in Figure 48. When the settlement ratio is positive, the shearing forces induced along the sides of the central prism of soil are directed downward and the load on the conduit is greater than the weight of the central prism. When the settlement ratio is negative, the shearing forces act upward and the load is less than the weight of the central prism.

The numerical magnitude of the product of the projection ratio and the settlement ratio, $r_{sd}p$, is an indicator of the relative height of the plane of equal settlement and, therefore, of the magnitude of the shear component of the load. The plane of equal settlement is at the top of the conduit when this product is equal to zero. There are no induced shearing forces in this case and the load is equal to the weight of the central prism.

It is not practicable to predetermine a value of the settlement ratio by estimating the magnitude of its various elements except in very general terms. Rather, it should be treated as an empirical factor. Recommended design values of r_{sd}, based on measured settlements of a number of actual conduits, are:

Type of Conduit	Soil Conditions	Settlement Ratio, r_{sd}
Rigid	Rock or unyielding foundation	+1.0
Rigid	Ordinary foundation	+0.5 to +0.8
Rigid	Yielding foundation	0 to +0.5
Rigid	Negative projecting installations	−0.3 to −0.5
Flexible	Poorly-compacted side fills	−0.4 to 0
Flexible	Well-compacted side fills	0

3. Embankment Soil Characteristics.—The load on a projecting conduit is influenced directly by the unit weight of the embankment soil. If the soil is to be compacted to a specified dry density, the corresponding wet density under normal moisture conditions should be used in calculating the load. A design value of not less than 120 or 125 lb/cu ft (1,900 or 2,000 kg/cu m) is recommended if specific information relative to unit weight of soil is not available.

The load also is influenced by the coefficient of internal friction of the embankment soil. Recommended values of the product $k\mu$ are as follows (also see Figure 49):

For a positive settlement ratio, $k\mu=0.19$,

For a negative settlement ratio, $k\mu=0.13$.

Example 5. Determine the load on a 48-in. diam reinforced concrete pipe installed as a positive projecting conduit under a fill 32 ft high above the top of the pipe. The wall thickness of the pipe is 5 in. and the fill weighs 125 lb/cu ft.

Assume the projection ratio is +0.5 and the settlement ratio is +0.6. Then $H=32$ ft; $B_c=4.83$ ft; $H/B_c=6.63$; $r_{sd}p=0.5\times0.6=0.3$; C_c (from Figure 49) $=10.0$; and $W_c=10.0\times125\times4.83^2=29,100$ lb/ft (43,400 kg/m).

(c) Negative Projecting Conduits and Imperfect Trench Conduits.

A negative projecting conduit (Figure 50) is one installed in a relatively shallow trench with its top at some elevation below the natural ground surface. The trench above the conduit is refilled with loose, compressible material, and the embankment is constructed to finished grade by ordinary methods.

Sometimes straw, hay, cornstalks, sawdust, or similar materials may be added to the trench backfill to augment the settlement of the interior prism. The greater the value of the negative projection ratio, p', and the more compressible the trench backfill over the conduit, the greater will be the settlement of the interior prism of soil in relation to the adjacent fill material. In using this technique, the plane of equal settlement must fall below the top of the finished embankment. This action generates upward shearing forces which relieve the load on the conduit.

An imperfect trench conduit (Figure 51) first is installed as a positive projecting conduit. The embankment then is built up to some height above the top and thoroughly compacted as it is placed. A trench of the same width as the structure next is excavated directly over the conduit down to or near its top. This trench is refilled with loose, compressible material, and the balance of the embankment is completed in a normal manner.

The formula for loads on negative projecting conduits is

$$W_c = C_n w B^2{}_d \dots\dots\dots\dots\dots\dots\dots\dots 8$$

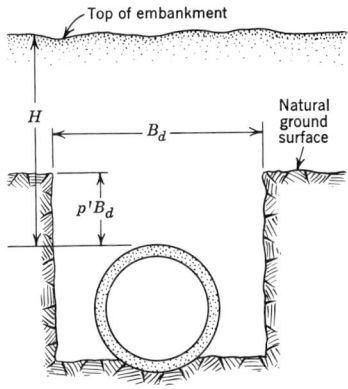

FIGURE 50.—Negative projecting conduit.

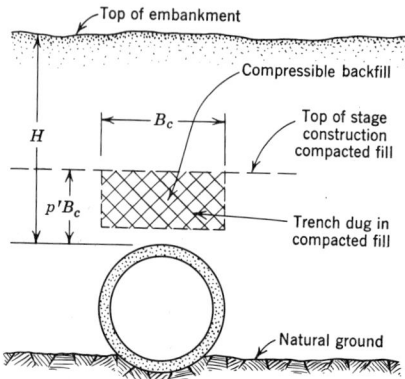

FIGURE 51.—Imperfect trench conduit.

in which W_c is the load on the conduit in lb/ft (kg/m); w is the unit weight of soil in lb/cu ft (kg/cu m); B_d is the width of the trench in ft (m); C_n is the load coefficient (Figure 52), a function of H/B_d or H/B_c, p', and r_{sd}; p' is the projection ratio; and r_{sd} is the settlement ratio as defined below.

In the case of imperfect trench conduits, B_c is substituted for B_d in Equation 8 in which B_c is the width of the pipe in ft (m), assuming the trench in fill is no wider than the pipe.

The projection ratio, p', is equal to the vertical distance from the firm ground surface down to the top of the conduit divided by the width of the trench, B_d, in the case of negative projecting conduits, or by the width of the conduit, B_c, in the case of imperfect trench conduits.

The settlement ratio, r_{sd}, for these cases is the quotient obtained by taking the difference between the settlement of the firm ground surface and the settlement of the plane in the trench backfill which was originally level with the ground surface (the critical plane) and dividing the difference by the compression of the column of soil in the trench. The formula for the settlement ratio is

$$r_{sd} = \frac{s_g - (s_d + s_f + d_c)}{s_d} \quad \dots \dots \dots 9$$

in which r_{sd} is the settlement ratio for negative projecting or imperfect trench conduits; s_g is the settlement of the firm ground surface; s_d is the compression of trench backfill within the height $p'B_d$ or $p'B_c$; s_f is the settlement of the bottom of the conduit; d_c is the deflection of the conduit, that is, the shortening of its vertical dimension; and $(s_d + s_f + d_c)$ is the settlement of the critical plane. The elements of the settlement ratio are shown in Figure 53.

Present knowledge of the value of the settlement ratio which may develop in these special cases is very meager. A design value of -0.3 is recommended temporarily.

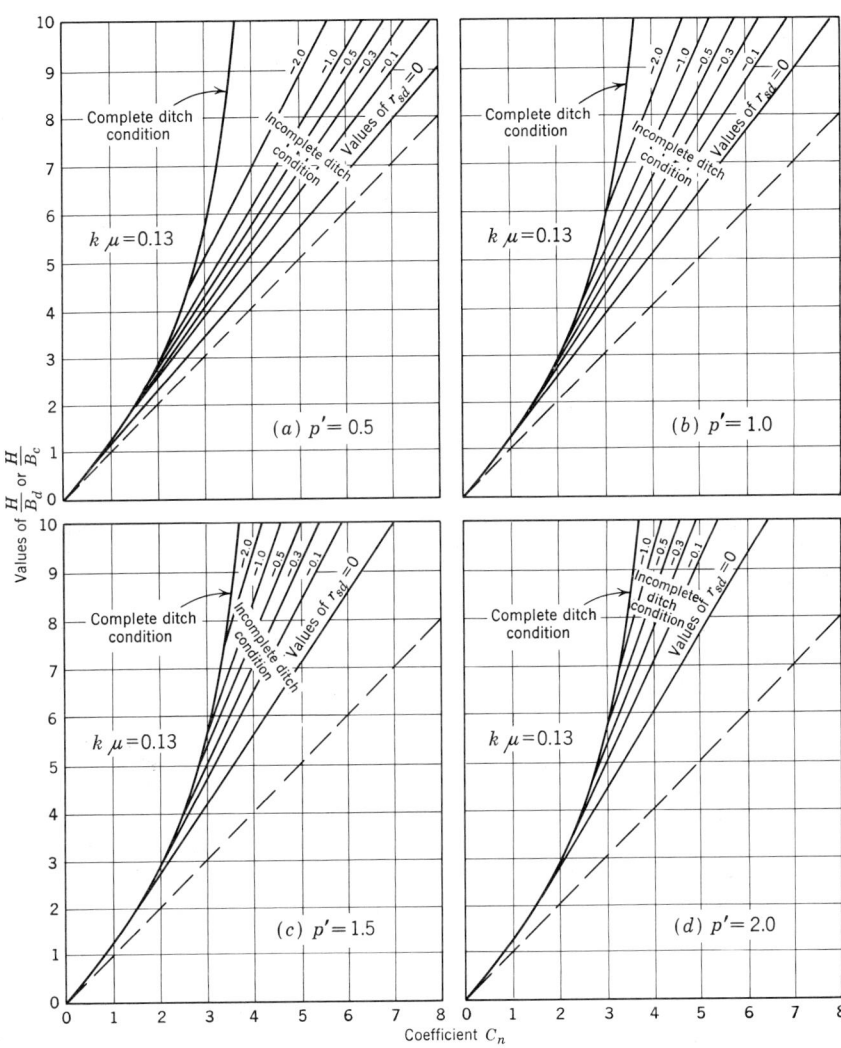

FIGURE 52.—Diagrams for coefficient C_n for negative projecting conduits and imperfect ditch conduits.

Example 6. Determine the load on the pipe of *Example 5* when installed as a negative projecting conduit in a trench whose depth is such that the top of the pipe is 7 ft below the surface of the natural ground in which the trench is dug. The width of the trench is 2 ft greater than the outside diameter of the pipe.

Assume the settlement ratio $= -0.3$. Then $H = 32$ ft; $B_d = 4.83 + 2 = 6.83$ ft; $H/B_d = 4.69$; $p' = 1.0$; C_n (from Figure 52) $= 3.0$; and $W_c = 3.0 \times 125 \times 6.83^2 = 17{,}500$ lb/ft $(26{,}100$ kg/m$)$.

Example 7. Determine the load on the pipe of *Example 5* when installed as an imperfect trench conduit with its top 2.5 ft below the elevation to

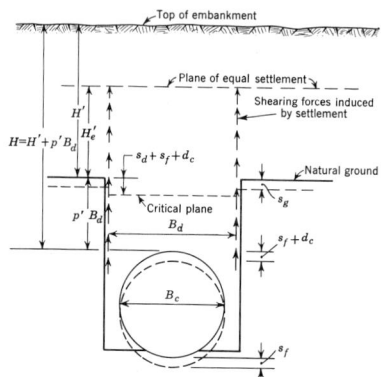

FIGURE 53.—Settlements that influence loads on negative projecting conduits.

which the soil is compacted thoroughly for a distance of 12 ft on each side of the pipe.

Assume the settlement ratio $= -0.3$. Then $H=32$ ft; $B_c=4.83$; $H/B_c = 6.63$; $p'=$ approximately 0.5; C_n (from Figure 52) $=4.8$; and $W_c = 4.8 \times 125 \times 4.83^2 = 14{,}000$ lb/ft (20,900 kg/m).

5. Loads for Jacked Conduits and Certain Tunnel Conditions

(*a*) **General.**—When the sewer is more than 30 or 40 ft (9 or 12 m) deep, or when surface obstructions are such that it is difficult to construct the sewer by cut and cover, it may be more economical to place the sewer by means of jacking or tunneling. The theories set forth herein usually will be appropriate for materials where jacking of the pipe is possible and for tunnels in homogeneous soils of low plasticity. Where a tunnel is to be constructed through materials subject to unusual internal pressures and stresses, such as some types of clays or shales which tend to squeeze or swell, or through blocky and seamy rock, the loads on the conduit cannot be determined from the factors discussed here. Reference should be made to Section B6.

The methods of constructing sewers by tunneling and jacking are described in Chapter 11. Tunnel supports carry the earth load until the conduit is constructed and the voids between the conduit and tunnel supports are filled. Jacked pipe (4) (5) is assumed to carry the earth load as it is pushed into place.

(*b*) **Load-Producing Forces.**—In the materials considered herein, the vertical load acting on the jacked pipe or tunnel supports, and eventually the pipe in the tunnel, is the resultant of two major forces. First is the weight of the overhead prism of soil within the width of the jacked pipe or tunnel excavation. Second is the shearing forces generated between the interior prisms and the adjacent material due to friction and cohesion of the soil.

During excavation of a tunnel, and varying somewhat with construction methods, the soil directly above the face of the tunnel tends to settle slightly in relation to the soil adjacent to the tunnel because of the lack of support during the period immediately after excavation and prior to the placing of the tunnel support. Also, the tunnel supports and the sewer pipe must deflect and settle slightly when the vertical load comes on them. This downward movement or tendency for movement induces upward shearing forces which support a part of the weight of the prism of earth above the tunnel. In addition, the cohesion of the material provides further support for the weight of the prism of earth above the tunnel. The resultant load on the horizontal plane on the top of the tunnel and within the width of the tunnel excavation is equal to the weight of the prism of earth above the tunnel minus the upward friction forces and minus the cohesion of the soil along the limits of the prism of soil over the tunnel.

Hence, the forces involved with gravity earth loads on jacked pipe or tunnels in such soils are similar to those discussed for loads on pipe in trenches except for the cohesion of the material. Cohesion also exists in the case of the loads in trenches and embankments but is neglected because the cohesion of the disturbed soil is of minor consequence and may be absent altogether if the soil is saturated. However, in the case of jacked pipe or in tunnels, where the soil is undisturbed, cohesion can be an appreciable factor in the loads and may be considered safely if reasonable coefficients are assumed. Jacking stresses must be investigated in pipe which is to be jacked into place.

(c) Marston's Formula.—When modified to include cohesion, Marston's formula may be used to determine the gravity earth loads on jacked pipe or pipe in tunnels through undisturbed soil (Figure 54). It takes the form:

$$W_t = C_t B_t (w B_t - 2c) \dots \dots \dots \dots \dots \dots 10$$

in which W_t is the load on the pipe or tunnel support in lb/ft (kg/m); w is the unit weight of the soil above the tunnel in lb/cu ft (kg/cu m); B_t is the maximum width of the tunnel excavation in ft (m) (B_c in the case of jacked pipe); c is the cohesion coefficient in psf (kg/sq m); and C_t is a load coefficient which is a function of the ratio of the distance from the ground surface to the top of the tunnel to the width of the tunnel excava-

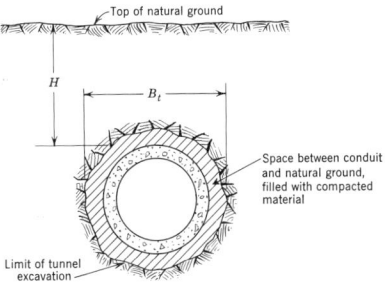

FIGURE 54.—Conduit in tunnel.

202 DESIGN, CONSTRUCTION OF SANITARY, STORM SEWERS

tion and of the coefficient of internal friction of the material of the tunnel.

The formula for C_t is identical to that for C_d (Equation 3), except that H is the distance from the ground surface to the top of the tunnel and B_t is substituted for B_d. The values of the coefficient for C_t for various ratios of H/B_t and various types of materials may be obtained from Figure 55.

An analysis of the formula for computing C_t indicates that for very

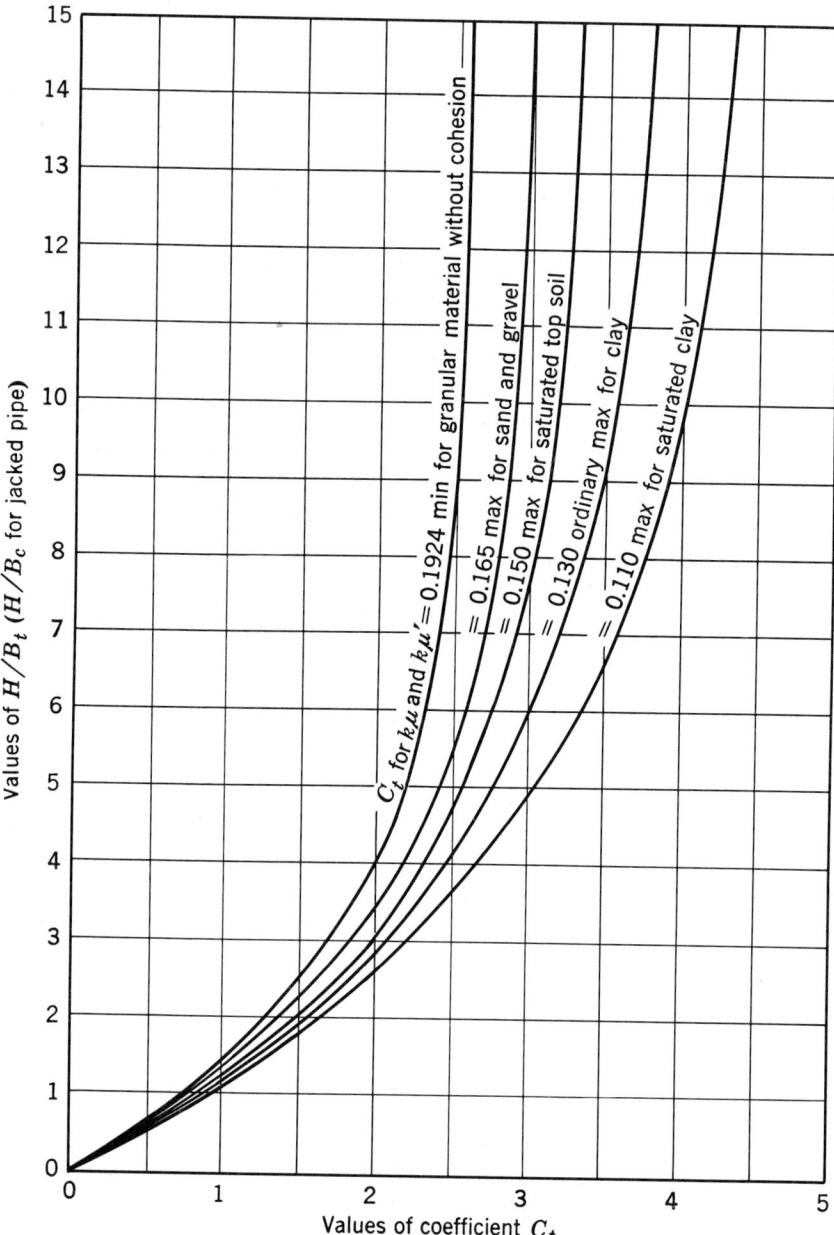

FIGURE 55.—Diagram for coefficient C_t for jacked pipe or tunnels in undisturbed soil.

high values of H/B_t, the coefficient C_t approaches the limiting value of $1/(2k\mu')$. Hence, where the tunnel is very deep, the load on the tunnel can be calculated readily by using the limiting value of C_t.

(d) **Tunnel Soil Characteristics.**—The discussion regarding unit weight and coefficient of friction for sewers in trenches applies equally to the determination of earth loads on jacked pipe or pipe in tunnels through undisturbed soil.

The one additional factor that enters into the determination of loads on tunnels is c, the coefficient of cohesion. An examination of Equation 10 shows that the proper selection of the coefficient c is very important; unfortunately it can vary widely even for similar types of soils.

It may be possible in some instances to obtain undisturbed samples of the material and to determine the value of c by appropriate laboratory tests; and this should be done whenever possible. It is suggested that conservative values of c be employed in order to allow for a saturated condition of the soil or for other unknown factors. Design values should probably be about 33 percent of the laboratory test value to allow for uncertainties.

Recommended safe values of cohesion for various soils (if it is not practicable to determine c from laboratory tests) are:

Material	Values of c (psf)
Clay, very soft	40
Clay, medium	250
Clay, hard	1,000
Sand, loose dry	0
Sand, silty	100
Sand, dense	300
Top soil, saturated	100

Values of $k\mu$ and $k\mu'$ are the same as those noted in Figure 45.

(e) **Effect of Excessive Excavation.**—Where the tunnel is constructed by a method that results in excessive excavation and where these voids above the pipe or tunnel lining are not backfilled carefully or packed with grout or other suitable backfill materials, saturation of the soil or vibration may eventually destroy the cohesion of the undisturbed material above the conduit and result in loads in excess of those calculated using Equation 10. If this situation is anticipated, it is suggested that Equation 10 be modified by eliminating the cohesion term. The calculated loads then will be the same as those obtained from Equation 2.

Example 8. Determine the loads on a 60-in. diam pipe in a tunnel 40 ft deep.

Assume the width of excavation, $B_t = 78$ in. $= 6.5$ ft; type of soil is silty sand ($k\mu' = 0.150$, $c = 100$ psf, and $w = 110$ lb/cu ft); and the depth of tunnel, $H = 40$ ft. Then $H/B_t = 40/6.5 = 6.15$; C_t (from Figure 55) $= 2.83$.

Employing Equation 10, $W_t = 2.83 \times 6.5 (110 \times 6.5 - 2 \times 100)$; or $W_t = 9,500$ lb (14,200 kg).

If the tunnel were very deep, $C_t = 1/(2k\mu') = 3.33$; and $W_t = 11,200$ lb (16,700 kg).

6. Loads for Tunnels

(*a*) **General.**—When the sewer is to be constructed in a tunnel through homogeneous soils of low plasticity, design should be based on theories set forth in Section B5 above. The design of tunnels through other types of materials is discussed in this section. The usual procedure in tunnel construction is to complete the excavation first and then place either a monolithic concrete liner or install pipe, grouting, or concreting it in place. The strength of such a section often is obtained by means of pressure grouting to strengthen the surrounding material instead of relying on the conduit itself. Tunnel loads therefore usually are determined for purposes of selecting supports to be used during excavation, and the pipe or cast-in-place liner designed primarily to withstand loads from pressure grouting.

A complete discussion of tunnels is not within the scope of this manual and the designer's attention is called to references listed at the end of this chapter (6) (7) (8) (9).

(*b*) **Load-Producing Forces.**—When the tunnel is to be constructed through soils which tend to squeeze or swell such as some types of clay or shale, or through blocky or seamy rock, the vertical load cannot be determined from a consideration of the factors discussed previously and Equation 10 is not applicable.

The determination of the rock pressures exerted against the tunnel lining is largely an estimate based on previous experience of the performance of linings in similar rock formations, although attempts at numerical analysis of stress conditions around a tunnel shaft have been made.

In case of plastic clay, the full weight of the overburden is likely to come to rest on the tunnel lining some time after construction. The extent of lateral pressures to be expected has not as yet been determined fully, especially the passive resistance which will be maintained permanently by a plastic clay in the case of flexible ring-shaped tunnel lining.

On the other hand when tunneling through sand, only part of the weight of the overburden will come to rest on the tunnel lining at any time if adequate precautions are taken. The relief will be due to the transfer of the soil weight immediately above the tunnel to the adjoining soil mass by shearing stresses along the vertical planes. In this case Marston's formula may be used for estimating the total load which the tunnel lining may have to carry.

Great care must be taken to prevent any escape of sand into the tunnel during its construction. Moist sand will usually arch over small openings and not cause trouble in this respect, but entirely dry sand, which is sometimes encountered, is liable to trickle into the tunnel through the smallest gap in the temporary lining. Sand movements of this kind destroy most

C. LOADS ON SEWERS DUE TO SUPERIMPOSED LOADS

1. General Method

Two types of superimposed loads are encountered commonly in the structural design of sewers and culverts. These two types are (*a*) concentrated load, and (*b*) distributed load. Loads on conduits due to both types of loading can be determined by application of Boussinesq's solution for stresses in a semi-infinite elastic medium through the convenience of an integration developed by Holl for concentrated loads and tables of influence coefficients developed by Newmark for distributed loads.

2. Concentrated Loads

The formula for load due to superimposed concentrated load, such as a truck wheel (Figure 56), is given the following form by Holl's integration of Boussinesq's formula

$$W_{sc} = C_s \frac{PF}{L} \quad \dots \dots \dots \dots \dots \dots 11$$

in which W_{sc} is the load on the conduit in lb/unit length (kg/unit length); P is the concentrated load in lb (kg); F is the impact factor; C_s is the load coefficient (Table XXVI), a function of $B_c/(2H)$ and $L/(2H)$; H is the height of fill from the top of conduit to ground surface in ft (m); B_c is the width of conduit in ft (m); and L is the effective length of conduit in ft

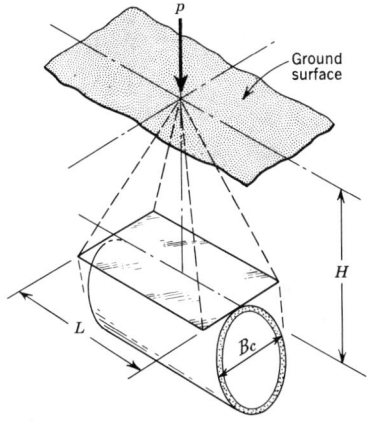

FIGURE 56.—Concentrated superimposed load vertically centered over conduit.

TABLE XXVI.—Values of Load Coefficients, C_s, for Concentrated and Distributed Superimposed Loads Vertically Centered Over Conduit*

$\frac{D}{2H}$ or $\frac{B_c}{2H}$	\multicolumn{14}{c}{$\frac{M}{2H}$ or $\frac{L}{2H}$}
	0.1	0.2	0.3	0.4	0.5	0.6	0.7	0.8	0.9	1.0	1.2	1.5	2.0	5.0
0.1	0.019	0.037	0.053	0.067	0.079	0.089	0.097	0.103	0.108	0.112	0.117	0.121	0.124	0.128
0.2	0.037	0.072	0.103	0.131	0.155	0.174	0.189	0.202	0.211	0.219	0.229	0.238	0.244	0.248
0.3	0.053	0.103	0.149	0.190	0.224	0.252	0.274	0.292	0.306	0.318	0.333	0.345	0.355	0.360
0.4	0.067	0.131	0.190	0.241	0.284	0.320	0.349	0.373	0.391	0.405	0.425	0.440	0.454	0.460
0.5	0.079	0.155	0.224	0.284	0.336	0.379	0.414	0.441	0.463	0.481	0.505	0.525	0.540	0.548
0.6	0.089	0.174	0.252	0.320	0.379	0.428	0.467	0.499	0.524	0.544	0.572	0.596	0.613	0.624
0.7	0.097	0.189	0.274	0.349	0.414	0.467	0.511	0.546	0.584	0.597	0.628	0.650	0.674	0.688
0.8	0.103	0.202	0.292	0.373	0.441	0.499	0.546	0.584	0.615	0.639	0.674	0.703	0.725	0.740
0.9	0.108	0.211	0.306	0.391	0.463	0.524	0.574	0.615	0.647	0.673	0.711	0.742	0.766	0.784
1.0	0.112	0.219	0.318	0.405	0.481	0.544	0.597	0.639	0.673	0.701	0.740	0.774	0.800	0.816
1.2	0.117	0.229	0.333	0.425	0.505	0.572	0.628	0.674	0.711	0.740	0.783	0.820	0.849	0.868
1.5	0.121	0.238	0.345	0.440	0.525	0.596	0.650	0.703	0.742	0.774	0.820	0.861	0.894	0.916
2.0	0.124	0.244	0.355	0.454	0.540	0.613	0.674	0.725	0.766	0.800	0.849	0.894	0.930	0.956

* Influence coefficients for solution of Holl's and Newmark's integration of the Boussinesq equation for vertical stress.

STRUCTURAL REQUIREMENTS 207

(m). The effective length of a conduit is defined as the length over which the average load due to surface traffic units produces the same stress in the conduit wall as does the actual load which varies in intensity from point to point. Little research information is available on this subject. Tentative recommendations are to use an effective length equal to 3 ft (0.9 m) for conduits greater than 3 ft long. Use the actual length for conduits shorter than 3 ft.

If the concentrated load is displaced laterally and longitudinally from a vertically centered location over the section of pipe under construction, the load on the conduit can be computed by adding algebraically the effect of the concentrated load on various rectangles each with a corner centered under the concentrated load. Values of C_s in Table XXVI divided by 4 equal the load coefficient for a rectangle whose corner is vertically centered under the concentrated load.

(a) Impact Factor

The impact factor, F, reflects the influence of dynamic loads caused by traffic. Suggested values for various kinds of traffic are:

Traffic Type	F
Highway	1.50
Railway	1.75
Airfield:	
Runways	1.00
Taxiways, aprons, hard stands	1.50

Example 9. Determine the load on a 24-in. diam pipe under 3 ft of cover caused by a 10,000-lb truck wheel applied directly above the center of the pipe.

Assume the pipe section is 2.5 ft long; the wall thickness is 2 in.; and the impact factor is 1.5. Then $B_c = 24 + 4 = 28$ in. $= 2.33$ ft; $L = 2.5$ ft; and $H = 3.0$ ft. Finally $B_c/2H = 2.33/6 = 0.39$; and $L/2H = 2.5/6 = 0.41$; the load coefficient is 0.240 from Table XXVI. Substituting in Equation 11,

$$W_{sc} = 0.240 \times \frac{10,000 \times 1.5}{2.5} = 1,440 \text{ lb/ft (2,140 kg/m)}.$$

3. Distributed Loads

For the case of a superimposed load distributed over an area of considerable extent, as shown in Figure 57, the formula for load on the conduit is

$$W_{sd} = C_s p F B_c \quad \ldots \ldots \ldots \ldots \ldots \ldots \ldots \ldots \ldots 12$$

in which W_{sd} is the load on the conduit in lb/unit length (kg/unit length); p is the intensity of distributed load in psf (kg/sq m); F is the impact factor; B_c is the width of the conduit in ft (m); C_s is the load coefficient, a function of $D/(2H)$ and $M/(2H)$ from Table XXVI; H is the height from the top of the conduit to the ground surface in ft (m); and D and M

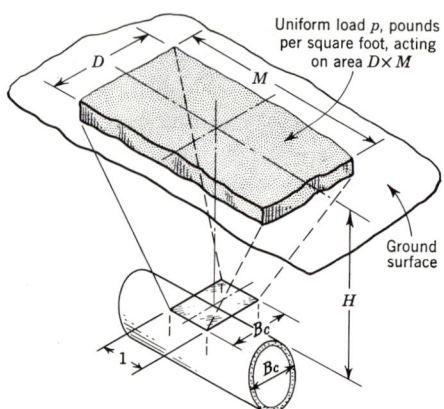

FIGURE 57.—Distributed superimposed load vertically centered over conduit.

are the width and length, respectively, of the area over which the distributed load acts, in ft (m).

Values of C_s can be read directly from Table XXVI if the area of the distributed superimposed load is centered vertically over the center of the conduit under consideration.

The load on the conduit can be computed by adding algebraically the effect of various rectangles of loaded area if the area of the distributed superimposed load is not centered over the conduit but is displaced laterally and longitudinally. It is more convenient to work in terms of load under one corner of a rectangular loaded area rather than at the center. Dividing the tabular values of C_s by 4 will give the effect for this condition.

4. Conduits Under Railway Tracks

The live load may be considered as a uniformly distributed load equal to the weight of locomotive driver axles divided by an area equal to the length occupied by the drivers multiplied by the length of ties when sewers are constructed under railroad tracks. In addition, 200 lb/ft (300 kg/m) should be allowed for weight of the track structure.

Example 10. Determine the load on a 48-in. diam concrete pipe under 6 ft of cover (bottom of ties to top of pipe) resulting from the Cooper E-70 railroad loading.

Assume the pipe wall thickness is 4 in., the locomotive load consists of four 70,000-lb axles spaced 5 ft center-to-center, the impact factor is 1.75, and the weight of track structure is 200 lb/ft. Then $B_c = 48 + 8 = 56$ in. or 4.67 ft; $H = 6$ ft; $D = 8$ ft; and $M = 20$ ft.

The unit load plus impact at the base of the ties is $\dfrac{4 \times 70{,}000 \times 1.75}{8 \times 20} + \dfrac{200}{8} = 3{,}085$ psf; $\dfrac{D}{2H} = \dfrac{8}{12} = 0.67$; and $\dfrac{M}{2H} = \dfrac{20}{12} = 1.67$; the influence coefficient is 0.641 (from Table XXV).

By Equation 12, $W_{sd} = 0.641 \times 3.085 \times 4.67 = 9{,}240$ lb/ft (13,800 kg/m).

5. Conduits Under Rigid Pavement

A method of computing the load transmitted to conduits under rigid pavement is given elsewhere (10).

D. SUPPORTING STRENGTH OF RIGID CONDUITS

The ability of a conduit to resist safely the calculated earth load depends not only on its inherent strength but also on the distribution of the vertical load and bedding reaction and on the lateral pressure acting against the sides of the conduit.

The inherent strength of a rigid conduit usually is specified by its resistance in the three-edge bearing test. This test is both convenient and severe but it does not reproduce the actual field load conditions. Thus, to select the most economical combination of bedding and pipe strength, a relationship must be established between calculated load, laboratory strength, and field strength for various installation conditions.

Field or supporting strength, moreover, depends on the distribution of the vertical load and the reaction against the bottom of the pipe. It also depends on the magnitude and distribution of the lateral pressure acting on the sides of the pipe. These factors, therefore, make it necessary to qualify the term "supporting strength" with a description of conditions of installation in a particular case as they affect the distribution of the load, the reaction, and the magnitude and distribution of lateral pressure.

As in the case of computing loads on the conduit, it is convenient when determining supporting strength to classify installation conditions as either "trench" or "embankment."

1. Laboratory Test Strength

Rigid pipe may be tested for strength in the laboratory by the three-edge bearing test. Methods of testing are described in detail in ASTM Specifications C301, C497, and C500 and USASI * Specification A60.2. ASTM and USASI specifications for pipe contain the minimum required strengths for three-edge bearing tests.

Laboratory strength, in the case of reinforced concrete pipe, may be expressed as the load per foot of length which causes the pipe to develop a 0.01-in. (0.025-cm) crack, or as the 0.01-in. crack load and the ultimate load which the pipe will withstand. The cracking load and the ultimate load, in the case of non-reinforced pipe, are essentially the same, and the cracking load is considered to be the ultimate strength of the pipe.

The strength of the pipe, in lb/ft (kg/m) at either 0.01-in. crack or ultimate, divided by the nominal internal diameter of the pipe, in ft (m), is called the D-load strength. Thus, if a 48-in. (1.22-m) diam reinforced concrete pipe has a three-edge bearing test strength at a 0.01-in. crack of

---

* American Society for Testing and Materials, Philadelphia, Pa.; United States of America Standards Institute, New York, N.Y.

8,000 lb/ft (11,900 kg/m) and an ultimate strength of 12,000 lb/ft (17,900 kg/m), the 0.01-in. crack strength is 2,000 D (9,800 D) and the ultimate strength is 3,000 D (14,700 D).

2. Pipe Bedding

The contact between a pipe and the foundation on which it rests is the pipe bedding. This has an important influence on the distribution of the reaction against the bottom of the pipe and therefore influences the supporting strength of the pipe as installed. In the case of bell and spigot pipe, a suitable excavation should be made to receive the pipe bells. It should be of sufficient width and depth to insure that the bottom reaction will act only on the pipe barrel and not on the bell.

Concrete cradle bedding for large diameter reinforced concrete pipe has been used advantageously in some areas. The concrete cradle provides positive uniform distribution of the reaction at the bottom of the pipe.

If the full benefit of the bedding method is to be achieved, the bottom of the trench or embankment must be stable. Ways of achieving this condition are discussed in Chapter 11.

3. Backfill

The soil at the sides of a pipe and above it is the backfill. It influences the supporting strength of the pipe by exerting lateral pressure against the sides.

4. Field Supporting Strength

The field supporting strength of a rigid pipe conduit is the maximum load in lb/ft (kg/m) which the pipe will support while retaining complete serviceability when installed under specified conditions of bedding and backfilling.

The "field supporting strength" should not be confused with the "safe supporting strength" or "working strength" which contains a factor of safety.

The field supporting strength, in addition to the inherent strength of pipe, is influenced by the distribution of the vertical load on the top, the distribution of the vertical reaction on the bottom, and the amount and distribution of effective lateral pressure against the sides of the pipe. It is greater than the three-edge bearing test strength because of the more favorable distribution of the load and reaction in a field installation and because of the complete absence of side pressure in the laboratory test.

5. Load Factor

The ratio of the strength of a pipe under any stated condition of loading and bedding to its strength measured by the three-edge bearing test is called the load factor. The relationship between field supporting strength, laboratory strength, and load factor is expressed as follows:

Field supporting strength = load factor × three-edge bearing strength

The load factor does not contain a factor of safety. Load factors have been determined experimentally and analytically for the commonly used construction conditions for both trench and embankment conduits.

6. Supporting Strength in Trench Conditions

(*a*) **Classes of Bedding.**—Four classes of beddings most often used for pipes in trenches are described as follows and illustrated in Figure 58.

1. *Class A—Concrete Cradle or Concrete Arch Bedding.*—This class of bedding may take either of two forms:

a. Concrete Cradle. The pipe shall be bedded in a monolithic cradle of plain or reinforced concrete having a minimum thickness of one-fourth the inside pipe diameter or a minimum of 4 in. (10 cm) under the barrel and extending up the sides for a height equal to one-fourth the outside diameter. The cradle shall have a width at least equal to the outside diameter of the pipe barrel plus 8 in. (20 cm).

Backfill above the cradle and extending to 12 in. (30 cm) above the crown of the pipe shall be compacted carefully.

b. Concrete Arch. The pipe shall be embedded in carefully compacted granular material having a minimum thickness of one-fourth the outside diameter between barrel and bottom of trench excavation and extending halfway up the sides of the pipe. The top half of the pipe shall be covered with a monolithic plain or reinforced concrete arch having a minimum thickness of one-fourth the inside diameter at the crown and having a minimum width equal to the outside pipe diameter plus 8 in. (20 cm).

The load factor for Class A concrete cradle bedding is 2.2 for plain concrete with lightly tamped backfill; 2.8 for plain concrete with carefully tamped backfill; and up to 3.4 for reinforced concrete with $p=0.4$ percent, in which p is the ratio of the area of steel to the area of concrete at the invert.

The load factor for Class A concrete arch type bedding is 2.8 for plain concrete; up to 3.4 for reinforced concrete with $p=0.4$ percent; and up to 4.8 for reinforced concrete with $p=1.0$ percent, in which p is the ratio of the area of steel to the area of concrete at the crown.

2. *Class B—First-Class Bedding.*—Class B bedding may be achieved by either of two construction methods:

a. Shaped Bottom with Tamped Backfill. The bottom of the trench excavation shall be shaped to conform to a cylindrical surface with a radius at least 2 in. (5 cm) greater than the radius to the outside of the pipe and with a width sufficient to allow six-tenths of the width of the pipe barrel to be bedded in fine granular fill placed in the shaped excavation. Carefully compacted backfill shall be placed at the sides of the pipe to a thickness of at least 12 in. (30 cm) above the top of the pipe. Shaped

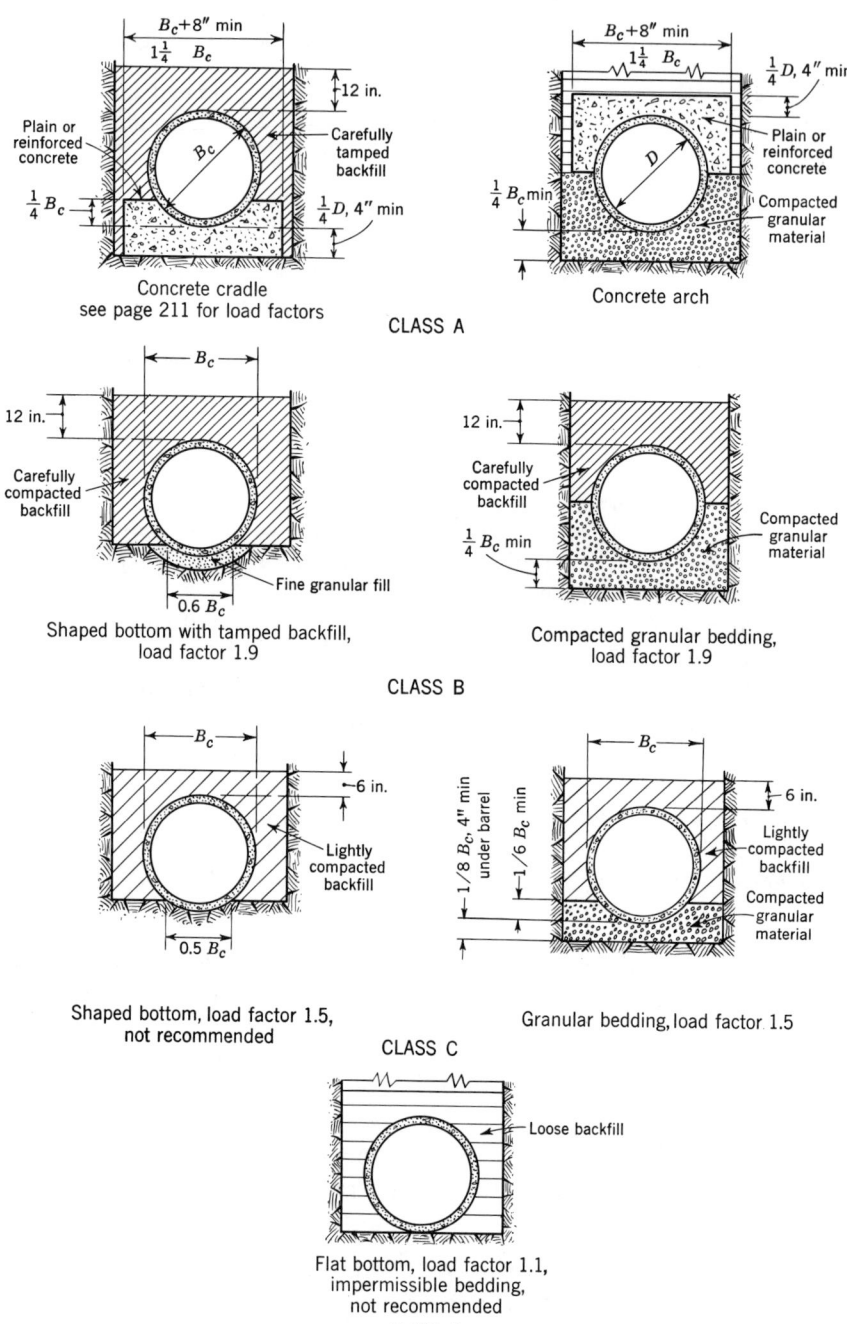

FIGURE 58.—Classes of bedding for conduits in trench. Note: In rock trench, excavate at least 6 in. (15 cm) below the bell of the pipe except where concrete cradle is used.

trench bottoms are difficult to achieve under current construction conditions.

b. Compacted Granular Bedding with Tamped Backfill. The pipe shall be bedded in compacted granular material placed on a flat trench bottom. The granular bedding shall have a minimum thickness of one-fourth the outside pipe diameter and shall extend halfway up the pipe barrel at the sides. The remainder of the side fills and a minimum depth of 12 in. (30 cm) over the top of the pipe shall be filled with carefully compacted material.

The load factor for either construction method is 1.9.

3. Class C—Ordinary Bedding.—Class C ordinary bedding may be achieved by either of two construction methods:

a. Shaped Bottom. The pipe shall be bedded with "ordinary" care in an earth foundation formed in the trench bottom by a shaped excavation which will fit the pipe barrel with reasonable closeness for a width of at least 50 percent of the outside pipe diameter. The side fills and area over the pipe to a minimum depth of 6 in. (15 cm) above the top of the pipe shall be filled with lightly compacted fill. The shaped bottom bedding is not recommended for pipeline construction because it is impractical and costly.

b. Compacted Granular Bedding with a Tamped Backfill. The pipe shall be bedded in compacted granular material placed on a flat trench bottom. The granular bedding shall have a minimum thickness of 4 in. (10 cm) under the barrel and shall extend one-tenth to one-sixth of the outside diameter up the pipe barrel at the sides. The remainder of the side fills and to a minimum depth of 6 in. (15 cm) over the top of the pipe shall be filled with lightly compacted backfill.

The load factor for Class C bedding is 1.5.

4. Class D—Flat Bottom Trench, Impermissible Bedding.—In this class of bedding the bottom of the trench is left flat, and no care is taken to secure compaction of backfill at the sides and immediately over the pipe. The load factor for Class D bedding is 1.1.

Class D bedding is not recommended for pipeline construction. Under present construction conditions, Class B or C bedding with a compacted granular bedding is generally a more practical and economical method of installation.

(b) Granular Material.—Granular material is used commonly to bed sewer pipes in lieu of shaping the trench bottom to fit the contour of the pipe. The granular bedding material, in addition to providing firm uniform support for the pipe, frequently also must stabilize the trench bottom. The pipe bedding material must remain firm and not permit displacement of the pipe either during pipe laying and backfilling or following completion of construction. Furthermore, the material should not have a tendency to flow when flooded nor when an excavation is opened through it.

The two general gradation classifications used for granular bedding materials are uniformly graded and well graded. Uniformly-graded materials, such as pea gravel, are one size materials with a low percentage of over- and under-size particles. Well-graded materials contain several sizes of particles, in stated proportions, ranging from a maximum to a minimum size.

Coarse sand, pea gravel, crushed gravel, gravel, crushed stone screenings, and crushed stone have been used for pipe bedding. Pea gravel and uniformly graded (one size) crushed stone of comparable size have been used commonly.

Fine materials, coarse sand, or screenings are not satisfactory for stabilizing trench bottoms and are difficult to compact in a uniform manner to provide proper pipe bedding.

Well-graded material is most effective for stabilizing trench bottoms and has less tendency to flow than a uniformly-graded material. However, uniformly-graded material is easier to place and compact around sewer pipes. Rounded granular material has a greater tendency to flow and allow pipe movement after laying than angular material.

Recent research (11) has shown that a well-graded crushed stone is most suitable for pipe bedding and that it is better suited for bedding than well-graded gravel. Both materials, however, are better suited for bedding than a uniformly-graded pea gravel.

It is considered that well-graded crushed stone or crushed gravel meeting the requirements of ASTM Designation C 33, Gradation 67 (¾-in. to No. 4) (1.9 to 0.48 cm) generally will provide the most satisfactory pipe bedding. A well-graded gravel meeting these same requirements also can be used.

Granular pipe bedding material when referred to in this manual generally means one of the foregoing well-graded materials. Any material used for pipe bedding should be compacted thoroughly as it is placed to provide uniform support for the pipe barrel and to fill completely all voids under and around the pipe.

(c) Rock or Other Incompressible Foundations.—Where ledge rock, compact rocky or gravelly soil, or other unyielding foundation material is encountered, the pipes should be bedded in accordance with the requirements of one of the foregoing classes of bedding, but with the following additions: The hard unyielding material should be excavated to the elevation of the bottom of the concrete cradle (Class A bedding) or below the bottom of the pipe and pipe bell (Class B, C, or D bedding) to a depth of at least 6 in. (15 cm). The width of the excavation should be at least five-fourths the outside diameter of the pipe and it should be refilled with granular material.

(d) Encased Pipe.—Total encasement of non-reinforced rigid pipe in concrete may be necessary where the required safe supporting strength cannot be obtained by other bedding methods.

A typical concrete encasement detail is shown in Figure 59 as used by the Department of Public Works, City of Los Angeles. The load factor

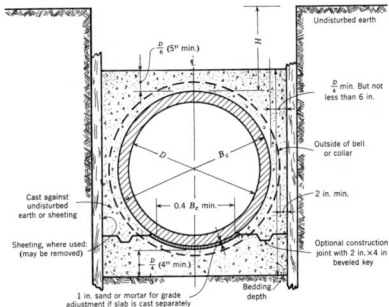

FIGURE 59.—Typical concrete encasement details.

for concrete encasement varies with the thickness of concrete and the use of reinforcing and may be greater than that used for concrete cradle or arch. The load factor for the encasement shown in Figure 59 is 4.5.

Concrete encasement also may be required for pipelines built in deep trenches in order to assure uniform support or for pipe lines built on comparatively steep grades where there is the possibility that earth beddings may be eroded by currents of water under and around the pipe.

7. Supporting Strength in Embankments

It is possible for the active soil pressure against the sides of a pipe placed in an embankment to be a significant factor in the resistance of the structure to vertical load. This factor is important enough to justify a separate examination of the supporting strength of embankment conduits.

(a) **Positive Projecting Conduits.**—The load factor for rigid pipes installed as projecting conduits under embankments or in wide trenches depends on the class of bedding in which the pipe is laid, the magnitude of the active lateral soil pressure against the sides of the pipe, and the area of the pipe over which the active lateral pressure is effective.

For projecting conduits of circular and elliptical cross section, the load factor is

$$L_f = \frac{A}{N - xq} \quad \ldots \ldots \ldots \ldots \ldots \ldots \ldots \ldots 13$$

in which L_f is the load factor; A is a pipe shape factor; N is a pipe bedding factor; x is a parameter dependent on the area over which lateral pressure effectively acts; and q is the ratio of total lateral pressure to total vertical load on the pipe.

Classes of bedding for projecting conduits are shown in Figure 60. The values of A for circular and elliptical pipe are:

Pipe Shape	A
Circular	1.431
Elliptical	
Horizontal elliptical	1.337
Vertical elliptical	1.021

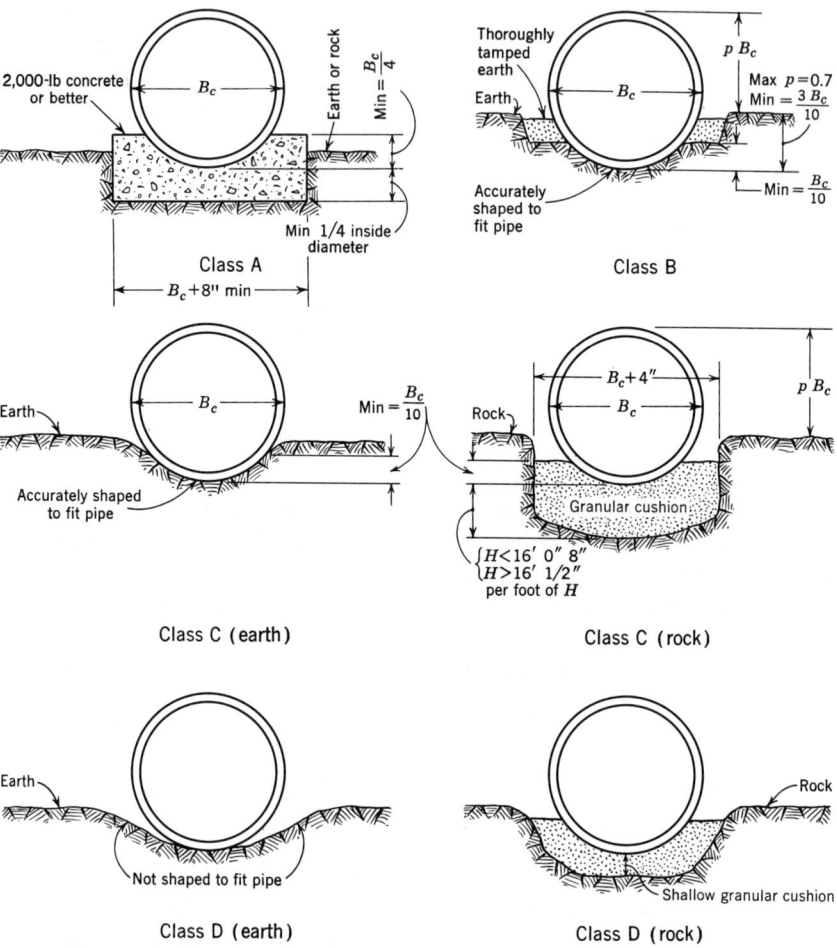

FIGURE 60.—Classes of bedding for projecting conduits.

Values of N for various classes of bedding are given in Table XXVII. Values of x for circular and elliptical pipe are listed in Table XXVIII.

The ratio, m, refers to the fraction of the pipe diameter over which lateral pressure is effective. For example, if lateral pressure acts on the top half of the pipe above the horizontal diameter, $m=0.5$.

The ratio of total lateral pressure to total vertical load, q, for positive projecting conduits may be estimated by the formula:

$$q = \frac{mk}{C_c}\left(\frac{H}{B_c} + \frac{m}{2}\right) \quad \quad \quad \quad \quad \quad \quad 14$$

in which k is the ratio of unit lateral pressure to unit vertical pressure (Rankine's ratio). A value of $k=0.33$ usually will be sufficiently accurate for use in Equation 14.

(b) Negative Projecting Conduits.—The load factor for negative projecting conduits may be the same as for trench conduits for the various classes of bedding as given in Section 6. These load factors for Class

TABLE XXVII.—Values of N

	Value of N		
	Pipe Shape		
Class of Bedding	Circular	Horizontal Elliptical	Vertical Elliptical
A (reinforced cradle)	0.421 to 0.505	—	—
A (unreinforced cradle)	0.505 to 0.636	—	—
B	0.707	0.630	0.516
C	0.840	0.763	0.615
D	1.310	—	—

TABLE XXVIII.—Values of x

Fraction of Pipe Subjected to Lateral Pressure, m	Class A Bedding	Other than Class A Bedding		
	Circular Pipe	Circular Pipe	Horizontal Elliptical Pipe	Vertical Elliptical Pipe
0	0.150	0	0	0
0.3	0.743	0.217	0.146	0.238
0.5	0.856	0.423	0.268	0.457
0.7	0.811	0.594	0.369	0.639
0.9	0.678	0.655	0.421	0.718
1.0	0.638	0.638	—	—

B, C, and D bedding do not take into account lateral pressures against the sides of the pipe for the reason that unfavorable construction conditions often prevail at the bottom of a sewer trench. However, in the case of negative projecting conduits, conditions may be more favorable and it may be possible to compact the side-fill soils to the extent that some lateral pressure against the pipe can be relied on. If such favorable conditions are anticipated, it is suggested that the load factor be computed by means of Equations 13 and 14, using a value of $k=0.15$ for estimating the lateral pressure on the pipe.

(c) Imperfect Trench Conduits.—Imperfect trench conduits usually are installed as positive projecting conduits before the overlying soil is compacted and the imperfect trench is excavated. Therefore, lateral pressures are effective against the sides of the conduit, and the load factor should be calculated by Equations 13 and 14.

E. SUPPORTING STRENGTH OF FLEXIBLE PIPES

1. General Method

Flexible pipes under earth fills derive their ability to support load from their inherent strength plus the passive resistance pressure of the soil as they deflect and the sides of the pipe move outward against the soil side fills. This type of pipe fails by excessive deflection and collapse or buckling rather than by rupture of the pipe walls as in the case of pipes made of brittle materials. Therefore, design of flexible pipes is directed toward determination of the deflection under load. A field supporting strength resulting in a deflection of five percent of the nominal diameter of the pipe is considered by many engineers to be a suitable criterion for design. Design criteria for buckling and longitudinal seam strength are suggested by Townsend (12).

A formula for calculating flexible pipe deflection under earth loading is

$$\triangle x = D_e \frac{KW_c r^3}{EI + 0.061 E' r^3} \quad \ldots \ldots \ldots \ldots \ldots 15$$

in which $\triangle x$ is the horizontal and vertical deflection of the pipe in in.; D_e is the deflection lag factor; K is a bedding constant dependent on the angle subtended by the pipe bedding; W_c is the vertical load on the pipe in lb/in.; r is the mean radius of the pipe in in.; E is the modulus of elasticity of the pipe material in psi; I is the moment of inertia per unit length of cross section of the pipe wall in in.^{4}/in.; $E' = er$ is the modulus of soil reaction in psi and e is the modulus of passive resistance of the enveloping soil in psi/in.

The deflection lag factor, empirically determined, compensates for the tendency of flexible pipes to continue to deform for some period of time after the full magnitude of load has developed on the pipe. Recommended values of this factor range from 1.25 to 1.50.

Values of the bedding constant, K, depending on the width of the pipe bedding are:

Bedding Angle (deg)	K
0	0.110
30	0.108
45	0.105
60	0.102
90	0.096
120	0.090
180	0.083

There is much yet to be learned about the modulus of passive resistance of soil, e, and its influence on flexible pipe deflection. Some recent research has indicated that this modulus is influenced strongly by the size of the pipe and that, for a given type of soil in a given state of compaction, the product of the modulus times the radius of pipe, E', is constant, that is, for the same soil, the modulus varies inversely with the pipe radius. Also, observations on a limited number of pipes in service, where sufficient information is available to make an estimate, indicate that the value of E' varies widely—from a minimum of 234 psi (17 kg/sq cm) in the case of an uncompacted sandy clay loam to a maximum of 7,980 psi (560 kg/sq cm) for a crushed sandstone soil which was compacted to maximum density. On five culvert installations where the soil was compacted (although not necessarily to maximum density), the values of E' ranged from 502 to 1,320 psi (35 to 93 kg/sq cm), the average being 765 psi (54 kg/sq cm). On the basis of these observations, a value of E' of 700 psi (49 kg/sq cm) is recommended in design if the side-fill soil is compacted to 90 percent or more of maximum density, AASHO T99, for a distance of two pipe diameters on each side of the pipe.

The first term in the denominator, EI, in Equation 15, reflects the influence of the inherent strength of the pipe on deflection; whereas the second term, $0.061 \, E'r^3$, reflects the influence of the passive pressure on the sides of the pipe. The second term may be excessively predominant in the case of large-diameter pipes, with the result that a very light-weight pipe may appear to be satisfactory. Since the pipe wall must have sufficient local strength in bending and thrust to develop and utilize the passive resistance pressure on the sides of the pipe, it is recommended as a practical measure that the value of EI should never be less than about 10 to 15 percent of the term $0.061 \, E'r^3$. Also, the gage of the metal must be sufficient to develop adequate strength of the bolted or riveted longitudinal joints of the pipe.

Almost the entire performance of a flexible conduit in retaining its shape and integrity is dependent on the selection, placement, and compaction of the envelope of earth surrounding the structure. For this reason, as much care should be taken in the design of the backfill as is used in the design of the conduit. The backfill material selected preferably should be of a granular nature to provide good shear characteristics. Cohesive soils can be used if careful attention is given to the moisture content.

If the material placed around the conduit is different from that used in the embankment or if for construction reasons fill is placed around the conduit before the embankment is built, the compacted backfill should cover the structure by at least 1 ft (0.3 m) and extend one diameter to either side of it.

2. Corrugated Metal Pipes

The most frequently used kind of flexible sewer pipe is constructed of corrugated metal. The sheets of pipes which are fabricated are of two general types, standard and structural plate. Standard corrugations are ½ in. (1.3 cm) deep and spaced 2⅔ in. (6.8 cm) center-to-center. Structural plate corrugations are 2 in. (5.1 cm) deep and spaced 6 in. (15.2 cm) center-to-center. The moment of inertia of the pipe wall for these two types of corrugations and various gage thicknesses of metal are shown in Table XXIX.

Equation 15 has been developed primarily for flexible conduits under embankments. Corrugated metal sewers that are to support a fill should not be placed directly on a cradle or pile bents. If such supports are necessary, they should have a flat top and be covered with a compressible earth cushion. Corrugated metal should not be encased in concrete.

TABLE XXIX.—Moments of Inertia of Corrugated Sheets

Gage	Thickness (in.)	Moment of Inertia (in.^{4}/in.)	
		Standard Corrugations, ½ in. × 2⅔ in.	Structural plate Corrugations, 2 in. × 6 in.
1	0.2690	...	0.16541
3	0.2391	...	0.14588
5	0.2092	...	0.12670
7	0.1793	...	0.10777
8	0.1644	0.00550	0.09610
10	0.1345	0.00450	0.07812
12	0.1046	0.00350	0.05455
14	0.0747	0.00250	...
16	0.0598	0.00200	...

Note: In. × 2.54 = cm.

For corrugated metal pipes installed in trenches, reference is made to manufacturers' handbooks for recommended gages and corrugations.

Example 11. A 60-in. diam, 12-gage, structural plate corrugated metal pipe is to be constructed under a 25-ft embankment. What is the long-term deflection which may be expected to develop in the pipe?

Assume $B_c = 5.33$ ft and $r_{sd}p = 0$. Then the load on the pipe is $5.33 \times 120 \times 25 = 16,000$ lb/ft (26,000 kg/m) or 1,333 lb/in. (260 kg/cm).

Also assume $E = 29,000,000$ psi; $E' = 700$ psi; $K = 0.100$; $D_e = 1.50$; and $r = 31$ in. The moment of inertia of 12-gage structural plate (Table XXIX) is $I = 0.05455$: and $EI = 1,581,950$.

Then, by Equation 15,

$$\Delta x = \frac{1.50 \times 0.1 \times 1,333 \times 31^3}{1,581,950 + 0.061 \times 700 \times 31^3} = 2.09 \text{ in. } (5.31 \text{ cm}).$$

3. Plastic Pipes

Several of the many types of plastics have been used for pipe of various classes. The wide range of physical properties of plastics makes possible the production of both flexible and rigid pipe, depending on the materials used. Some, such as plastic truss pipe, are semi-rigid. If semi-rigid pipe can deflect an allowable amount without failing, Equation 15 may be used in the strength calculations.

Plastic pipe technology is developing rapidly and new methods for structural design can be expected. For the present, in computing plastic pipe strength, a deflection lag factor of 1.50 and a bedding constant of 0.10 are suggested. A modulus of soil reaction, E', equal to 300 psi (21 kg/sq cm) is recommended when side fills are compacted by hand tamping to 65 percent of maximum density, AASHO T99, but an E' of 700 psi (49 kg/sq cm) can be used when the side fills are compacted properly by mechanical equipment to 90 percent or more of maximum density AASHO T99.

STRUCTURAL REQUIREMENTS

The suggested design values have been verified by laboratory and field-scale testing, but it must be remembered that, for a given pipeline installation, conditions may vary from one section to another. The deflection equation should be applied as a guide to design rather than an absolute rule, and the amount of deflection permitted in the installed pipe will depend on the factor of safety required by the design engineer. Several types of flexible plastic pipe are capable of undergoing parallel plate test deflections of more than 20 percent without cracking or other distress. Consideration should be given to allowable deflection so as not to impede cleaning operations in sewers or seriously impair flow capacity. For these types, a field supporting strength resulting in a deflection of five percent may be used.

F. FACTOR OF SAFETY

1. General

The term factor of safety is considered by most engineers, for any material, to be the amount the ultimate strength of the material is reduced to calculate the working strength used for design. Therefore, the factor of safety generally is independent of any technique used to determine the loads to be imposed on the material.

This definition is applicable to rigid sewer pipe where, for a given ultimate strength and a given factor of safety, the working strength remains constant regardless of any other conditions.

This definition is not, however, applicable to the design of flexible sewer pipe. Flexible sewer pipe may deflect as much as 20 percent of its original diameter before failure. However, a sewer pipe which has deflected to such an extent is no longer serviceable. In the case of flexible sewer pipe, the ultimate load which the installed conduit will support while retaining complete serviceability, under given conditions of bedding and backfilling, is considered to be the ultimate strength of the pipe. This load is termed the field supporting strength of the pipe. In the design of flexible sewer pipes, therefore, the factor of safety is the amount the field supporting strength is reduced to calculate the working, or safe supporting, strength of the pipe.

The factors of safety discussed herein are not to be used to compensate for poor inspection and construction. It is mandatory that design assumptions be realized in construction if pipe failures are to be prevented.

2. Rigid Pipes

Ultimate strengths of rigid pipe usually are measured in terms of the ultimate three-edge bearing strength for plain pipe, and of ultimate and 0.01-in. (0.025-cm) crack, three-edge bearing strengths for reinforced concrete pipe. Therefore, the specified minimum strength by the three-edge bearing method divided by the appropriate factor of safety gives the working strength in terms of three-edge bearing.

A factor of safety of at least 1.5 should be applied to the specified minimum ultimate three-edge bearing strength to determine the working strength for all rigid pipes.

3. Flexible Pipes

Flexible pipes are considered to have reached the limit of their serviceability when a deflection of five percent is attained. Therefore, the field supporting strength for flexible pipes is taken to be the load which produces the maximum deflection of five percent. A factor of safety of 1.25 should be applied to the field supporting strength to calculate the safe supporting strength of the flexible pipe.

G. DESIGN RELATIONSHIPS

1. Rigid Pipes

The various elements in the design of a rigid sewer have been discussed separately. Their combination into a safe and economical design may be expressed as follows:

$$\text{Safe supporting strength} = \frac{\text{field supporting strength}}{\text{factor of safety}}$$

in which field supporting strength equals the three-edge bearing strength times the load factor, or

$$\text{Safe supporting strength} = \frac{\text{three-edge bearing strength} \times \text{load factor}}{\text{factor of safety}}$$

and, since the three-edge bearing strength divided by the factor of safety is the working strength,

$$\text{Safe supporting strength} = \text{working strength} \times \text{load factor}$$

Also, since safe supporting strength is equal to the maximum allowable field load,

$$\frac{\text{Required three-edge}}{\text{bearing strength}} = \frac{\text{maximum allowable field load} \times \text{factor of safety}}{\text{load factor}}$$

Example 12.

(a) Refer to *Example 1* wherein the backfill load on a 24-in. diam pipe with 14 ft of cover was found to be 4,720 lb/ft. If vitrified clay pipe is to be specified and a factor of safety of 1.5 is selected, the design load will be $4,720 \times 1.5 = 7,080$ lb/ft (10,560 kg/cm).

The crushing strength requirement of 24-in. diam extra-strength clay sewer pipe (ASTM C200) by the three-edge bearing method is 4,400 lb/ft. Dividing this into 7,080 lb/ft (the design load), the minimum required load factor, $7,080/4,400 = 1.61$, is obtained.

Figure 58 indicates that a Class B bedding is required for this installation.

(b) Refer to *Example 3*. If a 24-in. diam reinforced concrete sewer pipe (one line of reinforcement near center of wall) and a factor of safety of 1.50 based on the minimum ultimate test strength are selected, the design load will be 1.50×5,750=8,620 lb/ft (12,900 kg/m).

If the minimum ultimate test strength of the pipe is 6,000 lb/ft (ASTM C76, Class IV—3,000 D), the required load factor will be 8,620/6,000 =1.44. According to Figure 58, this installation will require a Class C bedding.

Example 13. Assume the 48-in. diam pipe in *Example 5* is bedded on an unreinforced-concrete cradle, with $p=m=0.5$ and $k=0.33$. Using a factor of safety of 1.5 based on the ultimate three-edge bearing strength of the pipe, $\frac{H}{B_c}=6.63$; $C_c=10.0$; $x=0.856$; and $N=0.575$.

By Equation 14,

$$q=\frac{0.5\times 0.33}{10.0}(6.63+0.25)=0.114;$$

and, by Equation 13,

$$L_f=\frac{1.431}{0.575-(0.856\times 0.114)}=3.01$$

The required three-edge bearing strength at ultimate load is $\frac{29,100\times 1.5}{3.01}$

=14,500 lb/ft (21,600 kg/m) or 3,630 D (17,720 D in kg/m/diam). Use ASTM C76, Class V pipe.

2. Flexible Pipe

The combination of elements of design for a flexible pipe are stated below:

Field supporting strength = load producing five percent deflection

$$\text{Safe supporting strength}=\frac{\text{field supporting strength}}{\text{factor of safety}}$$

H. CHARTS FOR DETERMINING EARTH LOADS ON BURIED CONDUITS

Various tables and charts have been developed which allow direct and convenient determination of earth loads on buried conduits. It should be emphasized, however, that the designer should have a full understanding of the fundamental factors which determine the structural requirements of a sewer so that sound engineering judgment may be applied to the design.

One such method of computing the earth load on conduits involves the use of a set of curves of the type shown in Figure 61. These curves were developed for loads on reinforced concrete pipe buried in sand and gravel. Its practical value is illustrated in the example below:

Example 14. Determine the load on a 30-in. diam reinforced concrete pipe under 14 ft of sand and gravel cover in trench conditions if the width of trench at the top of conduit is 6 ft.

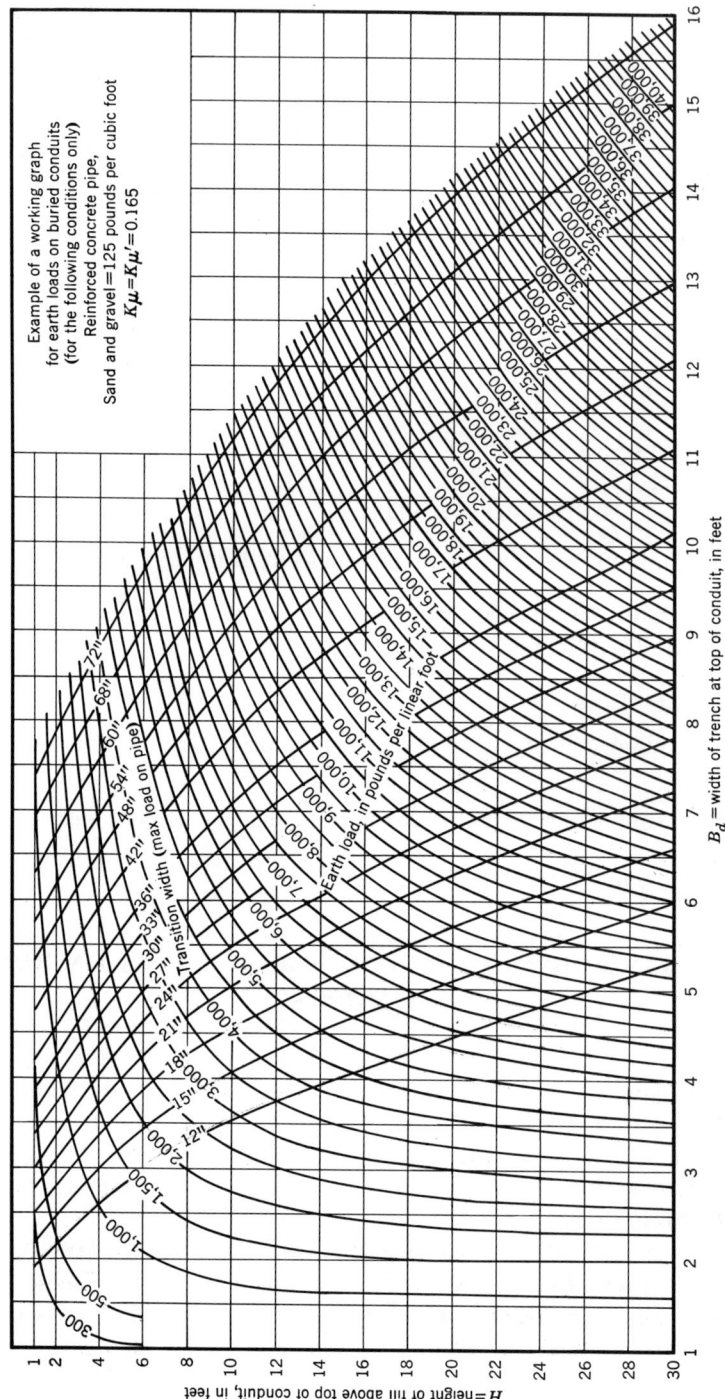

FIGURE 61.—Example of working graph for earth loads on buried conduits. (In. × 2.54 = cm; ft × 0.3 = m.)

Figure 61 gives the earth loads on conduits in lb/ft units for various trench widths and depths of sand and gravel backfill. Also plotted on these curves are the transition widths (trench widths at which further widening will have no effect on the load on the pipe) for each size of pipe.

For a fill height of 14 ft and a width of trench at the top of conduit of 6 ft, the load on the pipe is 7,250 lb/ft.

If a wide trench were used, the maximum load on the pipe would be 8,900 lb/ft at trench width (B_d) of 7 ft (7 ft is the transition width for 30-in. pipe with $H=14$ ft). Any further increase in the width of trench will not increase the load on the pipe.

The load per lineal foot determined from Figure 61 for the conditions given is the earth or dead load only. To this load must be added any live or superimposed load.

Following the determination of the total load and the selection of the proper factor of safety, the type of bedding and strength of pipe can be determined.

I. RECOMMENDATIONS FOR ATTAINING DESIGN LOAD SUPPORTING STRENGTH

1. Factor of Safety

The factor of safety against ultimate failure is generally at least 2.5 in the design of most engineering structures of monolithic concrete. The factor of safety of pipe sewers against ultimate collapse is considerably less. It is, therefore, important to guarantee that the loads imposed on the sewer are not greater than the design loads. To attain this objective the following procedures are recommended:

(a) Specifications should set forth limits for the width of trench below the top of pipe. The width limits should take into account the minimum width required to lay and joint pipe and the maximum allowable for each class of pipe and bedding to be used. Where the depth is such that a positive projecting condition will be obtained, maximum width should be specified as unlimited unless the width must be controlled for some reason other than to meet structural requirements of the pipe. Appropriate corrective measures should be specified in the event the maximum allowable width is exceeded. These may include provision for a higher class of bedding or concrete encasement. Maximum allowable construction live loads should be specified for various depths of cover if appropriate for the project.

(b) Construction should be observed by an experienced engineer or inspector who reports to a competent field engineer.

(c) Pipe testing should be under the supervision of a reliable testing laboratory and close liaison should be maintained between the laboratory and the field engineer.

(d) The field engineer should be furnished with sufficient design data to enable him to evaluate unforeseen conditions intelligently; and he should be instructed to confer with the design engineer if changes in design appear advisable.

(e) Where sheeting is to be removed, pulling should be done in stages, making certain that the space formerly occupied by the sheets is backfilled completely.

2. Effect of Trench Sheeting

Because of the various alternate methods employed in sheeting trenches, generalizations as to the proper construction procedure to follow to insure that the design load is not exceeded are risky and dangerous. Each method of sheeting and bracing should be studied separately. The effect of a particular system on the load on the conduit as well as the consequences of removing the sheeting or the bracing must be estimated.

It is difficult to obtain satisfactory filling and compaction of the void left by the pulling of wood sheeting. Sheeting driven alongside the pipe should be cut off and left in place to an elevation 1.5 ft (45 cm) above the top of the conduit.

If granular materials are used for backfill it is possible to fill and compact the voids left by the wood sheets if the material is placed in lifts and jetted as the sheeting is pulled. If cohesive materials are used for backfill, a void will be left by the pulling of the wood sheets and the full weight of the prism of earth contained between the sheeting will come to bear on the conduit.

Skeleton sheeting or bracing should be cut off and left in place to an elevation 1.5 ft (45 cm) over the top of the pipe if removal of the trench support might cause a collapse of the trench wall and a widening of the trench at the top of the conduit. Entire skeleton sheeting systems should be left in place if removal would cause collapse of the trench before backfill can be placed.

Where steel "soldier beams" with horizontal lagging between the beam flanges are used for sheeting trenches, efforts to reclaim the steel beams, before the trench is backfilled, may damage pipe joints. It is recommended that this type of sheeting be allowed when the beams are pulled after backfilling and the lagging is left in place.

Steel sheeting may be used and reused many times, and the relative economy of this type as compared with timber or timber and "soldier beams" should be explored. Because of the thinness of the sheets, it is often feasible to achieve reasonable compaction of backfill so that the steel sheets may be withdrawn with about the same factor of safety against settlement of the surfaces adjacent to the trench as that for other types of sheeting left in place.

References

1. Marston, A., and Anderson, A. O., "The Theory of Loads on Pipes in Ditches and Tests of Cement and Clay Drain Tile and Sewer Pipe." Iowa Eng. Exp. Sta., Bull. No. 31 (1913).

2. Marston, A., "The Theory of External Loads on Closed Conduits in the Light of the Latest Experiments." Iowa Eng. Exp. Sta., Bull. No. 96 (1930).
3. Schlick, W. J., "Loads on Pipe in Wide Ditches." Iowa Eng. Exp. Sta., Bull. No. 108 (1932).
4. "Jacked-in-Place Pipe Drainage." *Contractors and Engr. Monthly,* 45 (Mar. 1948).
5. "Jacking Reinforced Concrete Pipe Lines." Amer. Concrete Pipe Assn., Arlington, Va. (1960).
6. "Report of Test Tunnel." Part I, Vol. 1 and 2, Garrison Dam and Reservoir, Corps of Engineers, U. S. Army.
7. Procter, R. V., and White, T. L., "Rock Tunneling with Steel Supports." Commercial Shearing and Stamping Co.
8. "Soil Resistance to Moving Pipes and Shafts." *Proc. II Intl. Conf., Soil Mech. and Found. Eng.,* **7,** 149 (1948).
9. Von Iterson, F. K. Th., "Earth Pressure in Mining." *Proc. II Intl. Conf., Soil Mech. and Found. Eng.,* **3,** 314 (1948).
10. "Vertical Pressure on Culverts Under Wheel Loads on Concrete Pavement Slabs." Portland Cement Assn., Publ. No. ST-65, Skokie, Ill. (1951).
11. Griffith, J. S., and Keeney, C., "Load Bearing Characteristics of Bedding Materials for Sewer Pipe." *Jour. Water Poll. Control Fed.* **39,** 561 (1967).
12. Townsend, M., "Corrugated Metal Pipe Culverts—Structural Design Criteria and Recommended Installation Practices." Bur. Public Roads, U. S. Govt. Printing Office, Washington, D.C. (1966).

General References

Abernethy, L. L., "Effect of Trench Conditions and Arch Encasement on Load-Bearing Capacity of Vitrified Clay Pipe." Ohio State Univ. Eng. Exp. Sta. Bull. No. 158 (1955).

Civil Eng., **30,** 10 (1960).

Civil Eng., **30,** 12 (1960).

Frocht, M., "Photoelasticity," Vol. 1, John Wiley & Sons, Inc., New York.

Reitz, H. M., Spangler, M. G., White, H. L., Hendrickson, J. G., Jr., and Benjes, H. H., "Conduit Strengths and Trenching Requirements." Wash. Univ., Conf. Syllabus, St. Louis, Mo. (1958).

Schlick, W. J., "Concrete Cradles for Large Pipe Conduits." Iowa Eng. Exp. Sta. Bull. No. 80 (1926).

Schlick, W. J., "Supporting Strength of Concrete-Incased Clay Pipe." Iowa Eng. Exp Sta. Bull. No. 93 (1929).

Spangler, M. G., "Soils Engineering." 2nd. Ed., International Textbook Co., Scranton, Pa. (1960).

Spangler, M. G., "The Supporting Strength of Rigid Pipe Culverts." Iowa Eng. Exp. Sta. Bull. No. 112 (1933).

Spangler, M. G., "The Structural Design of Flexible Pipe Culverts." Iowa Eng. Exp. Sta. Bull. No. 153 (1941).

Spangler, M. G., "Underground Conduits—An Appraisal of Modern Research." *Trans. Amer. Soc. Civil Engr.,* **113,** 316 (1948).

Studley, E. G., and Aarons, A., "Current Sewer Design Practices in Los Angeles City." *Jour. Water Poll. Control Fed.* **38,** 10, 1656 (1966).

Swanson, H. V., and Reed, M. D., "Structural Characteristics of Reinforced Concrete Elliptical Sewer and Culvert Pipe." Publ. No. 1240, Highway Res. Bd., Washington, D.C. (1964).

Terzaghi, K., and Peck, R. B., "Soil Mechanics in Engineering Practice." John Wiley & Sons, Inc., New York (1966).

Timoshenko, S., "Strength of Materials—Part II." D. van Nostrand Co., New York (1948).

Tschibotarioff, G. P., "Soil Mechanics, Foundations and Earth Structures." McGraw-Hill Book Co., Inc., New York (1951).

"Reinforced Concrete Pipe Culverts—Criteria for Structural Design and Installation." Bur. of Public Roads, U.S. Govt. Printing Office, Washington, D.C. (1963).

"Trench Excavation." Nat. Safety Council, Data Sheet No. 254 (revised), Chicago, Ill.

CHAPTER 10. CONSTRUCTION CONTRACT DOCUMENTS

A. INTRODUCTION

The purpose of the contract documents is to portray clearly by words and drawings the nature and extent of the work to be performed and the conditions known or anticipated under which the work is to be executed. Most sewer construction projects are accomplished by contracts entered into between an owner and a construction contractor and the contract documents constitute the construction contract.

Frequently, the work is divided into various items, with either unit price or lump sum bids received for each item of work. The contract documents must clearly describe and limit these items to obviate all possible confusion in the mind of the bidder with regard to methods of measurement and payment. The subdivision of the work often is based on local customs or the customs and conventions of the designing engineer.

Lump sum bids have been applied most generally to special structures which are detailed completely and not subject to alteration or quantity changes during construction. A schedule of unit adjustment prices may be included in the proposal to provide a basis for payment in the event that changes are necessary in lump sum bid items.

Unit price bids have been used most generally where quantities of work are likely to be adjusted or varied during construction. Lineal feet of sewers or manholes, and cubic yards of rock excavation or concrete cradle are examples of such unit price items.

Lump sum bids may be taken for entire sewer construction contracts where the contract documents define the work with sufficient completeness to permit the bidder to make an accurate determination of the quantities of work. Such contracts may contain unit adjustment prices for items of work, such as rock excavation, piles, additional excavation, selected fill material, and sheeting requirements which cannot be determined accurately beforehand. This method may, however, lead to non-competitive quotations for unit adjustment prices. To prevent this, the amount of the unit adjustment price may be stipulated in the proposal or an appropriate quantity of the unit price work may be included for comparison of bids. The administration of the project, provided extensive changes are not made during construction, is simplified in the lump sum type of contract.

Plans and specifications are supplementary to each other, and all work portrayed in either is considered to be a part of the contract.

B. PLANS

1. Purpose

The purpose of the plans is to convey graphically to the bidders and later to the construction engineers and contractor the work to be done. All information which can best be shown by reference to drawings and their accompanying dimensions and notes should be shown on the plans. Lengthy specifications or word descriptions are best included only in the technical specifications, and should not be repeated on the plans.

2. Field Data

A survey of the route of the sewer is required to obtain information on the existing topography and underground utilities to be shown on the plans. The route may be mapped from data obtained by conventional ground surveys or by aerial photogrammetry. Survey work has been discussed in some detail in Chapter 2.

If the location of the sewer has been well defined by preliminary studies, it may be possible to run the ground survey directly on the centerline of the proposed alignment. This procedure will facilitate office plotting of field data. If the actual alignment is not established by field surveys, base lines or reference marks must be established in the field.

Aerial photogrammetry may be used to produce a strip map with or without contours. The sewer alignment is plotted on the strip map in the office and a field reconnaissance check made to confirm the alignment. Ground profile data along the centerline may be obtained from either survey method. The accuracies of the two survey methods must be considered for quantity surveys for payment. Both methods require horizontal and vertical control surveys along the sewer route.

3. Plan Preparation

Construction plans generally are drawn on a translucent medium in pencil or ink, to facilitate reproduction. Printed plan and profile sheets are available or may be specially printed to reduce drafting time. If the grid for the profile is printed on the back of the drawing sheet, grid lines will not be erased when corrections are made to the profile.

Topography for the sewer plan is plotted by the aerial photogrammetrist when aerial photography is used for the survey. Dates of aerial survey and mapping should be shown on the plans. This topography may be transferred directly to the sewer drawing sheet for use in sewer design. An alternate method is to plot the topography directly on the aerial photographs and to reproduce the resulting photomap in the plan strip. To complete the design, the sewer plan and profile must be plotted and underground and overhead utilities, which were not shown by the aerial survey, added.

A complete set of sewer plans should include a general map showing the extent of the project. Frequently it is possible to utilize an existing street or general sewer map, and add to it the overall project plan.

It is quite useful to adopt a standard set of symbols for elements of topography and the various items of the sewage works portrayed in the sewer plans. An appropriate legend should be included on the index or title sheet. An example of a legend for sewer plans is shown in Figure 62.

The scale of the sewer plan should be large enough to show all of the necessary surface and subsurface information without excessive crowding. A horizontal scale of 40 or 50 ft to the in. (4.8 or 6.0 m to the cm) is suitable for most sewer plans; in extremely crowded urban areas 20 ft to the in. (2.4 m to the cm) should be considered. In such areas, large-scale sections of street intersections are quite useful. A generally satisfactory vertical scale for the profile is 10 ft to the in. (1.2 m to the cm). Larger scales are used when necessary to develop details properly.

Commonly, plans are reduced approximately one-half scale and issued to bidders in this size. This practice requires careful preparation of the full-size drawings to assure clear and readable reductions. Reduced size plans should contain a note stating the magnitude of the size reduction.

4. Contents

(a) *Arrangement.*—The most logical arrangement for a set of plans develops the project from general views to more specific views, and finally to more minute details. The following paragraphs are arranged to follow this generally accepted order of plan presentation.

(b) *Index.*—Plans should contain an index which lists all of the drawings in the set by title and drawing number in order of presentation. It also is useful to include a sheet index on the general plan map to identify the sheets which show the details for each length of sewer on the general plan.

(c) *Location Map.*—Either on the cover page of the plans or on the page immediately following there should be a general location map showing the location of all work in the contract and its relationship to the community. This location map also may be used as an index map as outlined in the preceding paragraph.

(d) *Subsoil Information.*—Locations of soil borings made during the design phase of a project may be shown on the sewer plans. The boring logs may be shown on the plan or be made available to bidders as a separate document.

Plans and specifications should indicate where special construction is required because of known unfavorable subsoil conditions. In all other areas, it is the bidder's responsibility and risk to investigate soil conditions and make allowances, based on his interpretation of the soil, in his bid. Neither the owner nor the engineer can guarantee subsoil conditions as a known element of the contract agreement.

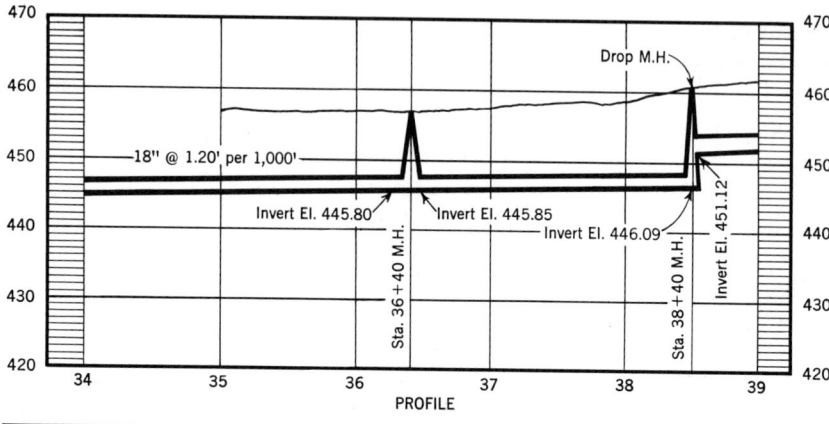

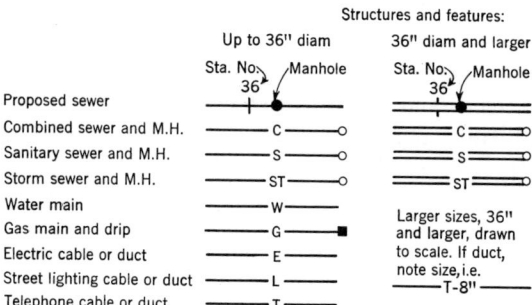

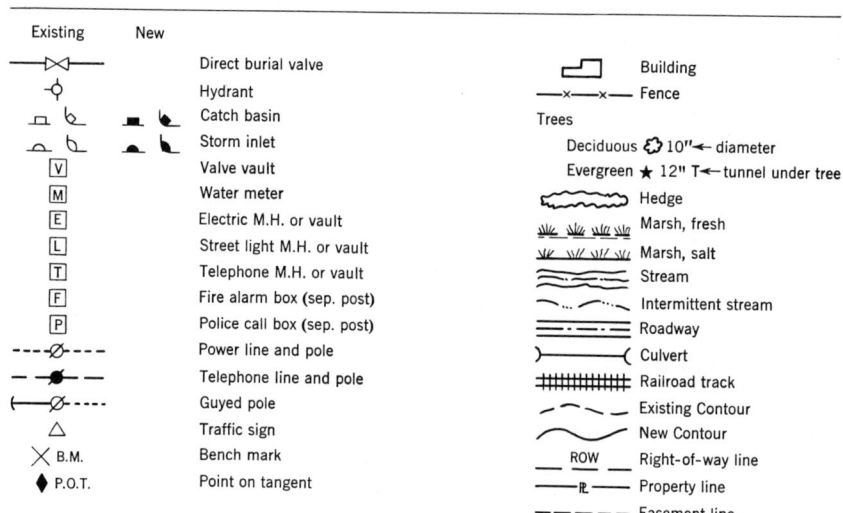

FIGURE 62.—Typical legend for sewer plans.

(e) *Datum Plane and Survey Reference Points.*—The datum plane used for determining elevations shown on plans must be defined and referenced to permanent bench marks. Survey points in street and property lines should be indicated wherever pertinent to the proposed work.

(f) *Sewer Plans.*—A continuous strip map, drawn directly above the profile, to indicate the plan locations of all work included as well as surface topography and existing improvements is an integral part of each sewer plan and profile drawing. Underground and overhead utilities along, across, or near the proposed construction route should be shown.

Efficient and economical drafting practice limits surface topography shown to that which directly affects construction or access to the work. A general note explaining the extent of the topography included on the strip map will make this approach to plotting the sewer plan known to the contractor.

Plans for sewers to be constructed in easements in private property should show survey and alignment data. Widths of temporary and permanent easements should be dimensioned.

Sewer plans generally are oriented so that the flow in the sewer is from right to left on the sheet and stationing is upgrade from left to right. Each sewer plan should include a north arrow. Match lines should be easily identifiable.

Any special construction requirement such as sheeting to be left in place should be shown on the plans.

Where interference with other structures is known to exist, explanatory cross sections should be shown. Such cross sections, often enlarged in scale, should be identified as to exact location and if practicable be placed on the plan and profile drawing where the section is cut.

A plan and profile for a sewer to be constructed in an extremely congested urban area is shown in Figure 63, and for a representative residential area in Figure 64.

(g) Sewer Profile.—Contract drawings should include a continuous profile of all sewer runs showing ground surface and sewer elevations and grades. The profile is also a convenient place to show the size, slope, and type of pipe, the limits of each size, pipe strength or type, and the locations of special structures and appurtenances.

The profile should be located immediately under the plan for ready reference. Stationing shown on the plan should be repeated on the profile. Stationing on construction drawings generally should be along the sewer centerline. Survey base-line stationing also may be shown. Indeed, both stationing notations may be shown, but stationing along the survey base line should not be substituted for stationing along the centerline of the sewer. Similarly, in the profile, ground surface along the centerline should be shown.

Stationing shown on construction drawings for location of manholes and Y-branches or house connections is to be considered

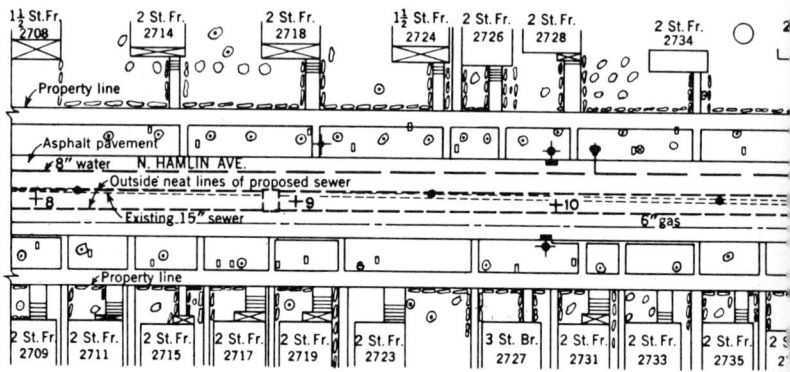

(a) Plan of existi

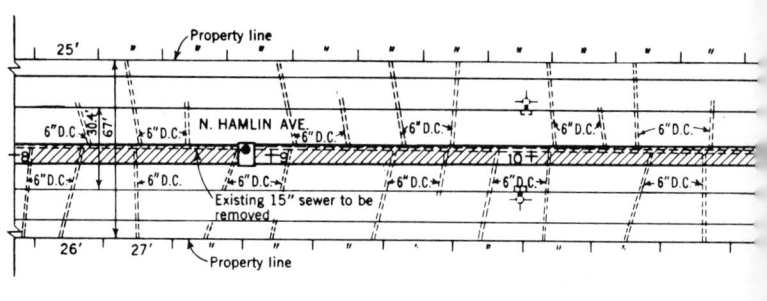

(b) S

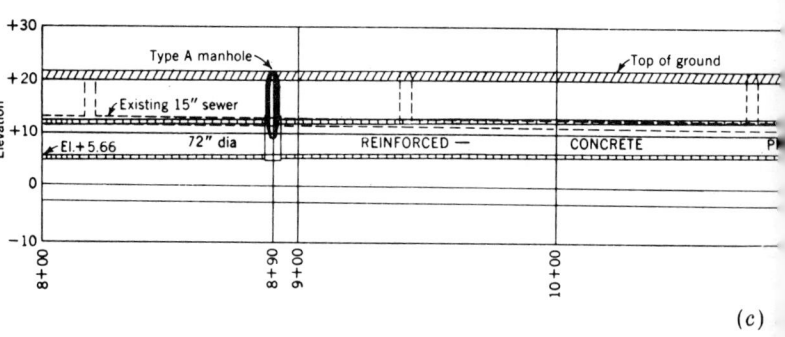

(c)

FIGURE 63.—Plan and profile f

CONSTRUCTION CONTRACT DOCUMENTS 235

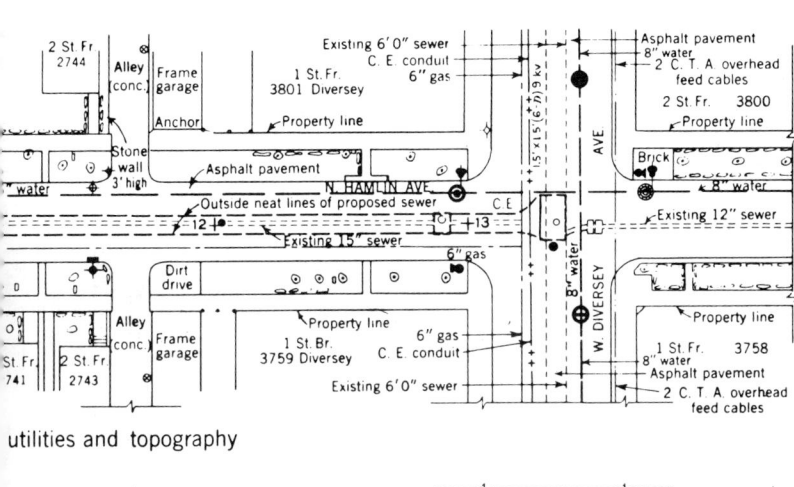

utilities and topography

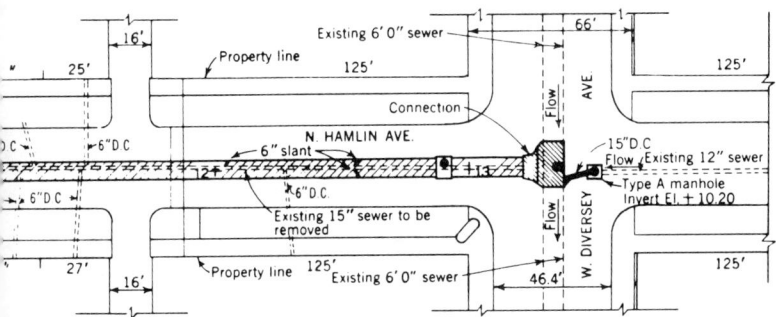

plan

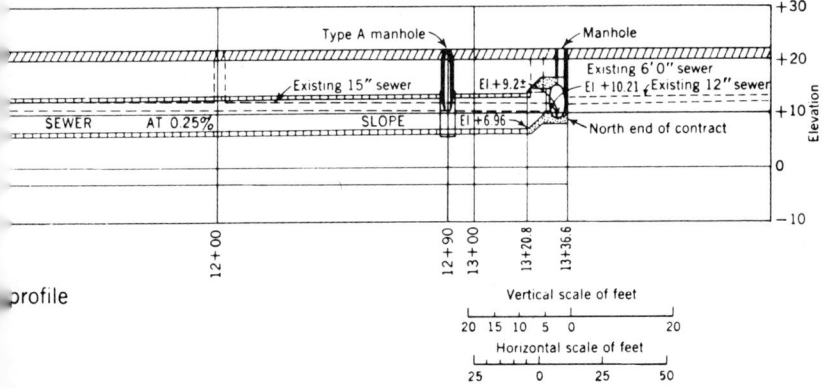

profile

or sewer in congested city street.

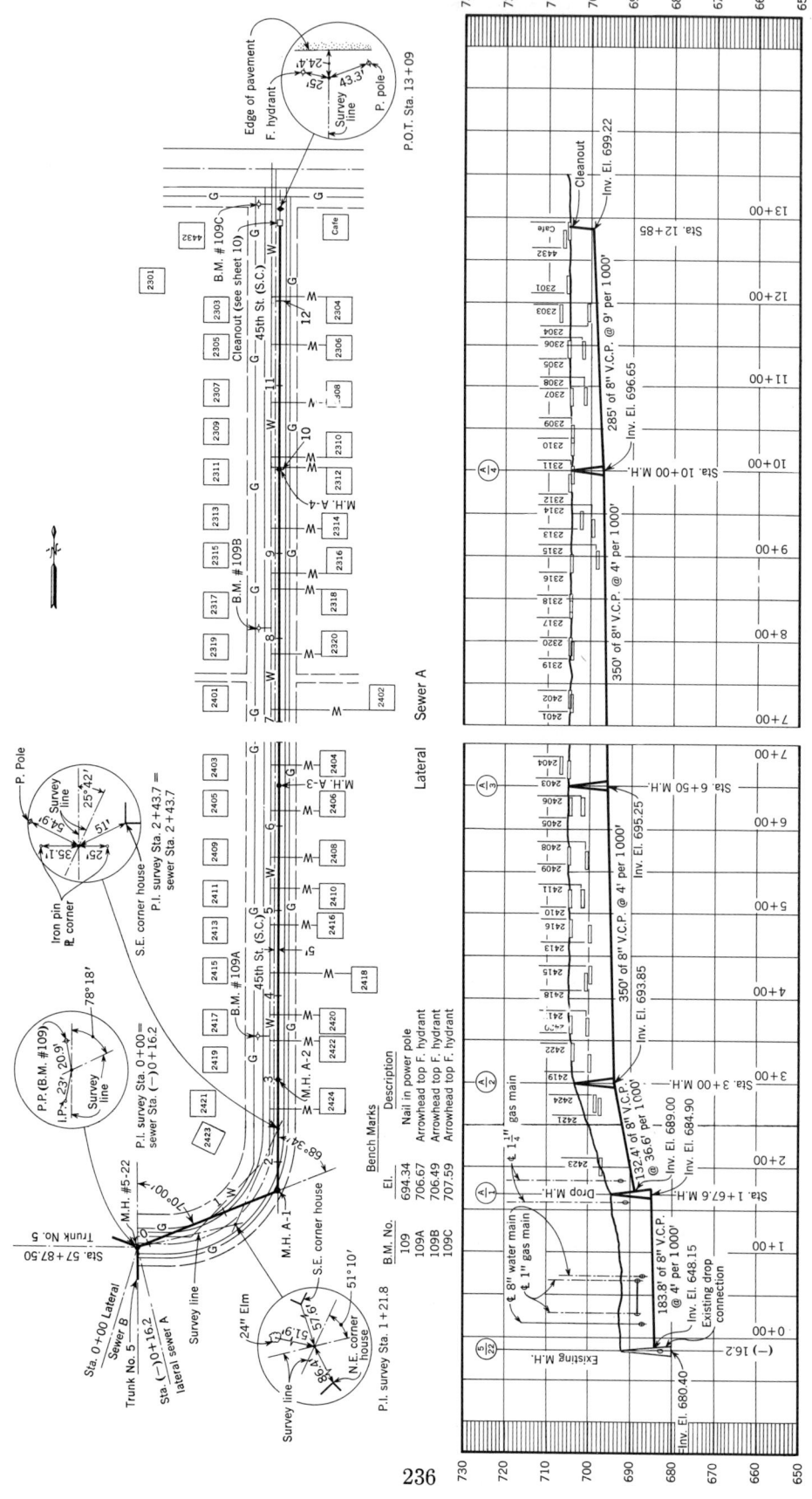

FIGURE 64.—Plan and profile for lateral sewer in residential street.

approximate only. Record drawings after construction must, however, give accurate locations of all features of a completed sewer system. Locations of junction structures must be held firm as shown on construction drawings.

(h) Sewer Sections.—When sewers consist of pipes of commonly known or specified dimensions, materials, or shapes, no sewer sections need to be shown.

For monolithic concrete sections, however, sections completely dimensioned and with all reinforcement steel shown should be included in the drawings.

(i) Sewer Details.—Separate sheets of sewer details including sewer bedding and special connections normally follow the plan and profile drawings.

(j) Appurtenances and Special Structures.—Standard casting items, such as manhole frames and covers and manhole steps, will be specified by reference to a manufacturer's catalog number in the specifications. If special details of such castings are required, they should be included in the detail sheet near the end of the set of plans.

Special structures not covered by standard details should be detailed fully to insure that the finished work is structurally sound and hydraulically correct.

(k) Standard Details.—Many details of sewer construction such as manhole bottoms and drop manholes find repeated use in sewer projects. It is helpful to develop standard details of these items which may be used repeatedly in sewer contracts.

5. Record Drawings

During construction of the sewer project, the engineer should measure and record the locations of all wyes, stubs for future connections, and other buried facilities which may have to be located in the future. All construction changes from the original plans, rock profiles, and other special classes of excavated material also should be recorded.

Construction plans should be revised to show this field information after the project is completed; and a notation such as "Revised According To Field Construction Records" should be made on each sheet. Record sets of such revised drawings should become a part of the owner's permanent sewer records.

C. SPECIFICATIONS AND RELATED DOCUMENTS

1. Purpose

Specifications and related documents set forth the details of the contractual agreement between the contractor and owner. They describe the work to be done, supplementing the information shown on the plans, and establish the method of payment. They also set forth the details for the performance of the work including necessary time schedules and requirements for insurance, permits, licenses, and other special procedures.

Specifications and related documents must be clear, concise, and complete. All portions should be written to avoid ambiguity in interpretation. Technical specifications should be understood easily and be devoid of unnecessary words and phrases, yet they must completely outline the requirements of the project. Reference to standards such as those of ASTM can be used to reduce the bulk of the specifications without detracting from completeness. A high degree of writing skill is needed to produce a good set of documents.

2. Arrangement

The arrangement of the contents of the specifications and related documents varies, depending to an extent on the requirements of the owner and the practices of the designing engineer. Furthermore, arrangement and division of the contents frequently are subject to local legal requirements. Preferably all parts of the specifications and related documents are bound in a single volume; but, in large jurisdictions or for extended programs of sewer construction, the part described as "Standard Specifications" often is bound separately and included by reference in the related documents.

The specifications and related documents can be divided most logically into four related categories defining various phases or functions in the overall administration and performance of the contract. Each category satisfies a definite requirement and information properly included in one portion should not be repeated in others.

The assembled volume should be prefaced with a table of contents. A convenient arrangement which organizes the parts in their chronological order of use is as follows:

(a) Bidding Requirements.
 1. Invitation to Bid.
 2. Instructions to Bidders.
 3. Form of Proposal.
(b) Contract Forms.
 1. Form of Contract.
 2. Wage Rates.
 3. Special Forms.
 4. Surety Bonds.
(c) General Conditions.
 1. Standard General Conditions.
 2. Supplement to General Conditions.
(d) Specifications.
 1. General Requirements.
 2. Technical Specifications.

3. Bidding Requirements

Bidding requirements should be clearly set forth in the first part of the volume of contract documents. This part covers all requirements, instructions, and forms pertaining to the submission of bids and chronologically spans the period from the date of invitation to immediately after the date of award. The various items in this portion of the contract documents are as follows:

(a) Invitation to Bid.

This is the preferred title for the section sometimes called "Advertisement for Bids," "Notice to Contractors," etc. The invitation to bid should be brief and simple, containing only the information essential to permit a prospective bidder to determine whether the work is in his line, whether he has the capacity to perform, whether he satisfies the pre-qualification requirements, whether he will have time to prepare a bid, and how to obtain bid documents. The basic elements are:

1. Identification of owner or contracting agency.
2. Name of project, contract number, or other positive means of identification.
3. Time and place for receipt of bids.
4. Brief description of work to be performed.
5. Details of submitting the bids including amount and character of any required bid deposit or bid security.
6. Means of making payment—cash, tax bills, or other method.
7. Reference to further instructions contained in the related documents.
8. Statement of owner's right to reject any or all bids.
9. Contractor's registration requirements.
10. Bidder's pre-qualification.
11. Method of obtaining bidding documents, amount of deposit for documents, and deposit refund, if any.
12. Reference to special federal or state aid financing requirements.

(b) Instructions to Bidders.

These instructions furnish prospective bidders with detailed information and requirements for preparing bids. Bidder's responsibilities and obligations, the method of preparation and submismision of proposals, the manner in which bids will be canvassed, the successful bidder selected, the contract executed, and other general information regarding the bid-award procedure are included in the instructions. The following list although not all inclusive contains the basic elements of instructions to bidders:

1. Instructions in regard to proposal form including: method of preparing, signing, and submitting same; instructions on alternates or options; data required with bids; etc.

2. Bid security requirements and form of bid bond if applicable.
3. Requirements for bidders to examine site of the work.
4. Use of stated quantities in unit price contracts.
5. Withdrawals or modification of proposal after submittal.
6. Rejection of proposals and disqualification of bidders.
7. Evaluation of bids.
8. Protection of owner from unbalanced unit adjustment prices to lump sum contracts.
9. Award and execution of contract.
10. Failure of bidder to execute contract.
11. Return of documents and bid security.
12. Instructions pertaining to subcontractors.
13. Instructions relative to resolution of ambiguities and discrepancies during bid period.
14. Taxes.
15. Insurance and bond requirements.

(c) Form of Bid Proposal.

The purpose of a stipulated form of bid proposal is to insure systematic submittal of pertinent data by all bidders in a form convenient for comparison. It must be so worded and prepared that all bidders will be submitting prices on a uniform basis.

The form of bid proposal is addressed to the owner of the proposed work and is to be signed by the bidder. The proposal contains spaces for insertion of unit prices or lump sum bid amounts and extensions for each item and a total bid price. The proposal may provide for taking bids on alternate materials or alternate methods of executing portions of the work. It may provide for combination bids on several contracts in the project. The bases for alternates and combinations must be described clearly in the Instructions to Bidders and set forth clearly in the proposal. Each proposal form may have space for an overall total bid price. An informal comparison of bids first may be made, based on the totals shown in the bidders' proposals. The formal bid comparison must be made after the extension of unit prices and totals of contract items have been checked and determined to be correct.

The time allowed for construction or completion may be a part of the proposal to be set by the bidder. Alternatively, the completion date or time may be set by the owner so that all bidders are submitting prices on the same time basis and the only variable is price.

If required, footnotes or paragraphs may be added at the end of the proposal form concerning certificates to be executed relative to receipt of addenda, the execution of "noncollusion affidavits," references to bid security submitted with the proposal, and any other supplementary certificates required. Affidavits often are appended to proposal forms which attest to the legal capacity of the bidder to submit a bid.

4. Contract Forms

The contract forms constitute the legal framework for the agreement between the contractor and owner. Once executed, each party is bound to fulfill its responsibilities as described in the contract documents. They comprise the following:

(a) Form of Contract.

There is no generally accepted Form of Contract since such documents are regulated by the laws of the local jurisdiction and state in which they are executed. The contract must, however, cover all items of work bid on in the proposal except those that have been eliminated as alternate items. It also must bind the contracting parties to conformity with the provisions of all the contract documents including both the specifications and the plans. These essentials must be included:

1. Identification of principal parties.
2. Date of execution.
3. Project description and identification.
4. Contract amount with reference to the proposal.
5. Contract time.
6. Liquidated damage clause if any.
7. Progress payment provisions.
8. List of documents comprising the contract.
9. Authentication with signatures and seals.

(b) Wage Rates.

Many public agencies specify minimum prevailing wage rates which must be made a part of the contract forms. All possible trades must be listed together with their hourly rate of compensation.

(c) Special Forms.

Included here are any special provisions such as anti-kickback regulations if they are to apply.

(d) Surety Bonds.

Acceptable bond forms should be included in the related documents and signing of the Form of Contract is contingent on prior receipt of the executed bonds. The first bond listed below is mandatory for all contracts; the others may or may not be included.

1. Performance bond.—The performance bond must be executed by a financially responsible and acceptable surety company. General practice is to require the performance bond to be in the amount of the contract bid price.

2. Payment bond.
3. Statutory bond.
4. Maintenance or guarantee bond.

5. General Conditions

This portion of the related documents is concerned with the administrative and legal relationships between the owner and the owner's representatives, the contractor and subcontractors, and the public and other contractors. General conditions should contain instructions on how to implement the provisions of the contract agreement and the technical specifications. They should not include detailed specifications for materials or workmanship.

General conditions often are standardized to apply to all public works contracts in a given city, town, or sanitary district. Some engineering offices also standardize on general conditions for projects of their design unless the owner has other requirements.

General conditions generally apply from contract to contract without need of revision or modification. If, however, amendments become necessary, the changes may be incorporated into a supplement. A supplement, for example, may be used to stipulate insurance requirements for the contract.

Model-type general conditions prepared by the American Society of Civil Engineers and the Associated General Contractors of America (1) may be used as a guide for preparation of this element of the contract forms.

6. Specifications

Specifications cover technical provisions of the contract as well as those non-technical items which are unique to the contract at hand. The two types of provisions should not be intermixed and may be kept separate by classifying one group as general requirements and the other as technical specifications.

(a) General Requirements.
Individual characteristics regarding conditions of the work, procedures, access to the site, co-ordination with other contractors, special scheduling, facilities available, and other details which are unique to the particular contract are placed properly in the general requirements.

(b) Technical Specifications.
Technical specifications set forth specific details regarding materials or workmanship applicable to the project. These specifications may be written especially for the contract or general technical specifications, called "Standard Specifications," may be prepared which are intended to apply to many contracts.

Commonly, materials are specified by reference to specifications of the American Society for Testing and Materials, United States of America Standards Institute, American Concrete Institute, American Water Works Association, and other similar organizations.

Standards of workmanship should be described in specific terms when feasible but specification of construction methods should be avoided. Nonetheless, parameters and limits often must be specified to insure construction procedures consistent with design assumptions.

For a major sewer project, technical specifications may be required for the following items of work:

1. Site preparation.
2. Excavation and backfill.
3. Concrete and reinforcing steel.
4. Miscellaneous and appurtenances.
5. Sewer materials and pipe laying.

7. Supplementary Information—Not a Part of Contract Documents

Frequently special studies, such as photogrammetric soil analyses and soil borings, are made during the design phase of a project. The results of such soil investigations and samplings with soil classifications should be made available to the prospective bidders during the bidding period. The location of soil borings made along the route of a sewer may be shown on the plans or may be indicated on a boring location map accompanying a separate soils report.

All available information pertinent to the work, especially with reference to subsurface conditions should be made available to the bidders. Bidders should be obliged to make their own interpretations of the subsoil information.

Arrangements should be made with the owner to permit bidders to inspect the soil samples and to make such additional soil borings as they may deem necessary prior to the bid date.

8. Standard Specifications

General technical specifications for workmanship and materials are intended to provide detailed descriptions of acceptable materials and performance standards which can be applied to all sewer contracts in a given jurisdiction. The description of acceptable construction procedures should be minimized in standard specifications and if such specifications are required, they should be included in the general requirements. Any time that construction procedures are specified, care should be exercised to prevent the substitution of fixed concepts for the contractor's initiative. In every case procedures must achieve, safely, specified final results.

General technical specifications usually are aimed at more than any one specific contract. They may be used on a group of similar contracts or even for larger groups of dissimilar contracts.

A supplement to the standard specifications may be written to include special requirements modifying the standard specifications for a particular contract. In this regard, care must be taken in using standard specifications so that the work involved in writing supplements is not greater than the work that would be required to write completely new specifications for the contract.

9. Sewer Specifications Check List

Technical specifications must cover, for example, concrete and reinforcing steel and other ordinary items of heavy construction. As these are common to much construction work the check list includes but does not further classify these items.

As a guide for determining completeness of construction specifications, with specific reference to the items of sewer construction discussed in this manual, the following check list of subjects is offered:

(a) Site Preparation.
    1. Protection of existing structures and plantings.
    2. Existing utilities.
    3. Tree removal.
    4. Pavement removal.

(b) Excavation and Backfill.
    1. General.
    2. Excavation classification.
    3. Trench excavation.

        a. Limitations on width of trench.
        b. Spoil placement.
        c. Preparation of trench bottom.
        d. Pipe bedding.

    4. Rock excavation.
    5. Excavation for appurtenances.
    6. Tunneling and/or jacking
    7. Casing installation.
    8. Sheeting and bracing.
    9. Dewatering.
    10. Construction along or across highways and railroads.
    11. Backfilling trenches.

        a. Quality.
        b. Placement.
        c. Compaction.
        d. Protection of pipe.

    12. Backfill for appurtenances.
    13. Pavement and sidewalk replacement.
    14. Surface restoration.
    15. Measurement and payment.

(c) Concrete and Reinforcing Steel.
(d) Miscellaneous Materials and Appurtenances.
1. Iron castings.
2. Manhole materials.
3. Miscellaneous metals.
4. Equipment for appurtenances.
5. Construction of appurtenanees.
6. Waterproofing appurtenances.
7. Measurement and payment.
(e) Sewer Materials and Pipe Laying.
1. Sewer pipe materials.
2. Pipe joints.
3. Pipe laying.

   a. Control of alignment.
   b. Control of grade.

4. Pipe jointing.
5. Service connections.
6. Connections to existing sewers.
7. Connections between different pipe materials.
8. Concrete encasement or cradle.
9. Sewer paralleling water main.
10. Sewer crossing water main.
11. Repair of damaged utility services.
12. Acceptance tests.

    a. Infiltration.
    b. Exfiltration.
    c. Smoke.
    d. Air.

13. Measurement and Payment.

Reference

1. "Form of Contract For Use in Connection with Engineering Construction Projects." Amer. Soc. Civil Engr. and Assoc. Gen. Contractors Amer., New York (1966).

CHAPTER 11. CONSTRUCTION METHODS

A. INTRODUCTION

The design and construction of sewers are so interdependent that knowledge of one is essential to the competent performance of the other. Indeed, the title of this manual bespeaks this relationship. It is fitting then, at this point, to examine and discuss some of the construction methods in common use. Local conditions, of course, will dictate variations; and the ingenuities of the owner, engineer, and contractor must be applied continually if construction costs are to be minimized and a quality job is to result.

Commencement of the construction phase normally introduces a third party, the contractor, to the sewer project and the division of responsibility and liability must be understood by all. The role of the engineer changes from active direction and performance during design to that of professional and technical observation during construction. Indeed, his duties during construction when properly defined and prosecuted assure the owner that the work when accepted is substantially in accordance with the contract documents. It is important to note, however, that the engineer's respresentative on the site is not expected to duplicate the detailed inspection of material and workmanship properly delegated to the manufacturer, supplier, and contractor.

Preconstruction conferences are helpful in deciding whether the contractor's proposed operations are not incompatible with contract requirements and whether they will result in finished construction approvable by the engineer and acceptable to the owner. These joint meetings of the owner, engineer, and contractor should culminate in definite construction plans and administrative procedures to be followed throughout the life of the construction contract and should include items such as progress schedules, payment details, method of making submittals for approval, and channels of communication. All of these aspects of construction should be settled before construction begins.

B. CONSTRUCTION SURVEYS

1. General

Base lines and bench marks for sewer line and grade control should be established along the route of the proposed construction by the engineer. All control points should be referenced adequately to permanent objects located outside normal construction limits.

2. Preliminary Layouts

Prior to the start of any work, the contractor should lay out rights-of-way, work areas, clearing limits, and pavement cuts to insure proper recognition of and protection for adjacent properties. Access roads, detours, bypasses, and protective fences or barricades also should be laid out and constructed as required in advance of sewer construction. All layout work must be completed and checked by the engineer before any demolition or construction begins.

3. Setting Line and Grade

The transfer of line and grade from control points established by the engineer to the construction work should be the responsibility of the contractor, with spot checks by the engineer as work progresses. The preservation of stakes or other line and grade references provided by the engineer is similarly the responsibility of the contractor. Generally a charge is made for re-establishing stakes carelessly destroyed.

The line and grade for the sewer may be set by one or a combination of the following methods:

(a) Stakes, spikes, shiners, or crosses set at the surface on an offset from the sewer center line.
(b) Stakes set in the trench bottom on the sewer line as the rough grade for the sewer is completed.
(c) Elevations given for the finished trench grade and sewer invert while sewer laying progresses.

The first method generally is used for small diameter sewers. The other methods are used for large sewers or where sloped trench walls result in top-of-trench widths too great for practical use of short offsets or batter boards.

In the first method, stakes, spikes, shiners, or crosses are set on the opposite side of trench from which excavated materials are to be cast at a uniform offset, insofar as practicable, from the sewer center line. A cut sheet is prepared (Table XXX) which is a tabulation of the reference points giving sewer station, offset, and the vertical distance from reference point to proposed sewer invert.

The line and grade may be transferred to the bottom of the sewer trench by the use of batter boards, tape and level, or patented bar tape and plumb bob unit.

Batter boards and batter-board supports must be suspended firmly across the trench and be adequate to span the excavation without measurable deflection. If the spanning member is to be the batter board, it is set level at an even foot (or other unit of measurement) above sewer grade and a nail is driven in the upper edge at center line of the sewer. Preferably, the spanning member is used as a support only and a 1-in. (2.5-cm) board is nailed to it with one edge in a true vertical plane at the center line of the sewer. A nail then is driven in the vertical edge of the batter board at an even foot above sewer grade. A string line is drawn taut across at

TABLE XXX.—Typical Cut Sheet

"A" street ....sewer
4th Street to 5th Street
Stakes 5 ft left

Sheet 1 of 1 Sheet
Notes book ....Page......
Prepared by ....Date......

Station	Size (in.)	Grade	Elevation		Cut		Remarks
			Invert	Stake	Feet	Hundredths	
0 + 00	12	0.0025	105.50	110.75	5	25	Existing manhole
0 + 25	..	0.0025	105.56	112.30	6	74	..
0 + 50	..	0.0025	105.63	109.70	4	07	Y-branch right
0 + 75	..	0.0025	105.69	110.35	4	66	..
1 + 00	12	0.0025	105.75	111.99	6	24	Etc.

Note: Ft × 0.3048 = m; in. × 2.54 = cm.

least three batter boards. The sewer center line is transferred to the trench bottom with a heavy plumb bob cord held lightly against the string line. Grade is transferred to the sewer invert with a grade rod equipped with a suitable metal foot to extend into the end of the pipe. For steep grades it is advisable to fasten a bull's eye level to the grade rod to assure that the rod is held plumb. For ease in reading, the grade rod may be marked at subgrade, finish grade, and invert grade with saw grooves, small nails, or colored bands. If nails or grooves are used, care must be taken in reading the rod to insure that the string line is not riding up or down on the reference devices. The line and grade of the string line should be checked by observation for possible error in cuts or in the establishing of the batter boards. Periodic inspection should be made during sewer laying to insure that the set line and grade have not been disturbed.

Another method of setting grade is from offset crosses or stakes or from offset batter boards and double string lines and the use of a grade rod with a target near the top. When the sewer invert is on grade, a sighting between grade rod and two or more consecutive offset bars or the double string line will show perfect alignment.

The transfer of surface references to stakes along the trench bottom is in some instances permitted; but the use of batter boards is preferred. If stakes are established along the trench bottom, a string line should be drawn between not less than three points and checked in the manner used for batter boards.

When trench walls are not sheeted but sloped to prevent caving, line-and-grade stakes are set in the trench bottom as the excavation proceeds. This procedure requires a field party to be on the work almost constantly.

A third method, applicable to large pipes or monolithic sections of sewers on flat grades, requires the line and grade for each pipe length or form sections to be set by means of a transit and level from either on top of or inside the completed conduit.

In the construction of large cast-in-place sewer sections in an open trench, both line and grade may be set at or near the trench bottom. Line points and bench marks may be established on cross bracing where such bracing is in place and rigidly set. When the concrete is placed in two or more lifts, grade stakes must be set for the bottom or invert placement. Later, alignment and grade of the wall or arch must be determined by checking the setting of the forms.

4. Tunnel Construction

Where tunnel construction is an extension of a sewer of sufficient size without change of alignment, the initial line and grade for the tunnel work may be established by extending lines and grades through and forward from the completed portion.

When tunneling begins from an isolated shaft, great care must be taken in transferring line and grade from the surface. If tunneling from any one shaft extends more than several hundred feet from that shaft, and especially if the alignment is curvilinear, it may be desirable to drive vertical line pipes to the elevation of the tunnel at intervals of 600 feet or less through which plumb lines can be dropped when excavation reaches those points. This will allow a check of both alignment and grade at each line pipe location. If tunneling is to be done under air pressure, these pipes must be capped at either the top or bottom at all times. All line pipes should be grout sealed on completion of construction.

A segment of surface alignment can be transferred to the bottom of the shaft with the aid of a pair of heavily weighted piano wire plumb lines. Plumb lines must be placed accurately on line and as far apart as size of shaft will allow. Sway of the wire plumb lines can be controlled by using a heavy cylindrical steel plumb weight immersed but freely suspended in a bucket of oil. Rotation of the plumb weight can be suppressed by adding vertical radial fins to the steel cylinder. The segment of alignment between plumb lines can be transferred by transit to spads set in the tunnel roof. As the heading advances, the alignment can be extended to additional roof spads and to line points on tunnel supports. Alignment may be checked from reference markers in finished sections and from line pipes as encountered.

Where the permanent tunnel lining is of cast-in-place concrete, small wood blocks can be attached to the forms on the approximate center line of the arch. On removal of forms, the blocks remain in the concrete arch. Spads then can be driven into them and an accurate point established for suspension of a plumb line.

In large tunnels, small platforms can be erected near the arch for the placement of a transit head which will be ready for use at all times.

Tunnel grade can be carried through the excavation on bench marks set in the tunnel wall or, in small bores, carried down from the bench marks set in the arch in conjunction with alignment spads.

As cleanup progresses in preparation for final lining operations, cross sections should be taken throughout the tunnel to insure that proper alignment and grade can be maintained. Tunnel geology and cross-sections should be recorded for incorporation in permanent construction records.

C. SITE PREPARATION

The amount of site preparation required varies from none to the extreme where the major portion of project costs is expended on items other than excavating for and constructing the sewer.

Operations which properly should be classified as site preparation are clearing; removal of unsuitable soils; construction of access roads, detours, and bypasses; improvements to and modification of existing drainage; location, protection, or relocation of existing utilities and pavement cutting. The extent and diversity of these operations make further discussion thereof impractical. Note, however, that the success of the contractor in keeping the project on schedule depends to a great degree on the thoroughness of the planning and execution of the site preparation work.

D. OPEN-TRENCH CONSTRUCTION

1. Trench Dimensions

With plans and specifications competently prepared it can be assumed that the location of the proposed sewers has been determined with proper regard for the known locations of existing underground utilities, surface improvements, and adjacent buildings. Barring unforeseen conditions, it becomes the objective of the contractor to complete the work as shown on the plans at minimum cost and with minimum disturbance of adjacent facilities. Because of load considerations as discussed in Chapter 9, the width of trench at and below the top of the sewer should be only as wide as necessary for the proper installation and backfill of the pipe or sewer section. The engineer must provide alternate designs or corrective measures to be employed if allowable trench widths are exceeded through overshooting of rock, caving of earth trenches, or over excavation. The width of trench from a plane 1 ft (30 cm) above the top of the sewer to the ground surface is related primarily to its effect on adjoining facilities such as other utilities, surface improvements, and nearby structures.

In undeveloped subdivisions and in open country, economic considerations often justify sloping the sides of the trench from a plane 1 ft (30 cm) above the top of the finished sewer to the ground surface. This eliminates placing, maintaining, and removing substantial amounts of sheeting and bracing unless safety regulations make some type of sheeting or bracing mandatory.

In improved streets, on the other hand, it may be desirable to restrict the trench width so as to protect existing facilities and reduce the cost of surface restoration. Available working space, traffic conditions, and economics will all influence the decision.

2. Excavation

With favorable ground conditions excavation can be accomplished in one simple operation; under more adverse conditions it may require several steps. In general, stripping, drilling, blasting, and trenching will cover all phases of the excavation operation.

(a) Stripping.—Stripping may be advantageous or required as a first step in trench excavation for a variety of reasons, the most common of which are:

1. Removal of topsoil or other materials to be saved and used for site restoration.
2. Removal of material unsatisfactory for backfill to insure its separation from usable excavated soils.
3. Removal of material having a low bearing value to a depth yielding material capable of supporting heavy construction equipment.
4. Reduce cuts to depths to which a backhoe can dig.
5. Make it easier to charge drill holes.

(b) Drilling and Blasting.—In some areas many lineal feet of sewers and house services must be installed in hard rock. In addition, some shales and softer rocks which may be ripped in open excavation will require blasting before they can be removed in confined areas as required in trenching or excavating for structures.

Normally the most economic method will involve preshooting, that is, drilling and shooting rock before removal of overburden. In some instances the occurrence of wet granular materials above the rock ledge will necessitate stripping before drilling as holes cannot be maintained through the overburden to permit placing of explosive charges.

For narrow trenches in soft rock, a single row of drill holes may be sufficient. One or more additional rows will be required in harder rock and for wider trenches. To reduce overbreak and improve bottom fragmentation, time delays should be used in blasting for trenches. In tight quarters, trench walls can be pre-split and the center shot in successive short rounds to an open face to produce minimum vibration.

It must be recognized that there will be a minimum feasible trench width varying with the rock formation and in the case of small sewers it may be necessary to design the conduit for the positive projecting condition.

(c) Trenching.—The method and equipment used for excavating the trench will depend on the type of material to be removed, the depth, amount of space available for operation of equipment and storage of excavated material, and prevailing practice in the area.

Ordinarily the choice of method and equipment rests with the contractor. However, various types of equipment have practical and real limitations regarding minimum trench widths and depths. The contractor is obligated, therefore, to utilize only that equipment capable of meeting trench width limitations imposed by pipe strength requirements or for other reasons as set forth in the technical specifications.

Spoil should be placed sufficiently back from the edge of the excavation to prevent caving of the trench wall and to permit safe access along the trench. With sheeted trenches, a minimum distance of 3 ft (1 m) from edge of sheeting to toe of spoil bank will provide safe and adequate access. Under such conditions the supports must be designed for the added surcharge. In unsupported trenches the minimum distance from the vertical projection of the trench wall to the toe of the spoil bank normally should be not less than one-half the total depth of excavation. In most soils, this distance will be greater in order to provide safe access beyond the sloped trench walls.

1. Trenching Machines.—This type of machine is not as widely used as it was 30 or 40 yr ago. With the development of the backhoe, or drag shovel, trenching machines have in some areas been relegated to trenches of moderate width and depth. For installation of sewers up to 24 or 30 in. (61 or 76 cm) in diameter, in cohesive soils which can be restrained by light sheeting and trench jacks or shores, the trenching machine can make rapid progress at low cost. Some contractors still prefer trenching machines and have used them for trenches up to 12 ft (4 m) deep.

2. Backhoes.—Backhoes (Figure 65) are available with bucket capacities varying from $\tfrac{3}{8}$ to 3 or more cu yd (0.3 to 2.3 cu m). They are convenient for the excavation of trenches with widths exceeding 2 ft (0.7 m) and to depths up to 25 ft (8 m) and are the most satisfactory equipment for excavation in rock. Minimum trench widths are compared with some common backhoe sizes in Table XXXI.

FIGURE 65.—Typical backhoe operation.

TABLE XXXI.—Trench Widths Associated with Various Backhoe Bucket Capacities

Bucket Capacity (cu yd)	Minimum Trench Width (in.)	
	Without Side Cutters	With Side Cutters
3/8	22	24–28
1/2	27	28–32
3/4	28	28–38
1	34	34–44
1 1/4	37	37–46
1 1/2	38	38–46
2	50	50–58

Note: in. $\times$ 2.54 = cm; cu yd $\times$ 0.76 = cu m.

The backhoe also is used with a cable sling for lowering pipe into the trench. By this means a single piece of equipment can maintain the pipe laying close to the point of excavation. Where the soil does not require sheeting and bracing, this method becomes a very economical one. When sheeting and bracing must follow the excavation closely, the combination of a back hoe for excavation and a crane for placement of pipe is a common practice.

3. Clamshells.—When the protection of other underground structures or soil conditions requires close sheeting and the use of vertical-lift equipment, the clamshell bucket is used. In very deep trenches where two-stage excavation is required, the backhoe is sometimes used in combination with the clamshell, with the backhoe advancing the upper part of the excavation and the clamshell following for the lower. Sheeting and bracing of the upper part are installed as required prior to the clamming of the lower part and the installation of the lower stage of sheeting.

4. Draglines.—In open country or in a wide right-of-way it may be feasible to do a large part of the excavation by means of a dragline, allowing the sides of the trenches to acquire their natural slope. In cases of very deep trench excavation, 30 to 50 ft (9 to 15 m), the dragline has been used for the upper part of the excavation, with a backhoe operating at an intermediate level. By rotating the backhoe the material thus excavated can be relayed to the dragline which then lifts it to the spoil bank or to trucks at the surface.

5. Front-End Loaders.—In wide, deep trenches the front-end loader has sometimes been used as an auxiliary to a backhoe or clamshell. In this arrangement the backhoe excavates the upper part of the trench and, perhaps, the center section of the lower part, leaving the bottom bench or benches for the front-end loader or dozer which completes the excavation, placing its spoil within the reach of the backhoe or clamshell.

3. Sheeting and Bracing

Trench bracing and sheeting should be adequate to prevent cave-in of the trench walls or subsidence of areas adjacent to the trench. Responsibility for adequacy of any required sheeting and bracing usually is stipulated to be with the contractor. The strength design of the system of supports should be based on the principles of soil mechanics as they apply to the materials encountered. Sheeting and bracing must always comply with applicable safety requirements.

In narrow trenches the simplest form is known as skeleton sheeting and bracing, and consists of pairs of planks set vertically with spreaders or trench jacks as shown in Figure 66. Horizontal spacing will vary with soil conditions. A similar but more satisfactory form of temporary

FIGURE 66.—Vitrified clay pipe installation in lightly braced trench.

support providing continuous sheeting involves the use of plywood panels prefabricated with plank stiffeners. The panels then are set vertically in the trench with spreaders, trench jacks, or hydraulic shores spaced as required.

For wider and deeper trenches a system of wales and cross struts of heavy timber shown in Figure 67 often is used. Sheeting is installed outside the horizontal wales as required to maintain the stability of the trench walls. Jacks mounted on one end of the cross struts maintain pressure against the wales and sheeting (Figures 68 and 69).

In some soil conditions it has been found economical and practical to use built-up units of sheeting called trench shields which are pulled forward as pipe laying progresses. Experience has indicated that the trench shield, in general, is a poor substitute for adequate sheeting and

FIGURE 67.—Concrete pipe installation in heavily sheeted trench, New York, N.Y.

FIGURE 68.—Corrugated metal pipe installation in a lightly braced trench.

where trench width is critical the shield should not be used. Care must be exercised in pulling shields forward so as not to drag or otherwise disturb the previously laid pipe sections or create conditions not assumed in calculating trench loads.

In non-cohesive soils combined with considerable groundwater, it may be necessary to use continuous steel sheet piling to prevent excessive soil movements. Such steel piling usually extends several feet below the bottom of the trench unless the lower part of the trench is in firm material. In the latter case the width of the trench in the upper granular material may be widened so that the steel sheet piling can toe-in to the lower strata.

In some soils, steel sheet piling (Figure 70) can be used with a backhoe operation for the upper part of excavation, but the piling usually needs to be braced before the excavation has reached its full depth and the remaining excavation then is performed by vertical lift equipment such as a clamshell.

Another means of trench sheeting occasionally adopted involves the use of vertical H-beams as "soldier beams" with horizontal wood lagging. This is sometimes advantageous for trenches under existing overhead viaducts where overhead clearances are low and spread footings lie alongside the trench walls. The holes for the vertical beams may be started by means of an earth auger. Then the soldier beams can be tilted into these holes and driven with the aid of an air hammer. As excavation progresses downward, the lagging is installed between adjacent pairs of

FIGURE 69.—Laying asbestos-cement pipe in a heavily sheeted trench.

soldier beams. For deep trenches with limited overhead clearances, the soldier beams can be delivered to the site in shorter lengths and their ends field-welded as driving progresses.

The removal of sheeting following pipe laying may affect the earth load on the sewer. This possibility must be considered during the design phase. If removal is to be permitted, appropriate requirements must be placed in the technical specifications. Whenever sheeting is removed it must be done properly, taking care to backfill thoroughly the voids thus created.

4. Dewatering

Trenches should be dewatered for concrete placement and pipe laying, and kept continuously dewatered for as long as necessary. Unfortunately, the disposal of large quantities of water from this operation in the absence of storm drains or adjacent water courses may present problems. Other

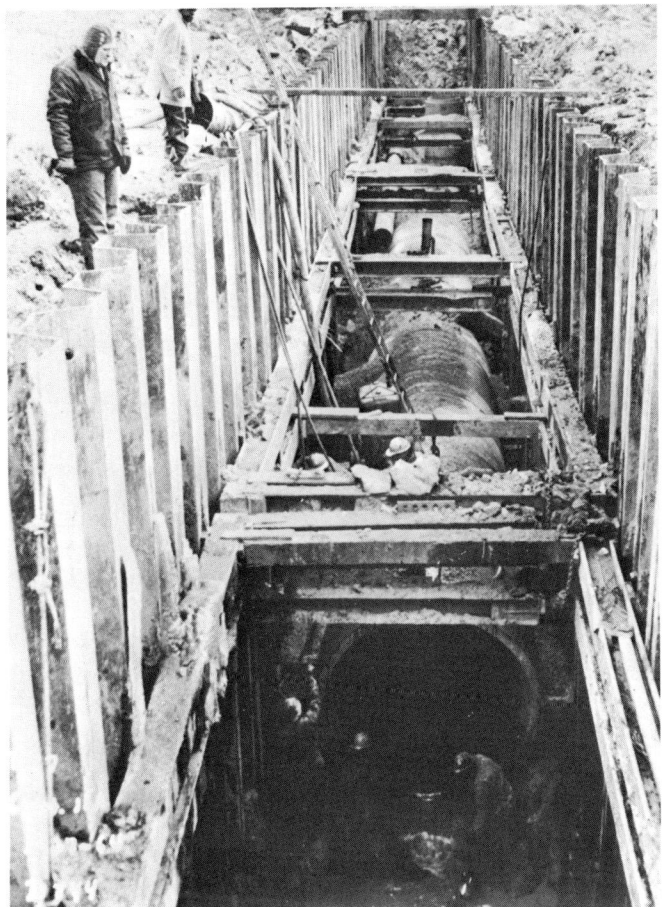

FIGURE 70.—Steel sheet piling, Chicago, Ill.

means of disposal being unavailable, the possibility of draining the water through the completed sewer to a permissible point of discharge should be considered, provided sufficient precautions are taken to prevent scour of freshly placed concrete or mortar. In all cases care must be exercised to insure that property damage, including silt deposits in sewers and on streets, does not result from the disposal of trench drainage.

Crushed stone or gravel may be used as a subdrain to facilitate drainage to trench or sump pumps. It is good practice to provide dams in the subdrain to minimize the possibility of undercutting the sewer foundation from excessive groundwater flows.

Monolithic sewers or structures cast in place which must be built in a dry excavation may make it necessary to install an underdrain leading to a pump sump. Again care must be taken to insure that this underdrain does not become clogged with silt from the excavation operation. The underdrain may consist of 4-in. (10-cm) diam or larger drain tile or perforated-metal pipe placed in the granular bedding of the sewer structure.

An excessive quantity of water, particularly when it creates an unstable soil condition, may require the use of a well-point system such as shown in Figure 71. A system of this type consists of a series of perforated pipes driven or jetted into the water-bearing strata on either side of the sewer trench and connected with a header pipe leading to a pump. The equipment for a well-point system is expensive and specialized. General contractors often seek the help of special dewatering contractors for such work. The cost of dewatering by well points is frequently more than offset by savings in the cost of sheeting and bracing.

Where excavation is in coarse water-bearing material, turbine well pumps may be used to lower the water table during construction. Chemical or cement grouting and freezing of the soil adjacent to the excavation have been used in extremely unstable water-bearing strata.

5. Foundations

(*a*) **Pipe Sewers.**—Firm cohesive soils provide adequate pipe foundations when properly prepared. The trench bottom must be shaped to fit the pipe barrel and holes dug to receive projecting joint elements. Careful trimming of the trench bottom is, however, rarely done today. Current (1968) practice includes over-excavation in depth followed by backfilling with granular material such as crushed stone, crushed slag, or gravel to provide uniform bedding of the pipe. Granular bedding is used increasingly because it is both practical and economical.

FIGURE 71.—Deep construction with special foundation, using well points.

In very soft bottoms it is necessary first, as a minimum, to overexcavate to greater depths and stabilize the trench bottom by the addition of gravel or crushed slag or rock. The stabilizing material must be graded to prevent movement of subgrade up into stabilized base and base into bedding material. The required depth of stabilization should be determined by tests and observations on the job.

In those instances where the trench bottom cannot be stabilized satisfactorily with a crushed rock or gravel bed and where limited and intermittent areas of unequal settlement are anticipated, a timber cribbing, piling, or reinforced-concrete cradle may be necessary.

Where the bottom of the trench is rock, it must be overexcavated to make room for an adequate cushion of granular material which will uniformly support the conduit. The trench bottom must be cleaned of shattered and decomposed rock or shale prior to placement of bedding.

In some instances, pipe sewers must be constructed for considerable distances in areas generally subject to subsidence, and consideration should be given to constructing them on a timber platform or reinforced concrete cradle supported by piling. The sewer's support should be adequate to sustain the weight of the full sewer and backfill. Piling in this case is sometimes driven to grade with a follower prior to making the excavation. This practice avoids subsidence of trench walls resulting from piledriving vibrations. Extreme care must be taken to locate all underground structures.

(*b*) **Cast-in-Place Concrete Sewer Sections.**—Foundation considerations for these sections are different from those for pipe sewers. A reinforced concrete section with both transverse and longitudinal steel reinforcement can bridge across intermittent variations in soil-bearing capacity. Long stretches of very soft bottom may dictate some stabilization of the subsoil and in extreme cases also may require pile supports. A moderate softening of the trench bottom by groundwater may be stabilized more economically by 1 or 2 in. (3 to 5 cm) of dry-mix concrete below the theoretical section, with the remaining invert concrete placed continuously on top of the additional thickness.

(*c*) **Slip-Form Pipe.**—Foundation conditions required for cast-in-place pipe are similar to those required for pipe sewers. Since the slip form slides along on the finished trench bottom, the accuracy of trench excavation and foundation preparation will determine the variations in invert grade of the finished sewer.

The contour of the trench bottom should match the outside bottom radius of the required section. In firm cohesive soils, shaping to the accuracy required may be accomplished with a rounded edge on the trenching machine or backhoe bucket. In rock trenches, overexcavation followed by shaping with selected backfill is required to support the slip form properly.

Wet trenches must be dewatered and the water table kept below the trench bottom while placing concrete. An effective method of water control is to overexcavate and refill to slip-form grade with a granular underdrain material.

6. Pipe Sewers

Proper sewer construction implies that quality materials and acceptable laying methods are to be used. Diligence in assuring both is required of all project personnel.

(*a*) **Pipe Quality.**—Pipe inspection is properly conducted by the manufacturer and by independent testing and inspecting laboratories. Moreover, with pipe sewers, transportation charges may constitute a substantial portion of material costs and inspection at the pipe plant, therefore, is usually desirable. Complete inspection consists of visual inspection of workmanship, surface finish, and markings; physical check of length, thickness, diameter, and joint tolerances; proof of crushing strength design with three-edge bearing; and material tests by compression, and absorption and acid resistance tests of representative specimens. On concrete pipe, both core and cylinder tests should be required as the correlation between core tests and cylinder tests has not been established for various pipe manufacturing processes. Pipe cores also permit checking tolerances on placement of reinforcing cages.

Pipe also should be checked visually at time of delivery for possible damage in transit and again as it is laid for damage in storage or handling.

(*b*) **Pipe Handling.**—Care must be exercised in handling and bedding all precast pipe, regardless of cross-sectional shape. As the pipe was designed to be loaded uniformly, all phases of construction should be undertaken to insure that insofar as practical, this will be the case. Pipe should be handled during delivery in a manner which eliminates any possibility of high impact or point loading, always taking care to protect joint elements.

(*c*) **Pipe Placement.**—Pipe should be laid on a firm but slightly yielding foundation true to line and grade, with a uniform bearing under the full length of the barrel of the pipe, without break from structure to structure, and with the socket ends of bell and spigot or tongue and groove pipe upgrade. Pipe should be supported free of the bedding or foundation during the jointing process to avoid disturbance of the subgrade. A suitable excavation should be made to receive pipe bells and joint collars where applicable so that the bottom reaction and support are confined only to the barrel of the pipe. Adjustments to line and grade should be made by scraping away or adding adequately compacted foundation material under the pipe and not by using wedges and blocks or beating on the pipe.

Extreme care should be taken in jointing to insure that the bell and spigot are clean and free of any foreign materials. Joint materials vary with the type of pipe used and are discussed in Chapter 8. All joints should be made properly using the jointing materials and methods specified. All joints should be sufficiently tight to meet infiltration or exfiltration tests.

In large sewers with compression-type joints, considerable force will be required to insert the spigot fully into the bell. Come-alongs and winches or the crane itself may be rigged to provide the necessary force (Figures 72 and 73). Inserts should be used to prevent the pipe from

CONSTRUCTION METHODS 263

(a) LAYING PIPE

(b) JOINING PIPE

FIGURE 72.—Laying reinforced concrete pipe.

FIGURE 73.—Laying asbestos-cement pipe.

being thrust completely home prior to checking gasket location. After the gasket is checked, the inserts can be removed and the joint completed.

Walking on small diameter pipe or otherwise disturbing any conduit after jointing must not be permitted. Interior pointing and exterior pointing or grouting of joints should not be done until sufficient pipe has been laid to insure against damage to finished joints. Moreover, interior pointing should be done after backfill is placed. Water must not be allowed to rise around the pipe until joints have cured adequately.

At the close of each day's work, or when pipe is not being laid, the end of the pipe should be protected by a close-fitting stopper with adequate precautions taken to overcome possible uplift.

If the pipe load factor is increased with either arch or total encasement, contraction joints should be provided at regular intervals in the encasement coincident with the pipe joints to increase flexibility of the encased conduit.

7. Cast-in-Place Concrete Sections

(*a*) **General Considerations.**—For sewer sizes 8 ft (2.5 m) in internal width or larger, cast-in-place reinforced concrete sections are often used. Figure 74 shows a sewer of this type. This size limit, however, will vary up and down, depending on the relative cost of pipe and cast-in-place sewers at the time and location of actual construction.

The concrete for cast-in-place sewers usually is cast in two lifts, generally designated as the invert and the arch. Large semi-elliptical sections sometimes are divided into three lifts—invert, side walls, and arch. In either case the arrangement of trench bracing must be planned to prevent movement of earth and the development of undue earth pressures on the partly constructed or partly cured sections.

FIGURE 74.—Two-barrel, 29-ft (9-m) section storm drain, St. Louis, Mo.

Sections with dished inverts and semi-elliptical arches are common. The curvature of the upper surface is made the maximum feasible without top forms. Occasionally semi-circular inverts are found necessary for hydraulic reasons, but such shapes are more costly because of the necessity of top forms and the requirements of special shaping of the trench bottom or excess concrete in the trench corners.

(*b*) **Invert.**—Installing invert reinforcement and placing the invert concrete should follow the trenching operation as closely as practical. The hardened invert can act as a continuous strut between the bottom ends of the trench sheeting and permit removal of temporary timber struts which would interfere with the placement and moving of the arch forms. The removal of the cross struts becomes mandatory if traveling steel forms are used for the arch section.

The top surface of the invert should have a trowel finish to produce a smooth surface with a low coefficient of friction. Smoothness is especially important in combined sewers where the dry-weather flow is proportionately small and the retardation of a rough invert might cause sludge and grit deposits.

(*c*) **Arch.**—Economy of construction usually dictates the use of collapsible traveling steel forms for substantial lengths of similar size and shape. Short runs may be formed entirely of wood or horizontal wood lagging supported by curved steel ribs. Forms must be kept clean and oiled to produce a smooth interior surface of the finished sewer section. The time of removal of forms and placement of backfill must be governed by a study of probable soil pressures and the ability of the partially cured concrete to resist them.

Steel reinforcement in the lower parts of the arch or side walls generally is placed before the positioning of the form. Crown steel is placed just prior to placement of the arch concrete.

It is usually more economical to use a flat slope on the top surface of the haunches of the arch than to steepen this slope so as to require top forms.

(*d*) **Slip-Form Pipe.**—Cast-in-place nonreinforced concrete pipe manufactured by slip-form methods is available in sizes from 12 to 120 in. (30 to 300 cm) diam.

As the casting or extruding machine is propelled along the trench, concrete is introduced into the hopper, extruded, and consolidated in intimate contact with the prepared trench. The exposed exterior arch is cured with a sprayed membrane, polyethylene film, or earth cover while the interior is cured by sealing all openings into the conduit. Forms normally are removed 4 to 6 hr after casting.

(*e*) **Quality of Concrete.**—The usual provisions for the mixing and placing of quality concrete in other structures should be even more rigidly adhered to in sewer construction. Sewers, in contrast to building and highway structures which usually can be protected from full design loading until thoroughly hardened, must be placed under full load within a few days after the concrete is placed. Arches, especially, must bear the brunt of this early loading and must be of high-quality concrete.

8. Backfilling

(*a*) **General Considerations.**—Backfilling of the sewer trench is a very important consideration in sewer construction and seldom receives the attention and inspection it deserves. The methods and equipment used in placing fill must be selected to prevent dislocation or damage to the pipe.

The method of backfilling varies with the width of the trench, the character of the materials excavated, the method of excavation, and the degree of compaction required.

(*b*) **Degree of Compaction.**—In improved streets, or streets programmed for immediate paving, a high degree of compaction should be required. In less important streets, or in sparsely inhabited subdivisions where flexible macadam roadways are used, a more moderate specification for backfilling may be justified. Along outfall sewers, in open country, it may be sufficient to mound the trench and, after natural settlement, return to regrade the area. Compaction results should be determined in accordance with current AASHO test procedure.

(*c*) **Methods of Compaction.**—

1. Cohesive Materials.—Cohesive materials with high clay content are characterized by small particle size and low internal friction. They have small ranges of moisture content over which they may be compacted satisfactorily and are very impervious in a dense state. Because of the strong adhesive forces of the soil particles, strong pressures must be exerted in order to shear the adhesive forces and remold the particles in a dense soil mass. These characteristics dictate the use of impact-type equipment for most satisfactory results in compaction. In confined areas, pneumatic tampers and engine-driven rammers may give good results. The upper portion of the trench can be consolidated by self-propelled rammers where trench widths are relatively narrow. In wide trenches sheepsfoot rollers may be used or if the degree of compaction required is not high, dozers and loaders may be used to compact the fill.

Regardless of equipment used, the soil must be near optimum moisture content and compacted in multiple lifts if satisfactory results are to be obtained. The trench bottom must be dry before placing the first lift of backfill.

If the material had a high moisture content at the time of excavation, some preparation of the material probably will be required before spreading in the trench. This may include pulverizing, drying, or blending with dry or granular materials to improve placement and consolidation.

2. Cohesionless Materials.—Cohesionless materials are granular with little adhesion, but with high internal friction. Moisture content at the time of compaction is not so critical and consolidation is affected by reducing the surface friction between particles, thus allowing them to rearrange in a more compact mass. Considering the characteristics of cohesionless material, the most satisfactory compaction is achieved through the use of vibratory equipment.

In confined areas, vibratory plates give the best results. For wider trenches vibratory rollers are most satisfactory. Again, if the degree of compaction required is not high, and if layers are thin, the vibration imparted by dozer or loader tracks may result in satisfactory consolidation.

In some areas water is used to consolidate granular materials. But unless the fill is saturated and immersion vibrators are used, the degree and uniformity of compaction cannot be controlled closely. With some materials, adequate compaction may be obtained by draining off, through drains constructed in manhole walls, water used to saturate or puddle fill. These drains are capped after the backfill has drained.

3. Borrow Materials.—Sometimes the material removed from the trench may be entirely unsatisfactory for backfill. Thus other materials must be hauled in from other sources.

Both cohesive and noncohesive materials are used, but an assessment must be made of the possible change in groundwater movement with the use of foreign materials. For example, the use of cohesive materials to backfill a trench in rock could result in a dam impervious to groundwater traveling in rock faults, seams, and crevices. On the other hand, granular materials placed in a clay trench could result in a very effective subdrain.

(a) Backfilling Sequence.—Backfilling should proceed immediately on curing of trench-made joints and after the concrete cradle, arch, or other structure gains sufficient strength to withstand loads without damage.

Backfill generally is specified as consisting of three zones, with different criteria for each: the first zone extends from the bedding material to 12 in. (30 cm) above top of pipe or structure, an intermediate zone generally containing the major volume of the fill, and the upper zone consisting of pavement subgrade or finish grading materials.

The first zone should consist of selected materials placed by hand or by suitable equipment in such manner as not to disturb the pipe and compacted to a density consistent with design assumptions. In some instances the material used for granular bedding is brought above the pipe to insure high density backfill with minimum compactive effort.

Gradation of the intermediate zone is not critical so long as the material is capable of consolidation resulting in a reasonably small number of voids. Good compaction generally can be obtained if maximum permissible particle size is limited to slightly less than the depth of the layer being compacted.

Depth and compaction of the upper zone are dependent on the type of finish surface to be provided. If the construction area is to be seeded or sodded, the upper 18 in. (45 cm) may consist of 12 in. (30 cm) of select material slightly mounded over the trench and lightly rolled, covered by 6 in. (15 cm) of top soil. If the area is to be paved, the upper zone must be constructed to the proper elevation for receiving paving and base course under conditions matching design assumptions for the subgrade. If the trench backfill is completed in advance of paving, the top 6 in. (15 cm) of the upper zone should be scarified and recompacted prior to paving. In such instance, it may be necessary to install a temporary surface to be replaced at a later date with permanent pavement.

Before and during the backfilling of a trench, precautions should be taken against the flotation of pipelines due to the entry of large quantities of water into the trench, causing an uplift on the empty or partly filled pipeline. An instance of this nature is shown in Figure 75. This photograph also illustrates poor selection of backfill material for the first zone as evidenced by large rocks bearing directly against the pipe barrel.

9. Surface Restoration

On completion of backfill the surface should be restored fully to a condition at least equal to that which existed prior to the sewer construction.

Portland cement or asphaltic concrete pavements should be saw cut and removed to a point beyond any caving or disturbance of the base materials prior to patching. If this results in narrow unstable panels, pavement should be removed to the next existing contraction or construction joint. Before replacing permanent pavement, the subgrade must be restored and compacted until smooth and unyielding.

The final grade in unpaved areas should match existing grades at construction limits without producing drainage problems. Restoration of grass, shrubs, and other plantings should be done in conformance with construction contract documents. Tree damage should be repaired in accordance with good horticultural practice.

FIGURE 75.—Sewer floated before backfilling was completed.

E. TUNNELING

1. General Classification

Tunneling is considered to be any construction method which results in the placement or construction of an underground conduit without continuous disturbance of the ground surface, and includes the various forms of jacking of prefabricated units from shaft or pit locations. Tunneling methods applicable to sewer construction can be classified generally as follows:

(a) Auger or boring method;
(b) Jacking of preformed steel or concrete lining; and
(c) Mining methods.

2. Auger or Boring

In sizes less than 36-in. (90-cm) diam, rigid steel or concrete pipe can be pushed for reasonable distances through the ground and the earth removed by mechanical means under the control of an operator at the shaft or pit location. Several types of earth augers are on the market; and some contractors specialize in this type of operation. In the case of thin-wall smooth steel pipe, the auger need not be larger than the inside diameter. In the case of the heavier wall concrete pipe, it may be necessary to use an auger with a special head, having a diameter equal to the outside diameter of the pipe being placed. The presence of boulders is a serious deterrent to this method of installation. If such obstructions are expected, particularly when pipes smaller than 36 in. (90 cm) are to be placed, it may be more economical first to install an oversize lining by conventional tunnel or jacking methods. The smaller pipe then can be placed within the liner pipe and the remaining space backfilled with sand, cement grout, or concrete.

3. Jacking

Although the limits will vary with geographic locations and soil conditions, finished interior diameters of 30 to 108 in. (75 to 275 cm) are the generally accepted limits for the jacking of permanent tunnel lining. Excavation and removal of the excavated material is by manual labor augmented with air spades, special knives, etc. The most commonly used materials for such jacking operations are reinforced concrete, corrugated-metal, or smooth steel pipe. The pipe selected for jacking must be strong enough to withstand the loads exerted by the jacking procedure.

The usual procedure is to equip the leading edge with a cutter or shoe to protect the pipe. As succeeding lengths of pipe are added between the leading pipe and the jacks, and the pipe jacked forward, soil is excavated by hand and removed through the pipe. Material is trimmed with care and excavation does not precede the jacking operations more than necessary. Such a method usually results in minimum disturbance of the natural soils adjacent to the pipe.

When jacking corrugated metal pipe, high frictional resistance of the corrugations can be overcome partially by providing a cutting edge or leading section of smooth external surface. In stable, cohesive soils, mining to a diameter slightly greater than the outside corrugations of the pipe generally will permit completion of jacking before subsidence of the earth and resulting increased resistance. Jacking of corrugated pipe is not very successful in non-cohesive materials due to lack of natural arch action and high frictional resistance.

When jacking, contractors have found it sometimes desirable to coat the outside of the pipe with a lubricant, such as bentonite, to reduce the frictional resistance. In some instances, this lubricant has been applied through pressure fittings installed in the wall of the leading pipe. Grout holes sometimes are provided in the walls of the pipes for use in filling outside voids. Protective joint spacers are used to prevent damage to pipe joints.

Because soil friction may increase with time, it is desirable to continue jacking operations without interruption until completed.

(*a*) **Alignment.**—In all jacking operations it is important that the direction of jacking be established carefully prior to the start of work. Guide rails must be installed in the bottom of the jacking pit or shaft. In the case of a large pipe, it is desirable to have such rails carefully set in a concrete slab. The number and capacity of the jacks used depend primarily on the size and length of the pipe to be placed and the type of soil encountered.

Backstops must be strong enough and large enough to distribute the maximum capacity of the jacks against the soil behind them. A typical installation for jacking concrete pipe is shown in Figure 76.

(*b*) **Continuous Sewers by Jacking.**—In some cases long sewer lines have been installed by jacking from a series of shaft locations spaced along the line of the pipe.

4. Mining Methods

Tunnels with finished interior dimensions of 5 ft (1.5 m) or larger in clay or granular materials are built ordinarily either with the use of tunnel shields, boring machines, or by open-face mining, with or without some breasting. Rock tunnels normally are excavated open-face by conventional mining methods or with boring machines.

(*a*) **Tunnel Shields.**—In very soft clay or in running sand, especially in built-up city areas, it may be necessary to use tunnel shields to make tunneling safe. Compressed air also may be required to control the entry of water into the tunnel.

With a shield it is necessary to install a primary lining of sufficient strength to support the surrounding earth and to provide a progressive backstop for the jacks which advance the shield. The lining may be installed against the earth, and the annular space between the lining and the earth filled with pea gravel and grout. Alternatively, the tunnel lining may be expanded against the earth as the shield is advanced. The

FIGURE 76.—Jacking 78-in. (198-cm) diam reinforced concrete pipe under railroad, Chicago, Ill.

latter method practically eliminates the need for grouting the annular opening.

Tunnel shields may be of the open-face type which provide a hood under which miners excavate the tunnel heading. Other shield designs have ports in the face through which the material being removed flows. Where the displaced material is allowed to flow through ports in the face of the shield, a continuous check must be maintained on the volume of materials removed against the linear displacement. Street surface elevations must be observed continually for vertical movement either up or down.

Shield alignment usually is maintained by varying the thrust of the jacks around the periphery of the shield.

(b) Boring Machines.—Boring machines, also called digger shields or mechanical moles, have been developed for tunnel excavation in clay and rock. They usually have cutters mounted on a rotating head which is advanced into the heading. A conveyor system moves muck to the muck cars. Machines may be braced against the walls of the excavation or against previously placed tunnel lining. Some machines also are equipped with shields. Machines have been used successfully in the construction of tunnels in clay up to 20-ft (6-m) diam.

Machines are most useful in fairly long runs through generally similar material. Difficulties have been encountered where the material to be excavated varies.

Soft ground tunnels in wet conditions also are being economically built with a permanent primary expanded lining of cast iron segments with the use of a shield. The cast iron is covered with a concrete lining in free air thus economizing on the more expensive compressed air labor. Such designs minimize street and adjacent building settlement as compared with temporary linings of steel liner plate.

(c) **Open-Face Mining without Shields.**—Where the ground allows the use of open-face mining methods (Figures 77 and 78) it is often more economical to use segmental supports of wood or steel for the sides and top of the tunnel only. The need for compressed air or breast boards in the tunnel heading will depend on the type of soil and amount of moisture or groundwater. Tunneling in soft clay in the presence of considerable groundwater usually would dictate the use of compressed air. Stiff clays which require the use of air spades for removal can be excavated normally by open-face methods without the use of compressed air.

(d) **Primary or Temporary Lining.**—Materials used for primary lining are usually steel, wood, or a combination of the two, but also may be segmental blocks of concrete or cast iron.

1. Timber and Timber and Steel Combination.—Tunnels have been timber supported with a continuous series of five-piece timber frames or cants kept close to the face of the tunnel. A combination of timber and steel often is used in which case rib sets fabricated from I-beams, H-beams, or channels support wood lagging (Figure 78). Rib sets may be fabricated full circle or have vertical legs with arch sections bent to the outside shape of the sewer. Ribs are set at required intervals with intermediate support supplied by wood lagging placed outside the ribs or just inside the outer flanges. Encroachment of ribs within the neat line of the tunnel lining section should not interfere with the placement of reinforcing steel and generally is limited to 2 to 3 in. (5.1 to 7.6 cm).

2. Steel.—Many tunnels use only steel for their primary lining. Such steel lining has been in several forms and combinations. In some instances, liner plates bent to conform with the outside shape of the sewer section are bolted together along longitudinal and circumferential seams. Curved I-beam ribs, conforming to the outside of the sewer section, may be placed at transverse joints between any two successive rings of liner plates. In other cases, I-beam ribs are placed inside the liner plates and the spacing of the ribs is varied according to pressure exerted by the soil on liner-plate sections. Deformed steel pans, or lagging plates, with flanges along two edges only may be used to provide intermediate support between rib sets similar to that of wood lagging described previously. The plates are 12 in. (30 cm) wide and vary in length from 24 to 60 in. (60 to 150 cm). Steel ribs are allowed to penetrate into neat lines of the concrete section 2 to 3 in. (5.1 to 7.6 cm). Concrete required to fill the

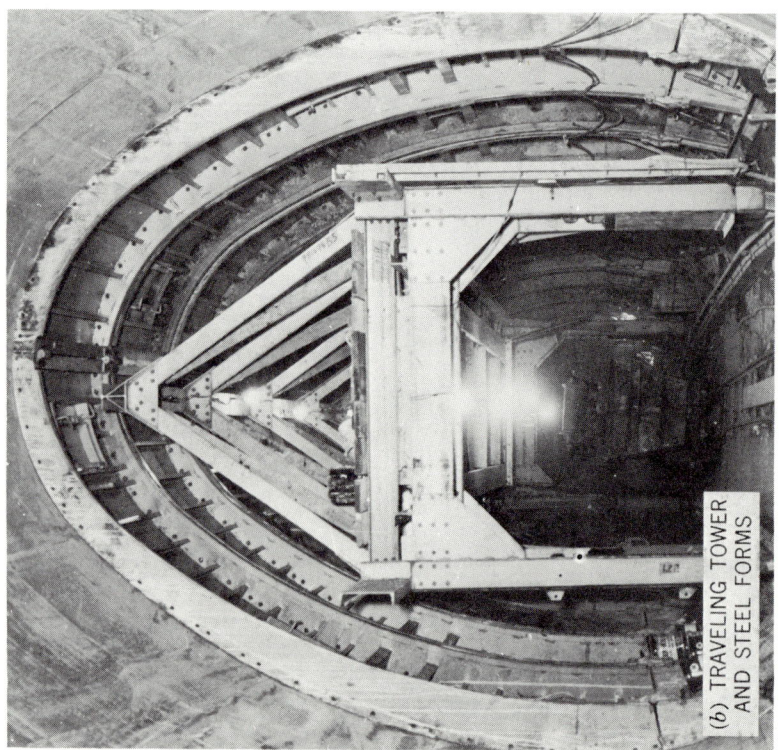

FIGURE 77.—Concrete sewer, 18- by 20-ft (5.5- by 6.1-m) cross section, Chicago, Ill.

(a) STEEL RIBS WITH WOOD LAGGING, 8-FT SEWER, CHICAGO, ILL.

(b) CIRCULAR STEEL LINING FOR SEWER, POCATELLO, IDAHO

FIGURE 78.—Primary tunnel linings of wood and steel.

space inside the flanges of the lagging plates is not considered a part of the finished sewer lining.

Some engineers and contractors prefer to use continuous rings of liner plates having sufficient section modulus to resist the earth pressures without use of special structure ribs or rings (Figure 78). A circular lining formed of such plates becomes a compression ring and has some inherent stability not equaled by horseshoe-shaped supports. Soil conditions and the contractor's preference determine the choice of such a lining. Some liner plates have shallow corrugations with flanges on all four edges for bolting to the adjacent plates. Others have deeper corrugations with flanges only on the circumferential edges. Lapped, bolted joints are used along longitudinal edges of the latter type. Nuts can be welded to the exterior of the plate at bolthole locations or the bolts can be held against rotation by special clips so that final tightening of the bolt can be accomplished from inside the tunnel. Steel liner plates frequently are used also as sheeting for circular shafts, either on or adjacent to the line of the tunnel.

If design of liner plate support is based on the assumption that plates will act as a compression ring, immediate grouting behind liner plates or immediate expansion of the lining is required to insure uniform loading. In any event, voids behind liner plates should be grouted or the lining expanded prior to subsidence of the overburden.

3. Precast Concrete.—Precast concrete segments for primary lining units have been used in several different forms. Construction of a 10-ft (3-m) diam sewer in Detroit was accomplished with seven-piece circular sets of precast blocks as seen in Figure 79. Rings of primary lining were placed with the aid of a hydraulic placer arm mounted within the trailing part of a tunnel shield. Hydraulic jacks, mounted on the shield, pushed against the completed rings of primary lining to advance the shield. A reinforced concrete lining was placed within the precast rings, resulting in a final section of approximately 12 in. (30 cm) of precast units and an additional 12 in. of concrete cast in place.

In a sewer of circular section in London, some precast concrete blocks formed with flanges and reinforcing ribs were used for primary lining. The finished lining was placed by guniting.

(e) Oval Precast Concrete Rings.—Another form of precast concrete tunnel lining consists of oval rings of short laying length. These sections are set with the long axis vertical. They are of such dimensions that they can be moved in a horizontal position through previously placed sections, then rotated in the heading of the tunnel, and jointed up with the last section placed. The oval rings are transported by a battery-operated locomotive specially designed for the purpose. Figure 80 showing installation of oval precast concrete rings does not indicate any primary lining. If soil conditions require initial primary lining, the lining should conform closely to the outside shape of the precast units. The space between precast units and primary lining must be filled with grout or other suitable material.

(a) RIBBED LINING BEFORE GUNITING, COLNE VALLEY, ENGLAND

(b) PRIMARY LINING OF SOLID CONCRETE BLOCKS BEFORE PUMPING CONCRETE, DETROIT, MICH.

FIGURE 79.—Precast concrete tunnel linings.

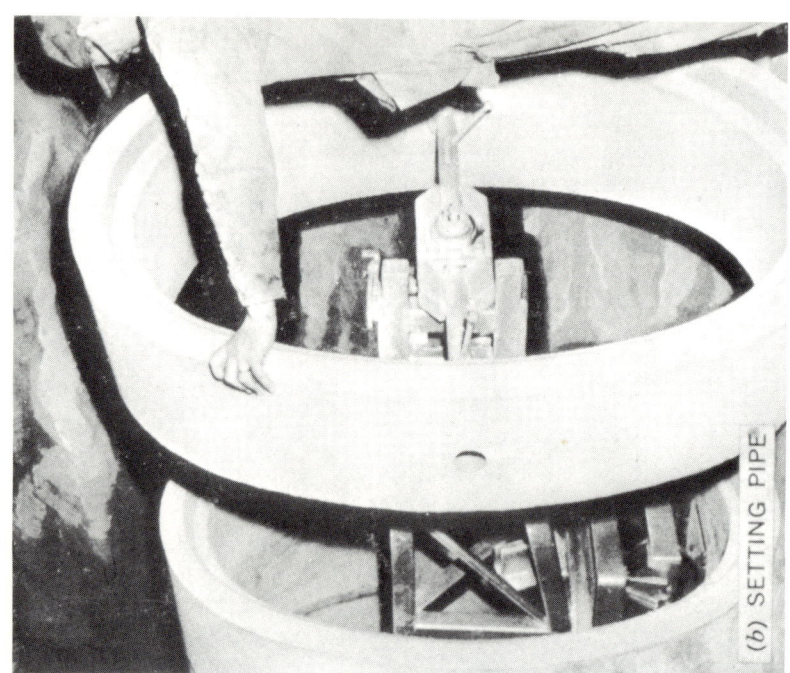

FIGURE 80.—Oval precast concrete rings. (a) DELIVERING PIPE. (b) SETTING PIPE.

(*f*) **Tunnel Excavating Equipment.**—The type of excavating equipment or tools used in sewer tunneling is the same as in tunneling for other purposes, and depends on the kind of material to be excavated and the work space available. Pneumatic spades and special knives are used widely in excavating clay. Drilling and blasting are necessary in rock tunnels. In the case of shale, undercutting machines like those used in coal mining have been used to advantage.

In a large sewer tunnel in Chicago, with a finished dimension of approximately 17 by 19 ft (5.2 by 5.8 m), an electric shovel with a ¾-cu yd capacity (0.6-cu m) dipper was used effectively. Excavated material usually is hauled from the face of the tunnel to the shaft location by means of small mucking cars pulled by electric locomotives.

(*g*) **Shafts.**—Where tunnels are of considerable length, one or more construction shafts need to be provided. On important thoroughfares these shafts are better located in an adjacent side street or vacant lot with access to the work through a short connecting entry tunnel.

Offset shaft locations are especially desirable when soil conditions require the use of compressed air. In such case only one air lock in the entry tunnel will be required. Shafts generally are located so that tunneling in both directions is possible. Construction shafts on long tunnels usually are spaced 1,200 to 2,500 ft (380 to 760 m) apart. Factors tending to extend this spacing are the need for compressed air, the size of tunnel, and the depth below ground.

Shafts should be large enough to permit the installation of an electric hoist. Such equipment should be used only for the handling of material, with separate facilities available for man lifts in deep shafts. Hoisting in shafts by means of a crane may be permitted when the length of the tunnel is short and safety precautions are taken to prevent engine exhaust from entering the shaft and tunnel.

Tunnel drainage also is discharged normally from the shaft. Therefore, the shaft must contain some form of collection sump and drainage pump.

(*h*) **Main Shafts and Emergency Exits.**—Workmen should not use the material shaft for entry except when a part of it is provided with an enclosed stairway. Usually a separate shaft, adjacent to the material shaft, equipped with a spiral stairway is used for entry by workmen. Where feasible, emergency exits should be provided whenever the distance from a heading to the nearest exit exceeds 1,000 ft (300 m). If the tunnel is under air pressure, then each exit must be equipped with an air lock.

(*i*) **Compressed Air Equipment and Locks.**—Compressed air equipment for tunneling should have sufficient capacity to maintain a pressure which will prevent movement of the soil as well as to supply the normal air demand.

The equipment includes compressors, air receivers, piping, control valves, air locks, main and emergency locks, bulkhead walls, gages, etc. Separate locks should be provided for materials and personnel. Generally, for long, large diameter tunnels, electrically operated compressors are used with two independent sources of power. Standby compressors in many cases are either diesel or gasoline powered.

The line supplying air to the working chamber is equipped with a pressure-reducing valve, a sufficient number of relief valves, and pressure-recording gages to insure a constant pressure in the chamber. Air locks are valved so that the lock may be operated either from within or without. Experienced lock tenders must be present at all times. Each lock door is equipped with a glass bull's-eye. Adequate signaling devices must be installed at each lock. A signal system for the compressor house, such as a siren operated from a mercoid switch on each of the pressure-reducing valves should be installed. Air supplied to the working chamber should be of proper temperature and contain no objectionable substance.

If air pressure exceeds 15 psi (1.05 kg/sq cm) all necessary safeguards should be taken, including a medical lock and stage decompression facilities, with a physician on call day and night at all times men are working.

(*j*) **Ventilating Air.**—In compressed air tunnels, air must be circulated in sufficient quantity to permit the work to be done without danger or excessive discomfort.

In free air tunnels, the ventilation rate must be adequate to clear the tunnel of gases in a maximum of 15 min if explosives are used. Rates also must be adequate to dilute exhaust of permissive diesel equipment to safe limits. In cold weather, it may be necessary to condition ventilating air to prevent excessive fogging at the heading.

F. SPECIAL CONSTRUCTION

1. Railroad Crossings

Sewers at times must be constructed under railroad tracks which may be at street level, on a raised embankment above street grade, or on an existing railroad viaduct.

Crossing of tracks at grade or on an embankment usually can be accomplished most economically by jacking, boring, tunneling, or combination thereof. Usually a casing pipe is inserted and the sewer pipe is placed inside.

When the distance from the base of rail to the top of the sewer is insufficient to allow jacking or tunneling (less than one diameter clearance), it is necessary, either to remove the tracks and interrupt service during an open-cut operation or to build a temporary structure for support of the railroad tracks after which the sewer may be constructed in open trench below that structure.

Construction of sewers under existing railroad viaducts involves wide variation in methods, depending on size of the sewer, its location in plan and elevation with respect to viaduct footings, type of footings, and the nature of the soil.

Where the soil is stable and the sewer is of sufficient size to allow the use of tunnel methods and located satisfactorily with respect to viaduct footings, tunneling may be both safe and economical.

When the proposed sewer does not meet these criteria, special methods of sheeting and bracing must be devised. To prevent subsequent movement of soil beneath the footings, all sheeting and bracing should be left in place.

2. Crossing of Principal Traffic Arteries

Residential and secondary traffic arteries usually can be closed to traffic during the construction of sewer crossings. But on heavily traveled streets and highways where public convenience is a major factor, it may be desirable to use tunneling or jacking methods for the crossing.

When required, limited traffic movements across open trenches can be accommodated by temporary decking. Trenches of narrow or medium width can be spanned with prefabricated decks placed on timber mudsills at the edges of the trench. Where the top of trench is wider than 16 to 20 ft (5.3 to 6.7 m), temporary piling for end and in some cases center support may be required. Where center supports pass through the sewer section, provision must be made for such piling to remain until the sewer is completed. On a project in Chicago, center piling of steel was set on the center line of a proposed twin-barrel sewer and later encased in the sewer section.

3. Stream and River Crossings

(*a*) **Sewer Crossing under the Waterway.**—Stream and river crossings may be constructed either in the dry through use of cofferdams and diversion channels or subaqueously. If constructed in the dry, planning and scheduling of construction should be such that completed portions of the line are not subject to damage in the event of cofferdam overtopping.

Concrete encasement, if required, should be placed with construction joints at 30-to 40-ft (9-to 12-m) intervals coincident with pipe joints. Pipe may be set conveniently to line and grade by supporting it on burlap bags filled with a dry-batched concrete mix. These bags also may be placed for construction of bulkheads in subaqueous concrete placements.

(*b*) **Sewer Crossing Spanning the Waterway.**—In Chapter 7 various methods of spanning obstacles are mentioned including hanging and fastening sewers to structural supports and the construction of sewer pipe beams. The latter type of construction consists of a manhole or other supporting structure on either side of the waterway and the spanning member itself. Where the crossing is of considerable width, intermediate piers or supports may be necessary. Figure 81 shows a 39-in (99-cm) diam prestressed concrete pipe beam section 80 ft (24 m) long, 72 ft (22 m) clear span, being lowered into final position over the west branch of the Housatonic River in Pittsfield, Mass.

4. Outfall Structures

(*a*) **Riverbank Structures.**—Outfall sewers and head walls may be located above or below surface water levels. When they are partly sub-

FIGURE 81.—Installing prestressed concrete pipe beam section, Pittsfield, Mass.

merged it is necessary to provide some form of cofferdam during construction.

In shallow water, an earth dike or timber piling may be sufficient to maintain a dry pit. In deep water, steel sheet piling cofferdams are desirable. Usually a single wall cofferdam with adequate bracing is sufficient, but in excessive depths at banks of main navigation channels, a double wall may be required. Standard practices of cofferdam design and construction should govern.

(*b*) **Ocean Outfalls.**—For long ocean outfalls there are two distinct phases of construction: the inshore section through the surf zone and the offshore section. The surf zone usually extends to a depth of 50 ft (15 m) but may be shallower or deeper depending on local ocean conditions. The inshore or surf zone section requires positive support and lateral restraint for the pipe. The inshore section usually requires a temporary pier for driving sheet piling to maintain the trench through the breakers and for pipe installation. If the shore is all sand, suitable piles must be driven for support and anchorage of the outfall pipe.

The offshore section of outfall pipe usually is laid from floating equipment as pictured in Figure 82 and often placed directly on the ocean floor provided the grade is satisfactory. Gravel or rock side fill to the spring line

FIGURE 82.—Construction of 102-in. (260-cm) diam diffuser for ocean outfall. Note diffuser ports, *a*. The pipe is attached to a "horse." The assemblage has swung clear of the barge and is being lowered for placement at a depth of 190 ft (58 m). When the "horse" has come to rest on the bottom, the man in the control capsule, *b*, will use hydraulic controls to maneuver the pipe into position.

of the pipe is added frequently to prevent lateral currents from scouring local potholes which might cause pipe movement.

Ocean outfall pipes have been made of cast iron, reinforced concrete, protected steel pipe, and a combination of these materials. Small pipelines may be assembled on shore and then pulled into position. Large lines must be laid in sections, although it is obviously advantageous to make the lengths as long as possible to minimize the number of underwater joints which must be assembled by divers. It is also desirable to select a joint type and construction procedure which will facilitate underwater connections. This is especially important if the outfall is in water more than 150 ft (50 m) deep because of the limited time divers can work at this depth. Flexible joint, ball-and-socket cast iron pipe laid with a cradle permits the jointing of pipe above water, thus eliminating the use of divers and underwater operations.

G. SEWER APPURTENANCES

Recent improvements in sewer pipe joints have made it possible to demonstrate that large quantities of extraneous flows are admitted at poorly constructed sewer appurtenances and service connections. Increased attention to this phase of sewer construction is essential if flow of surface and groundwaters in sanitary sewers is to be reduced further.

The sound principles of construction which apply to reinforced concrete and masonry structures must be applied also to sanitary sewer manholes.

Adherence to the following construction steps would eliminate most leakage in masonry manholes:

(a) Place manhole base slab on well prepared subgrade.
(b) Use clean water, free from silt and clay.
(c) Pre-wet masonry units with clean water to saturate surface prior to laying.
(d) Lay masonry units with fully mortared joints.
(e) Strike and tool inside joints.
(f) Parge (plaster) exterior of manhole with at least ½ in. (1.25 cm) of mortar.
(g) Moist cure masonry.
(h) Provide flexible joint at manhole wall or extend manhole base and cradle pipe to first flexible joint.
(i) Fit brick bats and spalls to pipe at wall openings to minimize curing shrinkage.
(j) Set manhole frame and cover at elevation to exclude surface water.

Sewer service connections should be permitted only at wye or tee branches or at machine cut, watertight-jointed taps. Fitting should be supported adequately during the making of the joint. Bell and spigot, compression-type flexible joints should be used at the junction of the house sewer and service tap.

Caps and plugs for any deadend branches or house service connections should be made as watertight as any other joint and be anchored to hold against internal pressure or external force.

H. OPEN CHANNELS

Open channels for conveyance of storm drainage may consist of natural streams, improved natural drainage ways, or entirely new structures. Improvement to existing channels may require special measures for protection of the new work and adjacent property from high flows during the construction period. Channels may be lined or unlined, depending on available space, magnitude of flow, and materials encountered or readily available from borrow sources.

The construction of large channels may involve the use of highly specialized equipment trains designed to excavate, shape, and line the waterway in a series of continuous operations. More normal procedure, however, is for excavation to be accomplished by conventional earthwork machinery followed by placement of impervious or non-erodible protective elements when required.

Observation of excavation generally is confined to that necessary to insure that planned grades and cross sections are being obtained. If the channel is to be lined, the engineer also must rule on the adequacy of foundations and order additional excavation to remove unsuitable materials encountered.

Construction of embankments must be observed and tested for proper gradation of materials, adequacy of mixing where blending of two or more materials is required, maintenance of proper moisture content, and attainment of proper degree of compaction. Compaction and moisture content of silty and clayey cohesive soils should be controlled by field density tests in conjuction with Proctor density determinations (AASHO T99). Compaction control of sandy and gravelly cohesionless soils should be by field density tests in conjunction with relative density determinations (ASTM D 2049). Borderline soils should be controlled by 95 per cent of Proctor maximum density or 70-percent relative density, whichever yields the highest unit weight.

Channel linings may be utilized to reduce seepage and possible waterlogging of adjacent land, control erosion, and improve hydraulic characteristics. Seepage control may be accomplished with blankets of relatively impervious cohesive soils placed and compacted in thin layers over the shaped channel. Construction and control in this event will be similar to that of embankment placement. When velocities are too great for compacted earth or sodded waterways, gravel, cobbles, or riprap may be placed to control erosion. Control of placement requires testing of gradation and observing for proper depth and density of blanket. Frequently an approved riprap panel is marked at the commencement of the project to indicate a standard of gradation and workmanship for visual comparison of remaining work. Riprap may be dumped from trucks progressing up the side slope or may be placed by clamshell buckets. Some hand manipulation after spreading followed by overcasting with rock spalls will result in a firm, well-interlocked blanket of stone.

When resistance to erosion and improved hydraulic characteristics are required, asphaltic or Portland cement concrete linings probably will be necessary. Construction procedures and control methods for continuous lining of this nature can be found in highway and street paving manuals and texts.

Whenever continuous linings are designed for uplift protection, the construction sequence and procedures also must be planned accordingly. Underdrains or other dewatering devices may be required to accomplish this purpose.

I. CONSTRUCTION RECORDS

It is the responsibility of the engineer to record details of construction as actually accomplished. These data should be incorporated into a final revision of the contract drawings to represent the most reliable record for future use.

Records should be sufficient to allow future recovery of the sewer itself, underground structures, connections, and services.

As-built invert elevations should be recorded for each manhole, structure, connection, and house service. In some instances, it may be advisable to set concrete reference markers flush with finished grade to facilitate future recovery.

Acknowledgments

The illustrations in this chapter have been used through the courtesy of the Dickey Clay Pipe Co.; Eyerman Studio, Cumberland, Md.; City of Chicago; American Concrete Pipe Assn.; New York City; Johns-Manville Co.; Armco Drainage and Metal Products, Inc.; Construction Methods and Equipment; Martin Marietta Co.; Metropolitan Sanitary District of Greater Chicago; Los Angeles County Sanitation Districts; and Mystic Photo, Medford, Mass.

General References

1. "Concrete Pipe Handbook." Amer. Concrete Pipe Assn., Arlington, Va.
2. "Concrete Sewers." Portland Cement Assn., Chicago, Ill., (1958).
3. "Installation of Concrete Pipe." Manual M9, Amer. Water Works Assn., New York (1961).
4. "Clay Pipe Engineering Manual," National Clay Pipe Inst. Crystal Lake, Ill.
5. "Recommended Practice for Installing Vitrified Clay Sewer Pipe." Std. C-12, Amer. Soc. Test. Mat., Philadelphia, Pa.
6. "Installation of Cast Iron Water Mains." Std. C-600, Amer. Water Works Assn., New York.
7. "Installation of Asbestos Cement Water Pipe." Std. C-603, Amer. Water Works Assn., New York.
8. "Installation Manual for Corrugated Metal Structures." National Corrugated Metal Pipe Assn., Chicago, Ill.
9. "Code for the Manufacture, Transportation, Storage and Use of Explosives and Blasting Agents." Publ. No. 495, Natl. Fire Protection Assn.
10. "Blasters Handbook." E. I. du Pont de Nemours and Company, Inc., Wilmington, Del. (1958).
11. "Manual of Concrete Inspection." Amer. Concrete Inst., Detroit, Mich. (1961).
12. "Manual of Accident Prevention in Construction." Associated General Contractors, Washington, D.C.
13. "Some Essential Safety Factors in Tunneling." Bur. Mines Bull. 439, U.S. Govt. Printing Office, Washington, D.C.
14. "Mine Gases and Methods of Detecting Them." Bur. Mines, Miner's Circ. 33, U.S. Govt. Printing Office, Washington, D.C.
15. "Protection Against Mine Gases." Bur. Mines, Miner's Circ. 35, U.S. Govt. Printing Office, Washington, D.C.
16. "Engineering Factors in the Ventilation of Metal Mines," Bur. Mines Bull. 385, U.S. Govt. Printing Office, Washington, D.C.
17. "Safety with Mobile Diesel Powered Equipment Underground," Bur. Mines Publ. RI 5616, U.S. Govt. Printing Office, Washington, D.C.
18. Tschibotarioff, G. P., "Soil Mechanics, Foundations, and Earth Structures." Ch. 8, 10, and 16, McGraw-Hill Book Company, Inc., New York.
19. Terzaghi, K. and Peck, R. B., "Soil Mechanics in Engineering Practice." Chapter 8, John Wiley & Sons, Inc., New York.

CHAPTER 12. WASTEWATER AND STORMWATER PUMPING STATIONS

A. INTRODUCTION

Location, arrangement, type of equipment and structure, and external appearance all are basic considerations in the design of any pumping station.

Most important is the proper location of the station. This means that a comprehensive study must be made to assure that the entire area to be served can be drained adequately. A detailed discussion covering this and related points is presented in Chapter 6. Careful attention should be given to probable future growth because the location of the pumping station will in many cases determine the future overall development of the area. In selecting the site good aesthetic judgment must be exercised so that the station will not be detrimental to the neighborhood.

Studies of the topography downstream from a proposed station should be made to determine whether the station can be eliminated feasibly by tunneling through or by providing conduits around high ground to obtain gravity drainage. All pumping stations require operation and maintenance, the cost of which when capitalized usually will justify a considerable first-cost expenditure to build a gravity sewer.

Wherever possible, stormwater pumping stations should be located in areas where water may be impounded without creating an undue amount of flood damage if the inflow should exceed the station capacity. Also, the ability to impound a part of the stormwater inflow will reduce the effect of peak runoff to the station and consequently the overall pumping capacity required for any given storm.

The depth of incoming sewers or drainage channels, and sometimes foundation conditions, will determine the depth of the station below ground level. Surface conditions, neighboring buildings, land use, and relative elevations with reference to flooding will determine the level of the operating floor, the type of superstructure, and exterior finish and trim. All stations must be designed to withstand flotation.

Where stations are located in built-up areas, the superstructure should be similar to that of adjacent buildings or structures. Descriptions of such designs may be found in the literature (1). When the location is isolated the type of structure may be left to the owner and engineer but the probable nature of future development in the vicinity of the site should not be overlooked. Isolated stations should be protected against vandalism and preferably not have windows or as few as architectural considerations will allow.

Underground stations may be either factory assembled or built in place and should conform to good design principles as outlined in this chapter. Locations beneath city streets should be avoided.

Dependability of equipment and power is mandatory in most pumping stations because failure of either can cause considerable damage. The types of drives and controls for pumping equipment should, for this reason, be weighed carefully. Electric stations, to be reliable, should have at least two independent incoming power lines with automatic switching equipment to transfer the load from the preferred line. Where this reliability cannot be obtained, standby engine-driven generators, standby engine-driven pumping units, or all engine-driven pumping units may be required.

Wastewater pumping stations in many instances can be located so that the wastewater may be overflowed if the power fails or the pumping capacity is exceeded.

A discussion of suggested devices to accomplish overflow is given in Chapter 7. Such bypassing can be permitted only in locations where a temporary overflow of wastewater will not be hazardous to health or injurious to property. Alarms should be provided to signal when the station is not functioning properly.

Careful attention should be given to the control of odors in the design of sewage pumping stations, especially when stations are located within 1,000 ft (300 m) of human habitation. However, if stale sewage is being handled, it is sometimes impossible to prevent odors emanating from the station unless the ventilating air is specially treated. The matter of sulfide control in sewers is discussed in Chapter 6.

Two differentiating terms which apply to both sewage and stormwater pumping stations have, by common usage, come to have certain general meanings: (a) lift (relift) station, in which the pumped liquid is released to atmospheric pressure a relatively short distance from the facility into a gravity sewer, open channel, or receiving body; and (b) pumping station, in which the liquid is pumped from the facility some distance, occasionally miles, in a pressure pipe. The term lift station is applied only in the first case while pumping station is applicable to both unless the two are being compared.

B. STATION CAPACITY

Methods used to determine present and future wastewater or stormwater inflow to the station are covered in other chapters of this manual. If the area is not developed fully, the designer will be obliged to establish an initial station capacity which probably will meet the requirements for a reasonable time, customarily a period of not less than 10 yr. The initial flows under these conditions may not be as great as allowed in the design. The effects of the minimum flow conditions must be estimated to be sure that retention of the sewage in the wet well will not create a nuisance and that pumping equipment will not operate too infrequently. Allowances must be made for future requirements so that additional or larger pumps can be installed as required.

Obviously, station capacity must be adequate to meet the maximum rate of flow. It also is mandatory that this capacity must not exceed the capacity of receiving conduits or sewers. This sometimes is overlooked because it is assumed that the pumps will operate intermittently and that there will be no downstream difficulties.

Probable minimum inflows to the station also must be taken into account because such flows will affect the design of screen flow-through channels and the size of the wet well. Average flow conditions are of interest in that they indicate the conditions under which the station usually will operate. To obtain the least operating expense, pumping equipment should be selected to perform at maximum efficiencies under average conditions.

C. WET-WELL DESIGN

Storage capacity usually is required for all wastewater and stormwater pumping stations where automatic controls and variable speed drives are not furnished to match pumping rates exactly with inflow rates. The selection of proper storage capacity is critical because it affects the time which the liquid will be retained in the station and the frequency of operation of the pumping equipment. The storage effect of incoming sewers and channels and stormwater overflows in low-lying areas all may be considered as part of the total station storage capacity, but usually is considered only in that portion of the capacity study relating to maximum flow conditions.

From a mechanical standpoint it is desirable to operate a pump for long periods if not continuously, but such performance is not compatible with the maintenance of aerobic conditions in the sewage when it results in long retention periods.

The shape of the wet well and the detention provided for sewage pumping stations should be such that deposition of solids is minimized and the sewage does not become septic.

Most design policies base detention on the average design rate of flow, but the maximum and minimum rates are the determining factors in sizing the wet well. The desired results can be attained with a minimum of objections except for large-capacity stations if the size of the wet well is such that with any combination of inflow and pumping, the cycle of operation for each pump will be not less than 5 min and the maximum retention time in the wet well will not average more than 30 min. Where large pumping units are involved, they should be operated continuously insofar as practical. It can be seen readily that to meet these requirements the design of the wet well must be coordinated with the selection of both the individual pumping units and the liquid levels at which the pumps are started and stopped.

It should be kept in mind that the longer the detention period, the greater the chance for the generation of objectionable odors from septic sewage and the accumulation of sludge in the bottom of the wet well, which in turn may increase the nuisance from odors and the frequency of pump stoppages. Accordingly, detention periods should be kept to the minimum compatible with proper operation of the pumping equipment.

Most engineers will accept the principles that the proper pump suction conditions should be maintained and that the deposition of solids should be minimized in the wet well. There is, however, great difference of opinion and variation in design practice as to how these ends should be accomplished. The principle that the level of sewage in the wet well should be above the level of the pump casings at all times during the entire pump cycle to insure a continuous positive prime is accepted by many engineers. It must be recognized, however, that this requirement entails a pump setting somewhat lower than the wet well with correspondingly increased structural costs. Other engineers follow the practice that the level of the liquid should be above the level of the pump casing only at the cut-in level of a given pump. The pump, therefore, operates under suction lift conditions during drawdown to its cutoff level, a normal capability of the non-clog sewage pump of the radial-flow type. This practice requires that a means of automatically repriming each pump after each drawdown be provided, for example, by an open venting line from the pump discharge nozzle discharging to the wet well. Such venting lines should use a minimum of bends, pass through the wet-well wall at a high level, be constructed of corrosion resistant tubing, and be of appropriate size to reduce clogging, but not less than ¾ in. (2 cm) in diam. Proof of pump output may be interlocked electrically with motor controls by means of a time delayed pressure switch on the unit discharge, or by a limit switch on the exterior arm of the check valve. Similarly, an air-bound unit may be shut down by a temperature switch mounted on the pump case. It generally is accepted that the pump control levels in the wet well should be such that the incoming sewers to the station will not be surcharged and that velocities will be maintained which will prevent deposition of solids and, thus, the formation of sulfide-producing slimes.

Probably the most controversial point in the design of sewage wet wells is the bottom slope needed to minimize deposition of solids. A relatively large number of state regulatory agencies call for a minimum bottom slope of 1 to 1 to the pump inlet. Designs occasionally are encountered where the slope is less; and some are steeper. States requiring a minimum slope of 1 to 1 to the pump inlet also require a minimum slope of 1 horizontal to 1.4 vertical at the bottom of Imhoff tank flow-through compartments. Considerable deposition and accumulation of solids have been encountered in Imhoff tank flow-through chambers having slopes of 1 horizontal to 1.5 vertical or flatter. A minimum accumulation of solids occurs where the slope is 1 horizontal to 1.75 vertical or greater, indicating that hoppers in wet wells should be constructed with slopes not flatter than 1 horizontal to 1.75 vertical.

A number of common pump-suction pipe arrangements in sewage wet wells are shown in Figure 83. Bellmouth inlets are far superior to the

WASTEWATER AND STORMWATER PUMPING STATIONS 291

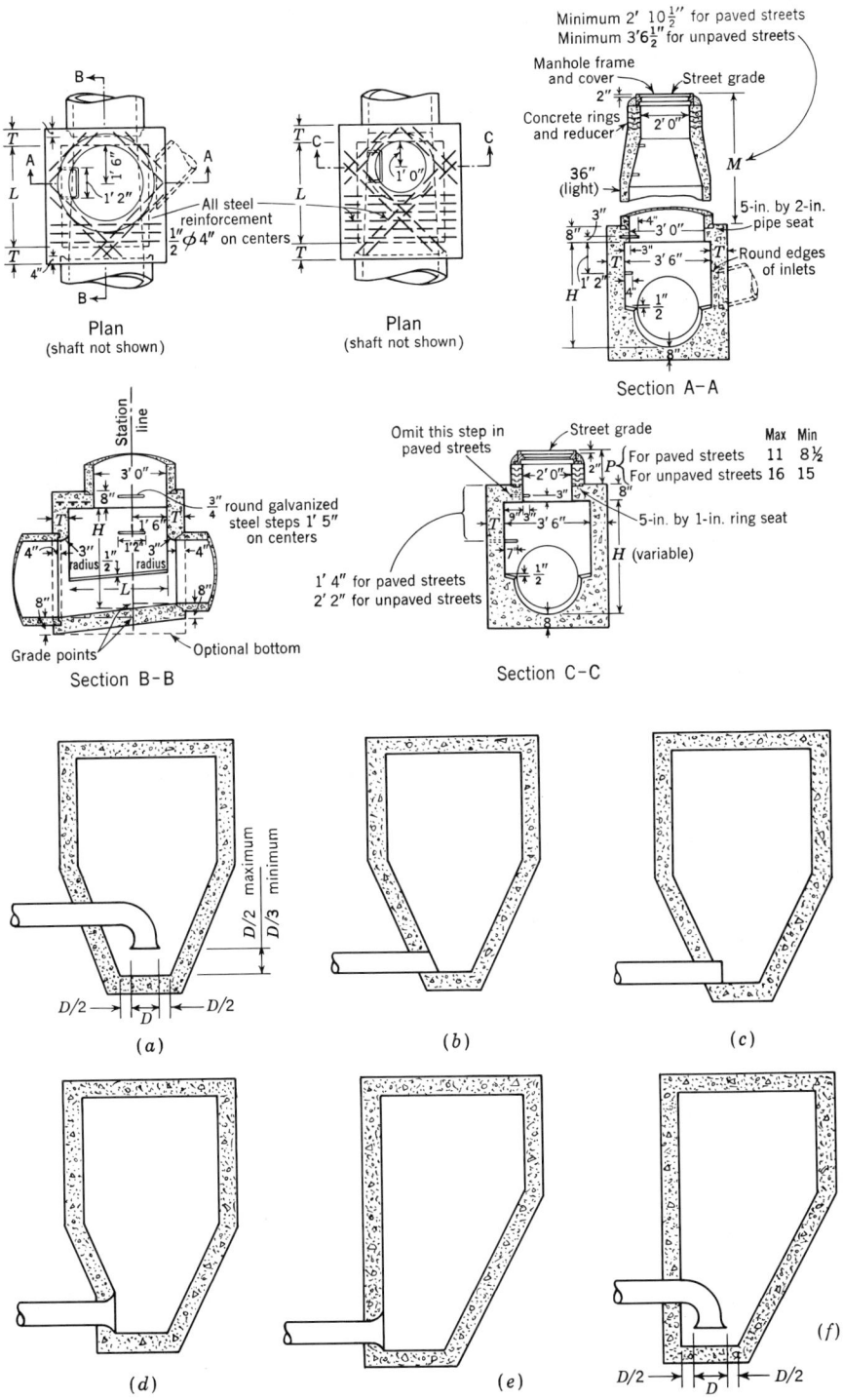

FIGURE 83.—Typical suction inlets for sewage pumps.

straight inlets shown in Figure 83 b and c. Flaring eliminates the sharp edges on which material may accumulate and also minimizes head loss. The inlets, Figure 83 a and f, are considered superior to any of the other arrangements illustrated since there is less possibility of vortices forming in the wet well.

Where turned-down bellmouth inlets are used, the bell should be not more than $D/2$ and not less than $D/3$ (in which D is the diameter of the suction pipe) above the floor of the wet well in order to obtain scouring inlet velocities and still obtain nearly optimum hydraulic entrance conditions.

The spacing between suction inlets in the wet well should be such that both hydraulic interference and deposition of solids between inlets are prevented insofar as possible. A spacing not more than two times the suction pipe diameter is required to prevent deposition of solids between suction inlets. However, this requirement cannot be satisfied, as a spacing of approximately four times the diameter of the suction inlets is required to prevent hydraulic interference.

The design of wet wells for stormwater pumping stations requires careful study where vertical axial-flow or diffusion-vane pumps are installed, taking suction directly from the wet well. These pumps are very sensitive to inflow arrangements, spacing between the units, and clearances from the bottom and side walls. The recommended clearances for installation are shown in Figures 84 and 85 (2), although variations sometimes are acceptable (3). The advice of pump manufacturers should be considered in the design and layout of wet wells. For large installations or unusual conditions, model tests may be advisable.

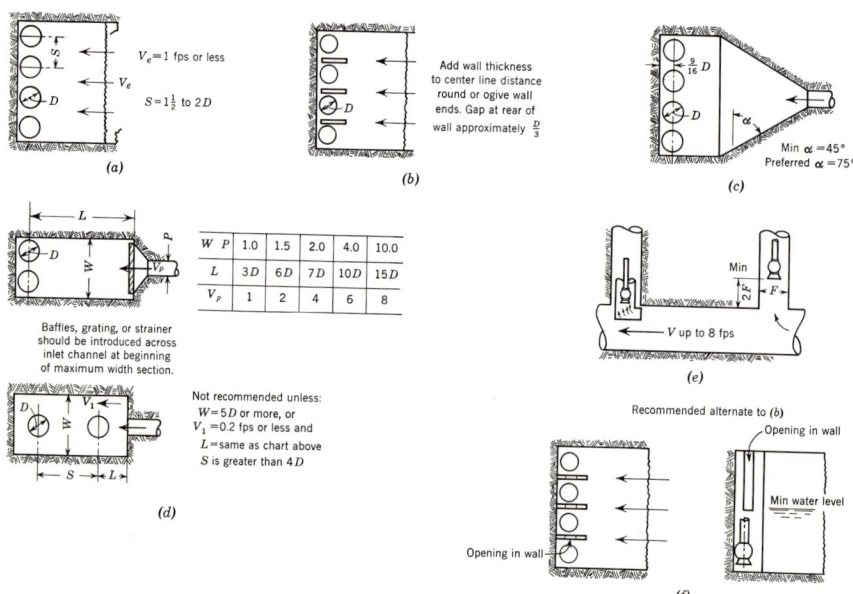

FIGURE 84.—Suction wells for diffuser-vane pumps.

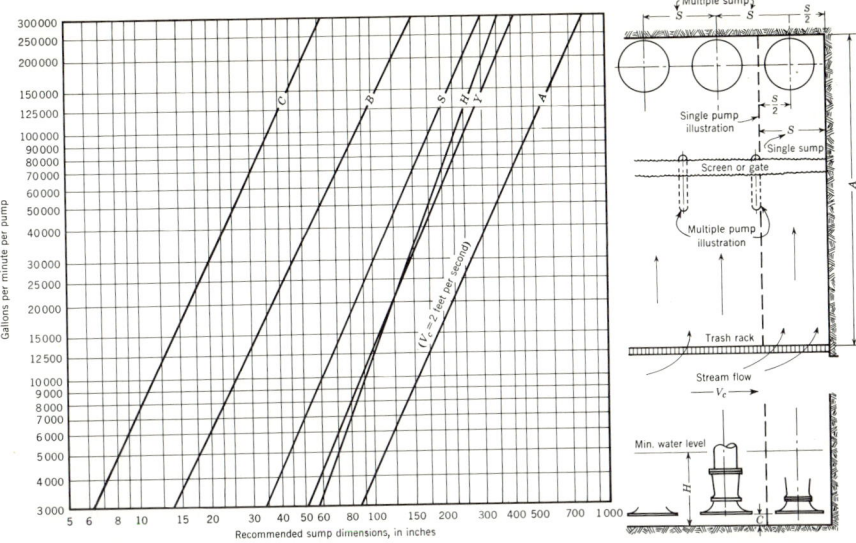

FIGURE 85.—Recommended sump dimensions for clear liquid applications.

Provision should be made for adequate access to wet wells for inspection and cleaning. Stairways and walkway platforms should be furnished in the larger stations. Proper lighting and ventilating facilities also should be available whenever the wet well is entered for any reason. Suitable access openings and facilities are needed for the removal of accumulated solid matter. This material may be in the form of sand, gravel, or other trash in pumping stations serving storm or combined sewers, and in many cases, sanitary sewers. In the case of sanitary sewage, there may be a collection of grease balls or scum on the water surface and grease accumulations on the wet-well walls. Thus, provisions generally are required for manual skimming.

Where wet wells are totally enclosed, adequate vents are needed to allow for entrance and exit of air as the liquid level in the sump rises and falls. Rooms located directly over wet wells, and having atmospheres common thereto, should be isolated from the rest of the station and be ventilated continuously by blowing air into them to provide a pressure slightly above atmospheric.

D. SCREENING DEVICES

1. Bar Screens

It is customary in major pumping stations to install bar screens to prevent large objects from entering the wet well which would clog or damage the pumps. Although numerous small and moderate-sized pumping stations are operating efficiently without screens, many engineers recommend that all sewage and stormwater pumps be protected by some method of screening.

Bar screens usually are made of iron or steel bars. The space between the individual bars varies: clear space of 1½ in. (3.8 cm) is customary for manually cleaned screens, and a 1-in. (2.5-cm) spacing for the mechanical type. Bar screens with relatively wide clearances are used for large pumping installations, depending on the ability of the pumps to pass solids.

At least 6 in. (15 cm) should be allowed in the hydraulics of a station for loss through the bar screen, and a greater loss should be allowed where manually cleaned screens are used, especially where they will receive relatively little attention. The floor of the screen channel should be placed at least 6 in. (15 cm) below the invert of the incoming sewers to allow for some accumulation of screenings without affecting the flow in the sewers.

Screenings should be disposed of by burying, burning, or shredding. In the last case, the screenings may be returned to the wet well. The handling and disposal of screenings are important design features. Moreover, inadequacies in this phase of pump station operation will create a nuisance.

A comprehensive presentation on screening theory and equipment is offered elsewhere (4).

(*a*) **Manually Cleaned Bar Screens.**—Manually cleaned screens are usually of the basket or bar-rack type. Basket screens are applicable only to smaller pumping stations and generally are in the form of a box with one open side facing the incoming sewer. Facilities are provided for hoisting the basket to the surface where the screenings may be removed.

Manually cleaned bar screens are placed in the entrance channel at angles varying between 30 and 45 deg from the horizontal. Because of the difficulty of cleaning by hand, manually cleaned bar screens are applicable only to channels of relatively shallow depths. Fabrication of the rack should be such that except for the toe all cross bracing is attached only to the rear edge of the bars. This provides an uninterrupted path for the rags, which tend to slide up the bars in the form of a wad when raked. Adequate head room is required to manipulate a hand rake of suitable length. A depressed pocket and drain should be provided in the operating floor adjacent to the top of the screen. Wet-well lighting should include at least one fixture located to illuminate directly the face of the screen.

Where appreciable quantities of screenings are to be removed, for example, immediately upstream of wastewater treatment processes, the bar screen should be sized so that the velocity through the clear openings will not exceed 2.5 fps (45 m/min) under all flow conditions. However, low velocities can have a detrimental effect in that rags tend to drift into the rack, rapidly blanketing the screen, and causing an undesirable surcharge in the influent sewer. Thus, where maximum screening efficiency is not the objective, higher velocities may be preferred in that they tend to reduce this blanketing effect by causing a single rag either to pass through or to stream out over a single bar. This method of operation provides the desired pump protection from oversized objects while at the same time maintains proper velocities in the sewer. The higher velocities are achieved by depressing the screen channel a minimum of 6 in. (15 cm) below the sewer invert, depressing the channel several inches again immediately downstream of the screen, and maintaining wet-well liquid levels below the toe of the screen.

(*b*) **Mechanically Cleaned Bar Screens.**—Mechanically cleaned bar screens are basically either front- or rear-cleaned, with the bars making an angle with the horizontal varying from 60 to 90 deg. Mechanical cleaning is accomplished by traversing the bar screen with a rake, although some manufacturers offer a rotary comb which not only traverses the bar screen but also shreds the screenings. The rake is moved either by means of endless chains or by cables which lower it to the bottom of the channel and then pull it vertically along the bars.

The front-cleaned type can be placed in a channel of almost unlimited depth. But since the cleaning mechanism is on the upstream side of the bars, objects may wedge between the cleaning mechanism and the bottom of the channel so that the machine is disabled. This disadvantage is overcome in the back-cleaned unit where all the cleaning mechanism is behind the bars. Cleaning is done by the teeth of the rake which project through the bar screen to the extent that they engage the screenings and elevate them to the surface. The back-cleaned bar screen does have one disadvantage: it can be installed only in channels with limited depths of flow because the individual bars in the screen must be supported from the bottom of the channel.

Close attention should be given to the resulting hydraulic effects of mechanically cleaned screens. Some equipment includes a gravel guard to protect the foot shaft assembly which causes a damming effect of up to 12 in. (30 cm) in depth. Similarly, some equipment requires a controlling weir downstream to insure a minimum water level for submergence of the water-lubricated foot shaft and sprocket bearings. These factors, together with head loss through the screen, must be considered in designing the channel to avoid undesirable surcharges upstream. Consideration of minimum velocities is important to prevent deposition of debris which would jam the mechanism, particularly during storm runoff when the screen is most needed.

In cold climates screens must be housed in heated spaces to keep solids from freezing to the rakes during off periods. In storm-sewer systems and large combined systems, it is customary to protect mechanical screens with coarse trash racks having openings of 3 to 6 in. (7.5 to 15 cm). Their purpose is to prevent logs, timbers, or other large objects from jamming and damaging the mechanism.

Where mechanically cleaned screens or comminutors are installed it is customary to provide at least two channels, both of which may be equipped with mechanical equipment. If only one channel is so equipped, a manually cleaned bar rack or basket screen usually is placed in the auxiliary channel. In any case, both should be provided with means for isolation, such as sluice gates, slide gates, stop plates, or wooden stop logs.

2. Comminutors

A comminutor is essentially a circular bar screen placed in the incoming channel which is traversed by a rotary rake that serves as a cutter to engage and shred the accumulated debris. Thus, screening and shredding are accomplished in one operation. Shredded solids are returned to the sewage flow, eliminating equipment and labor for separate handling, shredding, and disposal. Slot width of the screen usually varies from ¼ to ⅜ in. (0.6 to 1.0 cm). The life of the equipment can be extended if grit removal precedes it.

3. Water Supply

A water supply isolated by an air gap or other suitable means should be provided, if possible, at all screening installations in order to maintain cleanliness. Shredders for screenings require a water supply.

E. PUMPING EQUIPMENT

1. Types in General Use

Pumping equipment used in sewage and stormwater stations may be classified into two general types: centrifugal pumps and pneumatic ejectors. The latter are used only in the smaller installations where centrifugal pumps, if used, would be too large for the application.

2. Centrifugal Pumps

Centrifugal pumps fall into three general classifications:

(a) Axial-flow or propeller pumps;
(b) Mixed-flow or angle-flow pumps; and
(c) Radial-flow pumps (commonly referred to as centrifugal pumps).

The classification into which a pump falls usually can be determined by its specific speed (N_s) at the point of maximum efficiency. The specific speed of an impeller may be defined as the speed, in revolutions per minute (rpm), at which a geometrically similar impeller would run if it were of such size as to deliver 1 gpm (0.063 l/sec) against 1 ft (30.5 cm) of head. The formula for specific speed is as follows:

$$N_s = \frac{\text{rpm} \sqrt{\text{gpm}}}{H^{3/4}} \quad \dots \dots \dots \dots \dots \dots \dots \dots 1$$

in which H is in ft.

(a) *Axial-Flow Pumps.*—Axial-flow or propeller pumps develop most of their head by the propelling, or lifting, action of the impeller vanes on the liquid. They are characterized by a single inlet impeller with the flow entering axially and discharging nearly axially into a guide case and are customarily used for large, low-head installations in capacities greater than 10,000 gpm (630 l/sec) and for heads below 30 ft (9 m). This type of pump was developed primarily for flood control and irrigation work and is adapted particularly to stormwater pumping installations where the head-capacity relationship is within the range of this type of unit. The pumps are generally of the vertical wet-pit type, although they can be obtained for horizontal dry-pit installations. Axial-flow pumps have relatively high specific speeds, usually ranging between 8,000 and 16,000 rpm. It is desirable that vertical units have positive submergence of the impeller for proper operation. Units can be obtained that will operate with a suction lift.

(*b*) **Mixed-Flow Pumps.**—The head is developed by mixed-flow pumps partly by centrifugal force and partly by the lift of the impeller vanes on the liquid. This type of pump has a single inlet impeller with the flow entering axially and discharging in an axial and radial direction, usually into a volute-type casing. These units are applicable for medium-head applications of 25 to 50 ft (7.6 to 15 m) and for medium to large capacities. Mixed-flow pumps fall into a medium range of specific speeds, usually between 4,200 and 9,000 rpm. They generally require positive submergence, although, with proper selection of rotative speeds, they may be used for limited suction-lift applications.

(*c*) **Radial-Flow or Centrifugal Pumps.**—The head in radial-flow pumps is developed principally by the action of centrifugal force. Pumps of this type may be obtained with either single suction or double suction inlet impellers; the flow leaves the impeller radially and normal to the shaft axis. The majority of all pumps installed are of this type, and they may be obtained in almost any range of head and capacity. Centrifugal pumps are characterized by relatively low specific speeds, usually below 4,200 rpm for single inlet impellers and below 6,000 rpm for double suction units. Single suction pumps normally are used for sewage or stormwater pumping since they are less subject to clogging.

(*d*) **Relative Characteristics.**—A comparison of the characteristics of the axial-flow, mixed-flow, and centrifugal pumps is shown in Table XXXII.

(*e*) **Impeller Types.**—Impellers for centrifugal pumps may be classified into three general types:

1. Enclosed (front and back shroud);
2. Open (no shrouds); and
3. Semi-enclosed (back shroud).

Enclosed impellers with shrouds generally are specified for pumps which are to handle sanitary sewage. Not only are they less subject to clogging but they also pass stringy materials better than other types.

Open or semi-enclosed impellers sometimes are used in stormwater pumps, especially for handling large volumes and for intermittent service.

(*f*) **Casings.**—Two types of pump casings are common: the volute and the turbine or diffusion type. The primary difference is that with the volute case the velocity leaving the impeller is transformed to head by gradually increasing the area of the water passageway in a spiral-shaped scroll; whereas in the diffusion-vane case the velocity leaving the impeller is transformed to head by means of curved vanes. Dry-pit pumps generally are furnished with volute cases. Vertical wet-pit pumps used for stormwater pumping are usually of the turbine or diffusion-vane type.

TABLE XXXII.—Relative Characteristics of Centrifugal Pumps

Description	Axial Flow	Mixed Flow	Radial Flow
Usual capacity range	10,000 gpm	5,000 gpm	Any *
Head range	0 to 30 ft	25 to 50 ft	Any *
Shutoff head above rated head †	About 200 percent	165 percent	120 to 140 percent
Horsepower characteristic	Decreases with capacity	Flat	Increases with capacity
Suction lift	Usually requires submergence ‡	Usually requires submergence	Usually not over 15 ft
Specific speed	8,000 to 16,000	4,200 to 9,000	Below 4,200—single suction; below 6,000—double suction
Service	Used where space and cost are considerations and load factor is low	Used where load factor is high and where trash and other solid material are encountered	Used where load factor is high and high efficiency and ease of maintenance are desired.

* Heads of radial-flow sewage pumps may be limited.
† Maximum efficiency point.
‡ Lift limited.
Note: gpm × 0.0631 = l/sec; ft × 0.3048 = m.

Either type of casing may be split axially or radially to obtain access to the impeller for cleaning.

(g) Pump Construction.—Most pump casings are made of cast iron. Although for special applications where gritty or corrosive liquids are handled, other materials sometimes are specified. Pumps used for handling sanitary sewage usually are specified with cast iron impellers; trash pumps are furnished with this construction as standard by most manufacturers. Sleeves should be provided on shafting through stuffing boxes and preferably should be of chromium steel or other abrasion-resistant material. Where possible, clear water flushing should be provided for pump stuffing boxes. If this is not possible, provision should be made to apply grease under pressure to the stuffing box at all times when the pump is in operation. Mechanical seals have become quite popular with some engineers in recent years. Where they are used, a cooling and lubricating liquid must be applied to the seal. Brief operation without the lubricant can cause immediate damage. Where mechanical seals are used, spare seals should be kept available, since they cannot be repaired after they fail. Pump bearings may be either the anti-friction or sleeve type. Although advocates of each point out their relative merits, proper application of either type should result in successful operation.

(h) Non-Clog Pumps.—Actually, no pump has been developed that cannot clog. Experience shows that rope, long stringy rags, sticks, cans, rubber goods, and grease are objects most conducive to clogging. For pumps smaller than 10 in. (25 cm) in size which handle sanitary sewage, non-clog units are used almost exclusively. Pumps smaller than 4 in. (10 cm) in size are not recommended for this kind of service. Non-clog pumps differ from conventional units in arrangement, size, smoothness, and contour of channels and impellers to permit passage of clogging material. If efficiency is not the prime concern, passages are sometimes only one size smaller than the discharge pipe. Where coarse screening or comminution is done, the passages may be limited to approximately 1 in. (2.54 cm) greater than the clear openings between bars in the screen but not less than $2\frac{1}{2}$ in. (6.3 cm). The pump casing is a simple volute with a so-called end suction inlet to the impeller eye. The conventional non-clog impeller contains two blades and is fastened securely with a minimum overhang on a heavy shaft with at least one radial and one radial-thrust bearing. The radial bearing is placed as close to the stuffing box as is feasible, within a heavy supporting frame. Impeller blades should have smooth easy curves designed to prevent solids from agglomerating around the shaft between the impeller and back head or casing.

The leading edges of the impeller blades are rounded generously so that rags tend to slide off. Impeller wearing rings seldom are specified for pumps smaller than 10 in. (25 cm). Both case and impeller rings can be obtained for larger pumps where it is desired to maintain both axial and radial clearances. Most sewage pumps include a feature to allow field adjustment of the pump shaft to restore impeller end clearance without disassembly of the pump.

Some manufacturers offer a single blade (so-called bladeless) impeller, with a generally rounded passageway having good non-clog characteristics for lower capacity ratings. The balancing of this impeller is somewhat critical and is factory set at the selected design point. For this reason care should be exercised in its choice for systems where discharge head conditions may fluctuate in order to avoid vibration. It should be operated under submerged conditions as dynamic imbalance may be severe if it is operated dry. Another unit offered is the vortex (so-called torque-flow) type utilizing a semi-enclosed, single shroud impeller, in which the blades are recessed entirely into the curved shroud. The flow does not pass through the impeller, making the unit uniquely non-clog. Both of the special designs cited above normally achieve a lower efficiency than the standard non-clog impeller. But this disadvantage may be more than offset in smaller installations by the benefits of the larger passageways.

3. Pneumatic Ejectors

Pneumatic ejectors are used largely for lifting sewage from the basements of buildings and in small lift stations where their advantages outweigh their low efficiency, which is limited to about 15 percent. Their advantages are: (a) sewage is completely enclosed and consequently no sewer gases can escape except through the vent; (b) operation is fully automatic and the ejector goes into service only when needed; (c) the relatively few moving parts in contact with sewage require little attention or lubrication; and (d) ejectors are not clogged easily.

A pneumatic ejector consists essentially of a closed tank into which sewage flows by gravity until it reaches a certain depth. Then enough air under pressure is admitted into the tank to discharge the liquid. The inlet pipe check valve prevents sewage from leaving the tank except through the outlet pipe check valve, which also prevents backflow into the tank.

Ejectors and compressors should be installed in duplicate to assure that service will not be interrupted either for reason of mechanical failure or lack of regular maintenance.

The following is an empirical formula for the approximate capacity of air required to operate an ejector:

$$V = \frac{Q(H+34)}{250} \quad \dots \dots \dots \dots \dots \dots \dots \dots \dots .2$$

in which V is the volume of free air required in cfm, H is the total head in ft, and Q is the rate of sewage discharge in gpm.

To allow for expansion of air in the storage tank as the sewage is displaced, the volume of the air storage tank and the characteristics of the air compressors selected should be adequate to provide the proper volume of air at a pressure at least 40 percent higher than that required to raise all sewage the maximum computed lift.

4. Pump Selection

The number of pumps to be installed in the station will depend largely on the station capacity and range of flow. It is customary to provide a total pumping capacity equal to the maximum expected inflow with at least one of the largest units out of service. Two units sometimes are considered out of service in large stations in determining firm capacity. At least two pumps should be installed in any station.

In small stations two pumps are installed, normally with each unit having sufficient capacity to meet the maximum inflow rate. The size and number of units for larger stations should be selected so that the range of inflow can be met without starting and stopping pumps too frequently and without requiring excessive wet-well storage capacity. Variable-speed drives are provided in many cases to match the pumping rate with inflow rate. Pumping stations in close proximity to treatment works should be designed with consideration for the adverse effect of sudden or wide-ranging variations in flow.

Occasionally the capacity and depth of the wet well can be coordinated with the pumping units so that the rise and fall of the water level in the wet well will result in a variable pumping capacity which may approximate the inflow rates. This is possible since all centrifugal pumps have the inherent characteristic that, as the head increases, the capacity decreases. In stations where the pumping head is low, the normal variation of levels in the wet well will vary the pumping head so that a wide range of capacities may be obtained with each individual unit. This is most noticeably true for pumps having relatively flat characteristic curves.

Before describing the procedures for selecting pumping equipment it is appropriate to define various standard terms used to describe pump characteristics. The definitions which follow are given in the Standards of Hydraulic Institute (2):

(*a*) **Datum.**—All readings for suction lift, suction head, total discharge head, and net positive suction head are taken with reference to the datum which, in the case of horizontal shaft pumps, is the elevation of the pump center line and in the case of vertical shaft pumps is the elevation of the entrance eye of the suction impeller.

(*b*) **Suction Lift (h_s).**—Suction lift exists where the total suction head is below atmospheric pressure. Total suction lift, as determined on test, is the reading of a liquid manometer at the suction nozzle of the pump converted to feet of liquid and referred to datum, minus the velocity head at the point of gage attachment.

(*c*) **Suction Head (h_s).**—Suction head exists when the total suction head is above atmospheric pressure. As determined on test, it is the reading of the gage at the suction of the pump converted to feet of liquid and referred to datum, plus the velocity head at the point of gage attachment.

(*d*) **Total Discharge Head (h_d).**—Total discharge head is the reading of a pressure gage at the discharge of the pump, converted to feet of liquid and referred to datum, plus the velocity head at the point of gage attachment.

(*e*) **Total Head (H). Sometimes Referred to as Total Dynamic Head or TDH.**—Total head is the measure of the energy increase per pound of the liquid imparted to it by the pump and is therefore the algebraic difference between the total discharge head and the total suction head. Total head, as determined on test where suction lift exists, is the sum of the total discharge head and total suction lift, and, where positive suction head exists, the total head is the total discharge head minus the total suction head.

(f) Net Positive Suction Head (*NPSH*).—The net positive suction head is the total suction head, in feet of liquid absolute, determined at the suction nozzle and referred to datum, less the vapor pressure of the liquid, in feet absolute.

Graphic representations of pump-head relationships (5) are shown in Figures 86 and 87.

Pumps should be selected having head-capacity characteristics which correspond as nearly as possible to the overall station requirements. This can best be accomplished by the preparation of a system curve showing all conditions of head and capacity under which the pumps will be required to operate. The head-capacity curve is developed using standard hydraulic methods for determining friction losses. Several available methods for calculating pipe friction losses are described in most standard hydraulic textbooks. Friction losses in straight piping usually are calculated using the Hazen-Williams formula.

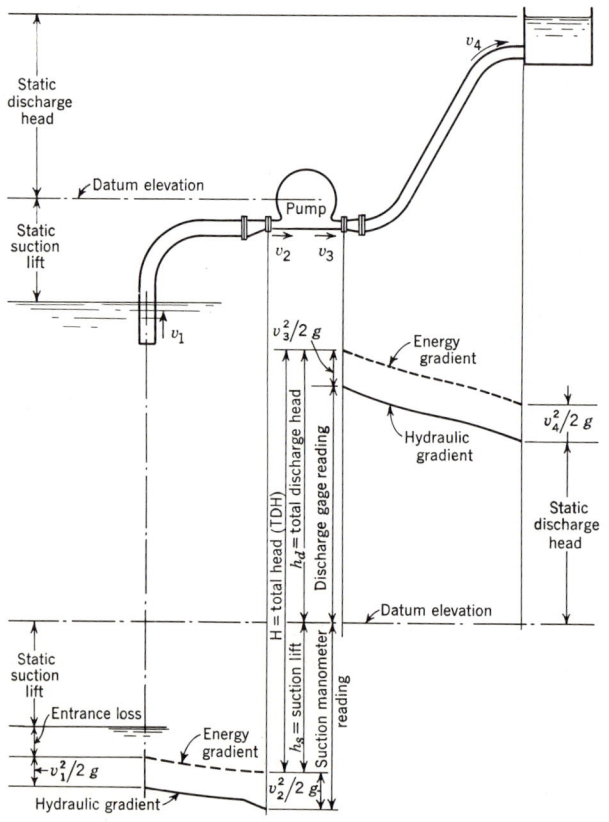

FIGURE 86.—Pump-head relationships with suction lift.

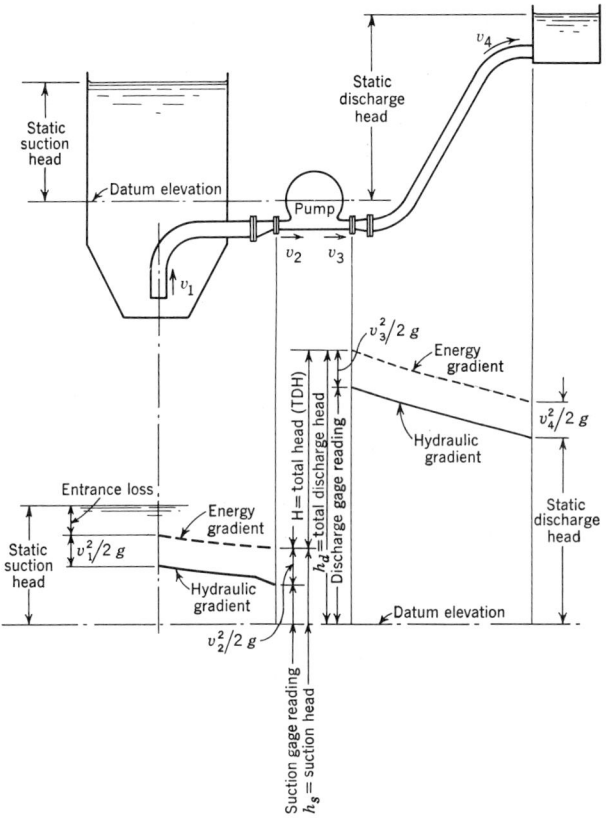

FIGURE 87.—Pump-head relationships with suction head.

They also may be estimated by extracting loss of head values directly from tables (6).

Losses of head in valves and fittings (excepting increasers) may be approximated using the following formula:

$$H_l = K \frac{V^2}{2g} \quad \ldots\ldots\ldots\ldots\ldots\ldots\ldots\ldots\ldots\ldots 3$$

in which H_l is the head loss in ft; V is the velocity in the fitting in fps; g is acceleration due to gravity; and K is the resistance coefficient from Table XXXIII, or from appropriate manufacturers' tables.

Loss of head due to increasers may be computed using the following relationship:

$$H_l = K \frac{V_1^2 - V_2^2}{2g} \quad \ldots\ldots\ldots\ldots\ldots\ldots\ldots 4$$

in which H_l is the head loss in ft; and K is the resistance coefficient from Figure 88.

These coefficients represent average losses in fittings and valves 4 in. in diam (10 cm) and larger. Where head loss in fittings represents a major part of the total system loss, especially in rating pumps, an analysis

TABLE XXXIII.—Approximate Values of Resistance Coefficient, K, for Valves and Fittings

Type of Fitting	Coefficient, K
Standard elbow	0.30
Long radius elbow	0.20
45-deg elbow	0.20
90-deg bends in flumes	1.50
Tees:	
Straight through	0.20
Branch to line	0.50
Line to branch	0.80
Entrance:	
Inward projecting	0.78
Sharp cornered	0.50
Bellmouth	0.04
Gate valves	0.20
Check valves:	
Swing	2.50
Tilting disk	0.34
Butterfly valve	0.30

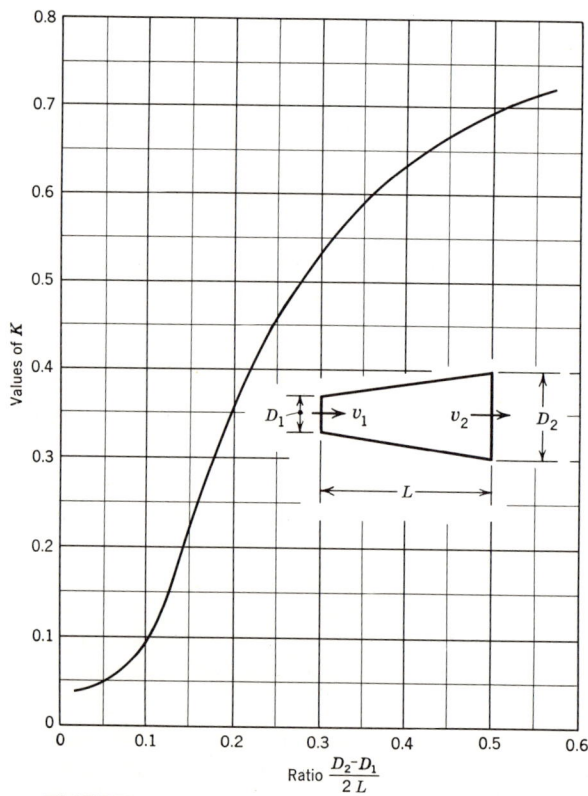

FIGURE 88.—Values of K for use with Equation 4.

should be made to determine the minimum and maximum possible using more accurate methods. In such cases a hydraulics handbook (7) and manufacturers' data on losses through valves should be consulted.

The procedure outlined above for estimating loss of head in valves and fittings may be described as the "resistance coefficient" method. Another procedure known as the "equivalent length" method may be sufficient where loss through fittings is a relatively small component of total dynamic head. There are available in hydraulics texts and manufacturers' literature for the various types and sizes of valves and fittings, an equivalent length in ft (or in pipe diameters) of pipe of the type normally used with the particular valve or fitting. The equivalent lengths so determined are added to the actual pipeline length and losses through them determined concurrently as a part of the overall system loss.

In the design of long force mains, it is difficult to predict accurately the changes in total friction loss that may occur over an extended period of time. It has been reported from some localities that friction losses in sanitary sewage force mains do not increase substantially with time, regardless of the type of pipe material used. Where this is observed as a normal local condition, the designer frequently uses a Hazen-Williams C of 130 or 140. In areas where the friction loss is known to increase with time, it is customary to use the more conservative Hazen-Williams C of 100. These considerations will affect materially the designer's selection of pumps and pipe sizes, as well as the capacity of the pumping units and their successful operation. System curves, for this reason, should be developed to show the possible maximum and minimum friction losses to be expected in the pipe line during the lifetime of the pumping units. Chapter 5 presents a detailed discussion of pipeline hydraulics.

When a system head study is being considered in which two or more pumps will discharge into a common discharge header and pressure main, it is normal practice to construct a system head curve on which may be superimposed the head-characteristic curves of some specific pumps being considered. The system head curve is constructed, for the full range of desired station capacity, by plotting against pumping rate the sum of the static head and the hydraulic losses in the system for the given pumping rate. Since the wet-well levels fluctuate considerably, it is usual to plot two system head curves, one for minimum level in sump (maximum static head) and one for the maximum level (minimum static head). Further, additional system head curves may be plotted using different friction factors. Only the losses in that part of the system common to all pumps normally are included in the system head curve computation. This provides a modified system curve reflecting performance at the station header (including static head) but temporarily excluding the variable of losses in individual pump suction and discharge piping, which losses depend solely on the flow rate through the individual pump.

The individual pump curves next are superimposed on the system curve. However, the losses in individual pump piping previously deleted from the system curve now are considered. They are computed for each pump from the suction inlet to the station header, for varying pumping rates, and subtracted from the pump characteristic curve. The resulting data are plotted as a modified pump curve. Combined modified pump curves then are plotted by adding the capacities (from the individual modified pump curves) for points of equal head to determine the output capacity of multiple pump operation in parallel.

Figure 89 shows a typical set of system curves together with representative individual pump characteristic curves, modified pump curves, and combined modified pump curves showing multiple pump operation in parallel (5). The intersection of the modified pump curves and the combined modified pump curves with the system head capacity curves shows the station pumping capacity for the several conditions of operation. Figure 89 shows four system curves, two for a Hazen-Williams C of 100 and two for C of 140 for maximum and minimum water levels in the sump. These coefficients usually can be considered as the maximum and minimum which will obtain in sewage force mains.

It is good practice to select pumps which will deliver the station capacity at the maximum head. However, this capacity and head will not necessarily be the point of maximum pump efficiency. Pumps should be selected with maximum efficiency at average operating conditions. In the case of Figure 89, assuming that the total station capacity is to be obtained by operating Pumps 1, 2, and 3 in parallel, the total head required at the station discharge header would be approximately 51 ft (15.5 m), with maximum sump level and $C=100$ in the discharge line. If this point

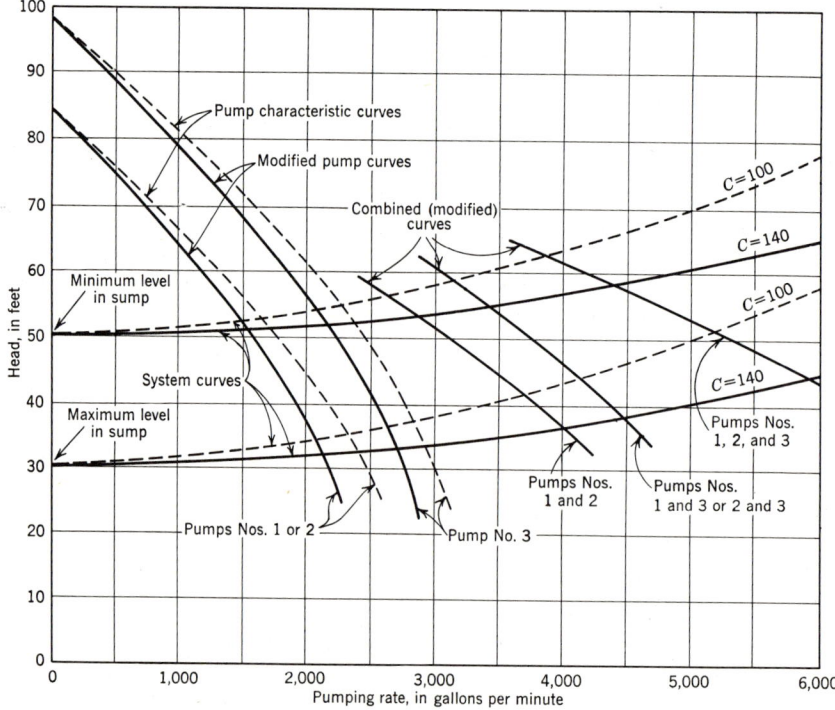

FIGURE 89.—Head capacity curves for parallel operation.

is projected horizontally to the individual modified pump curves and thence vertically to the pump characteristic curves, the required head for Pumps 1 and 2 would be 54 ft (16.5 m), and for Pump 3, approximately 57 ft (17.4 m). The difference between the head required at the station header and the head required for each pump is the head loss in the suction and discharge piping to each individual pump and is portrayed graphically as the vertical distance between the pump characteristic curve and the modified pump curve. The minimum head at which each individual pump will operate also is indicated in Figure 89.

From the intersection of the modified pump curve and the system head curve for maximum sump level and $C=140$, a vertical projection to the pump characteristic curve shows the minimum head in the case of Pumps 1 and 2 is approximately 39 ft (11.9 m). For Pump 3 it is approximately 42 ft (12.8 m). These minimum heads are important and should be furnished to the pump manufacturer since they usually will determine the maximum brake horsepower required to drive the pump and the maximum speed at which the pump may operate without cavitation.

The relatively large passageways of the non-clog impellers somewhat limit the practical heads against which they can operate. For this reason, the series operation of two standard sewage pumps occasionally has been employed to meet the requirements of some installations having higher than normal heads. The designer should examine closely, with the manufacturer, all factors relating to such installations, including the ability of the pump cases to handle the interior pressures. The performance of the series operation may be approximated by plotting the sum of the heads of the individual pumps for points of equal capacity.

It must be remembered that the capacity of a centrifugal pump is a variable which depends on the total head at which the unit operates. When a pump is referred to as having a certain capacity, this capacity applies only to one point on the characteristic curve and will vary depending on the actual head conditions.

Pump sizes usually are designated by the size of the pump discharge nozzle. The size of a pump for a given set of performance characteristics will vary with individual manufacturers. There has been a tendency in some cases to decrease the size of the suction and discharge nozzles to the extent that excessive nozzle velocities are obtained. It is, therefore, reasonable in purchasing pumps to specify the minimum size of suction and discharge nozzle which will be considered. For smaller installations it is recommended that no sewage pump be installed less than 4 in. (10 cm) in size or unable to pass a 2½-in. (6.3-cm) sphere. It has been found that excessive clogging occurs in pumps smaller than 4 in. (10 cm) when handling sewage. This means that a pump having a greater capacity than actually is required will sometimes be installed; satisfactory operation is more important than matching actual station requirements. Pneumatic ejectors are probably the best solution for installations where flows are such that a 4-in. (10 cm) pump would be too large.

The maximum speed at which a pump should operate is determined by the net positive suction head available at the pump, the quantity of liquid being pumped, and the total head. The limiting suction requirements for pumping units (2) are shown in Figure 90. When specifying pumps, especially those which are to operate with a suction lift, the speed at which the pumps will operate should be checked against these charts, which are based on sea level conditions and a temperature of the liquid of 85° F (29.4° C). Conditions different from these must be taken into account when selecting the proper speed. For a given pump, damaging cavitation will occur if it is installed where the permissible maximum suction lift (or minimum suction head) determined from Figure 90 is exceeded. For a given head and capacity, a pump of low specific speed will operate safely with a greater suction lift than one of higher specific speed (2).

It generally is not good practice to operate sewage or stormwater pumping units at speeds over 1,800 rpm. Although sewage pumps are operated between the general limits of 200 rpm, for large capacity low-head units, and 1,800 rpm, for small capacity high-head units, there is a tendency among some engineers and owners to keep operating speeds as low as practical. This tends to increase considerably the life span of the unit.

Sewage pumps should be of the dry pit type. Wet pit pumps are difficult to maintain because they are not accessible to the operator except by removing the pump which at most station locations creates both a nuisance and a health hazard. The general trend for stormwater pumping units, especially in the larger sizes, has been toward the vertical diffusion-vane type since the pumps require a minimum of space and in the larger sizes are considerably more economical. Close coupled, submerged pumping units are used in many cases for pumping stormwater from small drainage areas, such as highways or railroad underpasses and where it is desirable to keep pumping stations below ground level.

Sewage or volute-type pumps for stormwater pumping are available in both horizontal and vertical units. Most installations now being built use vertical pumps since they require less space. And, where the pump pit may be flooded, the pump drives can be mounted high enough to keep them out of the water. The determination of whether or not to use vertical or horizontal units must be governed by the individual conditions as no general rule can be established. Horizontal units are basically more stable than close-coupled vertical units, since the motors for the former are mounted separately from the pumps. Extended shafting for vertical units presents problems of alignment and maintenance, especially in large installations, and must be weighed against the disadvantage of horizontal units with the drives usually mounted in the pump pit and consequently subject to flooding. For small and medium-sized vertical units with motors mounted on the upper floor, the most popular shafting is the hollow type, having universal joints at the pump, motor couplings, and the intermediate steady bearings, and including a slip spline at the pump end. This combination allows the shaft to be disconnected quickly and swung aside for pump servicing. Minor shaft misalignment is compensated for automatically when proper installation is used. It is good practice to mount the vertical motor on a high ring base, rather than on the flat motor

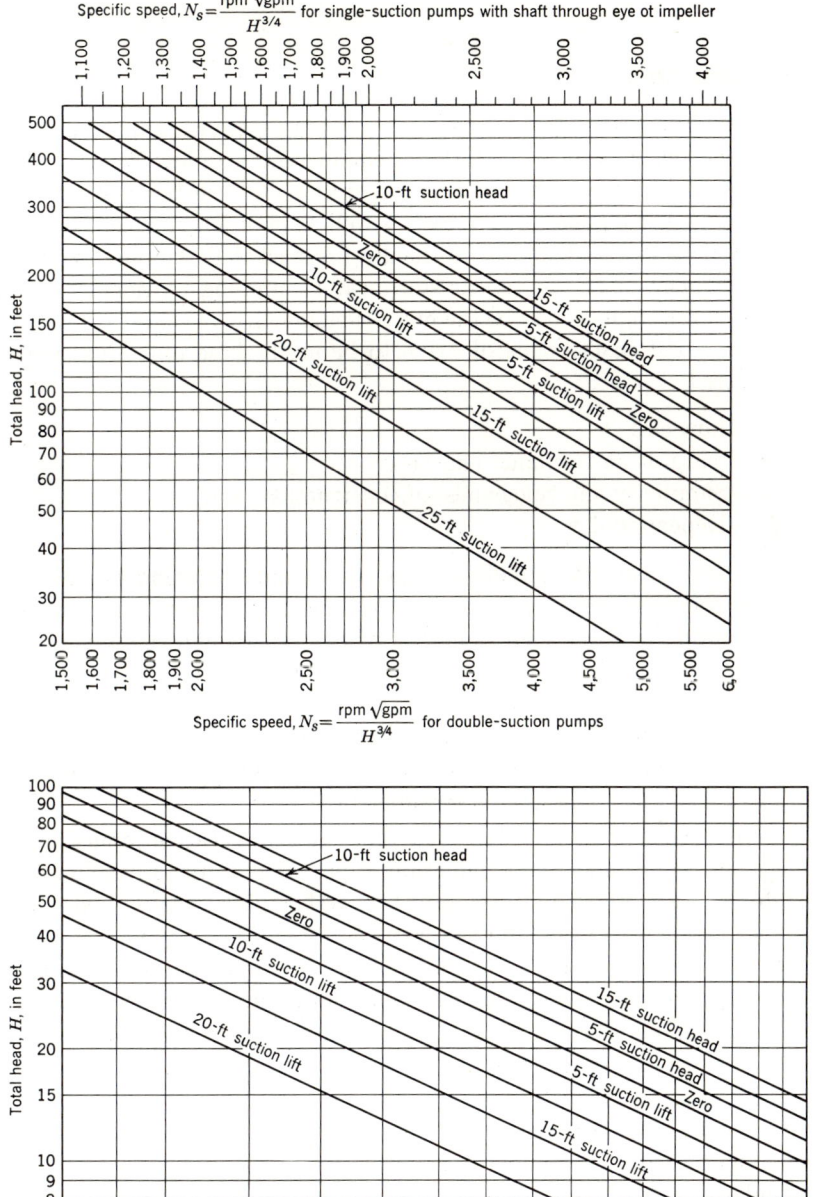

FIGURE 90.—Upper limits of specific speeds for clear water pumps at sea level and 85°F (29.4°C).

room floor, and to provide 1 to 3 in. (2.54 cm to 7.6 cm) of grout under the base.

5. Pump Drives

In modern sewage and stormwater pumping stations, pumps are driven by either electric motors or internal combustion engines. Where uninterrupted power and continuous duty are anticipated, electric motors usually prove more economical in first cost as well as in maintenance. Where firm power is not available or where the duty is such that the pumping will be at infrequent intervals (as in stormwater pumping stations), gasoline, diesel, or gas engines may be the most economical.

In deciding on the type of drive to be used, these objectives must be considered:

(a) Low cost;
(b) Suitable performance characteristics;
(c) Simplicity and ruggedness of construction; and
(d) Dependability.

The cost of a drive involves both initial and operating expense. The former comprises the purchase price and the cost of installation, including all mounting features such as floor supports, wiring and conduit, and special rigging. Operating costs, on the other hand, include the energy costs, demand charges and power-factor billing penalties, repair and maintenance costs, and extra operational labor expense, if any (8).

Electric motor drives may be classified into three general groups: constant-speed, multi-speed, and variable-speed.

Constant-speed motors may be either the squirrel-cage induction, synchronous, or wound-rotor induction type. The squirrel-cage induction motor is the most widely used and for most applications will cost less to install. It also has the simplest controls of the three types listed and is the simplest to maintain. For those drives where the motor rating exceeds more than one hp/rpm, the synchronous motor may have lower initial cost.

A comparison of constant-speed drive characteristics for electric motors (9) is shown in Table XXXIV.

The initial costs of the three general types of constant-speed motors and their controls are compared in Figure 91. Two sets of costs are shown: one, the cost of the bare motor only; second, the cost of these motors with standard controls. All are based on 440-v rating. Included in the control cost for induction motors is a standard primary starter both across the line and reduced voltage. For the wound-rotor and synchronous motors the cost includes both primary and secondary controls including the exciter for the synchronous motors. Costs shown are based on a 1965 level and should be used for a comparative study only. Quotations from manufacturers should be secured for final analysis. Since unusual circumstances, other than cost, would be needed to justify the use of synchronous or wound-rotor motors instead of the induction type for drives below 100 hp, cost comparisons below 100 hp have not been shown in Figure 91.

TABLE XXXIV.—Comparison of Constant-Speed Motor Characteristics

Type of Motor	Full-Load Eff.	Full-Load Power Factor	Simplicity	Initial Cost
Squirrel-cage induction	1–3% less than synchronous	75–90% lagging	Simplest	Lowest
Synchronous	Highest	Unity or 80% leading	Least simple	Second
Wound-rotor induction	1–3% less than synchronous	75–90% lagging	Less simple	Highest

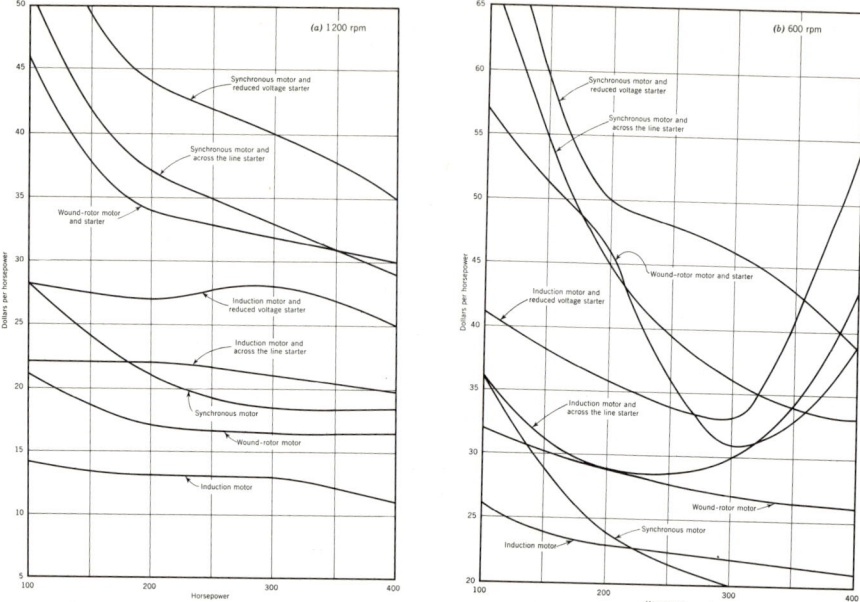

FIGURE 91.—Comparison of initial cost for constant-speed motors (1965 price level).

Multi-speed motors are a compromise between constant-speed and variable-speed drives. Because of pump characteristics, their application usually is limited to only two speeds. Both speeds, of course, are restricted to the available synchronous speeds, which may limit hydraulic capability. A two-speed drive costs approximately twice as much as a single-speed motor of the same type and maximum ratings (8). But multi-speed motors have an advantage over variable-speed drives in that they will operate at a higher efficiency at lower speed than any form of variable-speed drive.

Variable-speed electric drives also may be classified into three general groups: wound-rotor induction, brush-shift, and slip-coupling combinations.

The wound-rotor induction motor is the most popular variable-speed drive used for pumping equipment. Brush-shift motors are limited at present to horsepower ratings of 75 and less. Slip couplings used on pumping equipment are of either the hydraulic or magnetic type. Speed adjustment is obtained with the hydraulic coupling by changing the clearance between the vanes in the coupling or by changing the vane angles. Speed adjustment is obtained in the magnetic coupling by adjusting the excitation of the coupling. For those stations employing variable-speed pumps and in addition requiring standby generation, consideration should be given to the use of wound-rotor motors to limit the size of the generating set required. Standard wound-rotor controls can be arranged to limit the starting or inrush current to a value of 100 percent or less of normal full-load running current, thereby allowing in most cases a generator set of much smaller capacity than that required for other types of motors.

Many new type speed controllers are available or under development. A number of them may find practical application following adequate testing and operating experience (9).

A comparison of the characteristics of adjustable-speed drives (9) is shown in Table XXXV.

In general, it may be said that for each one percent of speed change, one percent of drive efficiency (based on the brake horsepower at the reduced speed) is lost no matter what type of variable-speed drive is selected. However, the variation in pump efficiency may tend to compensate for the drop in drive efficiency at reduced speeds and consequently, one may cancel the other. For this reason, the overall pump and drive efficiencies at the various operating speeds should be studied together with other methods of matching inflow requirements before deciding whether variable-speed or constant-speed drives are the better selection from an economic standpoint.

Enclosures for motor drives on sewage or stormwater pumps are usually of the open drip-proof type, although in some installations protected or enclosed motors may be required. Motors and starting equipment for pumps which operate infrequently well may be equipped with heaters to prevent condensation during periods of idleness. For small- and medium-sized installations it is good practice to select motors for the maximum load anticipated under adverse conditions.

Two general factors usually will determine the type of engine best suited to drive sewage or stormwater pumps; first, the type of fuel available at the most economical price, and second, service to which the engine will be subjected—that is, continuous duty or infrequent service.

High-speed gasoline engine drives are probably the most economical from a first-cost standpoint for horsepower ratings of 500 or less. However, they are not used ordinarily for continuous duty because of high maintenance expense. Low-speed diesel or gas engines are best for continuous duty, considering all factors, including first cost and operating and maintenance expense. One advantage of engine drives is that variable speed may be obtained by controlling engine speed by a direct mechanical linkage from float controls. By this method, pumping rates can be varied over rather wide limits to match the sewage flow rate. Remote control of engines is more complex than remote control of motors.

The proper selection of a drive for pumping equipment requires a careful study of all factors concerned, including but not necessarily limited to:

(a) First cost;
(b) Availability and dependability of electric power;
(c) Pump characteristics as to horsepower, torque, and speed under all operating conditions;
(d) Total annual power cost considering not only the energy rate but also demand charges;
(e) Maintenance costs; and
(f) Dependability of equipment.

TABLE XXXV.—Comparison of Characteristics of Adjustable-Speed Drives

Item	Type of Drive	Efficiency	Power Factor (at full load)	Simplicity	Initial Cost
1	Wound-rotor induction motor with cam switch	High	75–90% lagging	Simplest	Lowest
2	Wound-rotor induction motor with liquid rheostat	High	75–90% lagging	Simplest	Higher
3	Brush shift	High	90% lag to unity	Less simple	High (except in small ratings)
4	Slip coupling with synchronous motor	2–4% less than Item 1	Unity or 80% lead	Less simple	Higher than Item 1
5	Slip coupling with squirrel-cage induction motor	2–4% less than Item 1	75–90% lagging	Less simple	Higher than Item 1

F. PIPING AND VALVES

Suction, discharge, and header piping in the station should be sized to handle the flows adequately. Proper sizing of piping is usually a matter of economics. Ordinarily, piping is sized so that the velocity in the suction line will not exceed 5 fps (1.5 m/sec), and in the discharge piping, 8 fps (2.4 m/sec). Piping less than 4 in. (10 cm) in diam should not be used for conveying sewage. Valves should be provided on the suction and discharge side of each pump to allow proper maintenance of the unit. Where sewage check valves are used, they should be installed so that they are readily accessible for cleaning or repair without removal from the line. Check valves should be of the outside lever, spring- or weight-loaded type and be installed only in horizontal lines. Pivoted valves, where the disk shaft is not clear of the flow, should not be used. Studies should be made and facilities provided to prevent excessive surges due to water hammer when starting and stopping the pumps where stations discharge into long force mains. This may require automatic valves coordinated with the pump motor controls (2) (11).

Velocities previously mentioned should not be exceeded through valves, which usually require that increasers and decreasers be placed on each side of the pump. Reducers on the suction side of pumps should be eccentric and installed with top flat to prevent entrapment of air in the suction line. Sectionalizing valves should be installed in the station header to the extent that the firm capacity of the station may be maintained if it should prove necessary to take any of the station piping out of service. Pump discharge piping should not connect to the header piping from the bottom since solids will have a tendency to settle out from the header into any vertical riser. Suction piping should be arranged so that it can be dismantled readily for cleaning. Each pumping unit should have a separate suction line from the wet well.

Flexibility is essential in laying out flanged piping. It may be provided by using hub-end joints, flexible couplings, or other means so that tolerances in such piping may be taken up in the flexible joints. Cast-iron flanged pipe, flanged specials or flanged fittings should not be encased in concrete because a failure in the flange would be very difficult to repair; rigid connections should not be provided between flanged piping and concrete walls or floors.

It is essential to provide proper hangers, braces, and supports for all piping to assure that no undue strains are induced. Particular attention must be paid to proper blocking and tying of pipe where hub-end or flexible joints are used. Small drain valves should be installed at all low points and air relief valves at all high points in the piping.

G. ELECTRICAL EQUIPMENT

1. General

The basic requirements for the electrical equipment at pumping stations are adequacy, reliability, and safety. Adequacy of the major equipment is determined largely by the continuous current requirements of the station loads and the available short-circuit capability of the power supply. The reliability of the equipment concerns the capacity of the electrical system to deliver power when and where it is required under abnormal, as well as normal, conditions. Safety involves the protection of plant personnel as well as the safeguarding of equipment under all conditions of operation and maintenance. None of these three requirements should be sacrificed for the sake of initial economy. The electrical system should be designed with enough flexibility to permit one or more components to be taken out of service at any time without interrupting the operation of the station.

The type of electrical equipment that is chosen for quite similar pumping stations may vary considerably in size, cost, and complexity because of the following differences: voltage, short-circuit capability, regulation, anticipated changes and reliability of the power source, the initial and ultimate station capacity, and the general appearance of the station.

2. Voltage Selection

A proper selection of voltages in the station electrical system is one of the most important decisions affecting overall system characteristics and plant performance.

The station main-bus voltage should be at the level most suitable for the pump motors which constitute the major part of the load. If a choice is available in the voltage level of a purchased power supply, the desired voltage rating for the largest pump motors will dictate how the choice should be made.

From the viewpoint of service voltage, the purchased power supply for a user will come from one of the following:

(a) Substation owned and operated by the power company at the station site to provide a service voltage satisfactory for the pump motors;

(b) Incoming voltage to the site satisfactory for the pump motors; and

(c) Main substation owned and operated by the user to produce a voltage satisfactory for the pump motors.

The user has design control of the main-bus voltage for supplying the pump motors under the last kind of service arrangement. For the other two kinds of service, the user should make known to the power company the service voltage level which best suits his purposes. In residential areas, the power company may prefer to supply 220-v service for small pumping facilities with aggregate loads of 15 hp or lower, so that portions of the residential area may be served from the same transformer bank. In this case, a penalty charge may be assessed to provide higher voltages. For aggregate loads between 15 and 100 hp, either 220- or 440-v service should be available without penalty. Some users, for safety or policy reasons, prefer service voltages of 220 for facilities up to an aggregate loading of 100 hp; others prefer to retain 220-v service until the largest single motor is rated 100 hp or larger. The user preferences may be maintained in face of the knowledge that equipment installed costs probably are higher at 220 than at 440 v.

For these reasons the selection of service and equipment voltage involves economic factors, user preference, and utility company policy. It should be resolved only after close coordination among the designer, power company, and user.

Figure 92 is a general guide for preferred motor voltages (8), but it is based on equipment installed costs alone. The recommendations allow for the effects produced in other components for typical system arrangements. The heavy vertical lines extending across two voltage levels indicate the horsepower ratings at which total system costs will be about the same for either voltage.

The economic voltage is still 440, down to about 1 hp, for small motors below the 10-hp minimum shown in Figure 92. Fractional horsepower motors can profitably be single-phase machines at 115 or 230 v.

3. System Selection

The equipment for a particular pumping station depends on the type of distribution system selected. These types vary not only in arrangement but also in operating characteristics and cost, depending on the following:

(a) Quality of service required;
(b) Total load;
(c) Magnitude of individual loads;
(d) Location of station; and
(e) Size and arrangement of station.

System types are classified according to the arrangement of primary supply and secondary interconnections. Several basic types are shown in Figure 93.

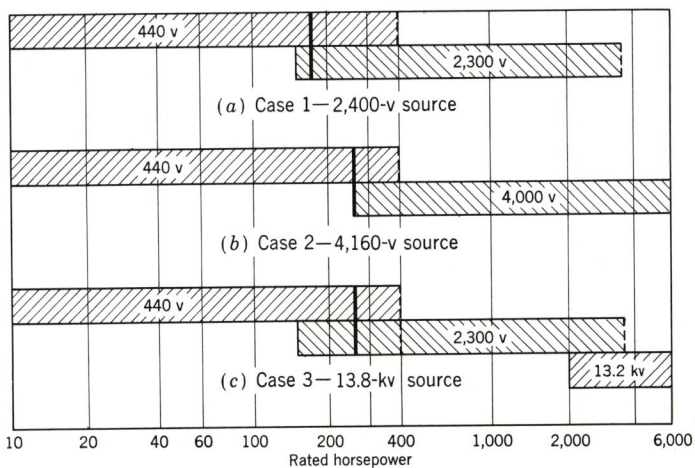

FIGURE 92.—Economic motor ratings by voltage classification.

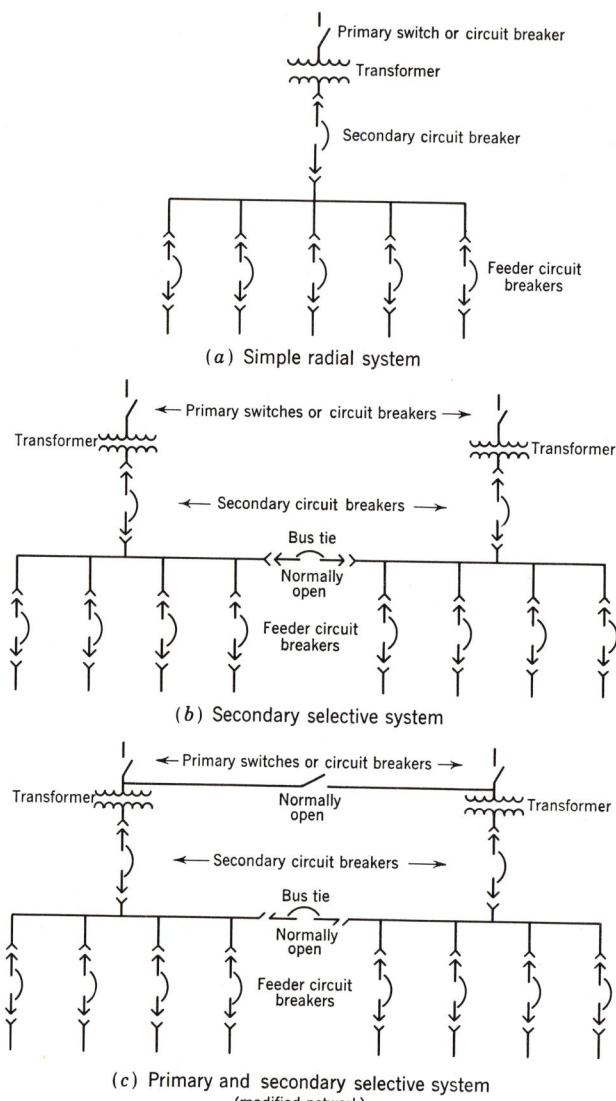

FIGURE 93.—**Typical electrical systems.**

The simple radial system is the least expensive and most common. The secondary selective system is more expensive and may be provided with two full-capacity transformers so that either the primary feeder or its associated transformer may be out of service while continuous service to the station is maintained. Still more flexible and reliable yet even more expensive is the primary and secondary selective system, which permits either transformer or primary feeder to be out of service simultaneously while still maintaining continuous service.

Generally the equipment selected for the primary and secondary selective system will require the highest interrupting capacity of the three systems described. The interrupting capacity of the equipment selected for the secondary selective network will be greater than for the simple radial system.

Another characteristic of the system that establishes certain equipment requirements is the grounding design. The National Electric Code and various local ordinances require systems to be grounded up through certain voltages; they do not, however, preclude grounding of higher voltage systems. Important benefits are derived from grounded neutral operation. Destructive transient overvoltages of several times normal are eliminated in a grounded system. Grounding systems basically fall into three classifications: solidly grounded, resistance grounded, and reactance grounded.

Small, low voltage systems, when grounded, are as a rule solidly grounded; larger systems, including those where relay protection is used or generating equipment is employed, will require other grounding methods. System grounding should not be confused with equipment grounding. Indeed, equipment always should be grounded solidly, regardless of whether the system is grounded.

Attention should be given to the protection of all systems from lightning and surges. Destructive overvoltage surges can occur on lines feeding a station even in the absence of a lightning storm. Lightning and overvoltage equipment is a form of insurance, in that, the greater the investment in station equipment, the greater should be the investment in protective equipment. In general, average-sized stations, and especially those fed from overhead lines, should have lightning arrestors and surge capacitors connected between the main bus or lines and ground for minimum protection. The application of more extensive protection should be studied for each particular application as no general rule is conclusive for all cases.

4. Transformers

Transformers are either dry self-cooled, liquid-immersed self-cooled, or forced-cooled. Dry transformers are limited in size and generally are chosen of necessity for lower weight, elimination of liquid, or improved appearance. Dry units compare unfavorably with liquid-immersed transformers with respect to impulse strength. Their ability to withstand surge voltages is only about one-half that of a liquid-filled transformer of the same voltage rating. The use of the dry-type should, therefore, be avoided where the system may be hit by lightning. Liquid-filled units are available in nearly all sizes and are considered to possess the highest degree of reliability of any of the main components of a power system. Two general types of liquid-filled transformers are the oil filled and the non-flammable liquid filled. Oil-filled transformers may be used only outdoors or in special transformer vaults. Non-flammable, liquid-filled transformers are for indoor use or for outdoor installation at locations where fire hazards must be reduced.

5. Switchgear

The functions of switchgear in a distribution system include normal and fault switching operations and equipment protection. Motor-starting functions sometimes are vested in switchgear, but only when the required frequency of starting and stopping is low or in applications where the motors are of such magnitude that no other equipment is practical. The basic classifications of switchgear equipment are made according to their rated voltage and their rated-fault interrupting capacity. Successful fault handling extends beyond the selection of devices which will interrupt satisfactorily an abnormal overcurrent; it also is important to prevent or minimize the damage, and perhaps hazard, that may appear in a cable, a motor controller, or other device during the clearing operation, even though the device itself was not initially in trouble. The problem is to choose equipment to withstand the duration and magnitude of faults and to operate selectively so that only a minimum of the system is disconnected in the event of trouble.

Outdoor metal-clad switchgear similar to the indoor type is used commonly for both medium and low voltages. All features of indoor switchgear are retained in the outdoor type, such as compartmentation, insulation levels, and removable and interchangeable breakers. Other added features include extra heavy base construction, planned ventilation, heaters, illumination, and weather-proof enclosures with weather resisting outdoor finishes.

6. Unit Substations

Transformers and switchgear or controls may be built integrally to combine into unit substations in modern distribution systems. Electrical and mechanical coordination of components provide a variety of benefits that assist in the planning, purchase, installation, and operation of the system. The total installed cost of unit substations is almost always lower than the total installed costs of the piecemeal components.

7. Motor-Starting Equipment

The full-voltage type starting equipment always is preferred from the standpoint of cost, floor space, and reliability. Experience has shown that full-voltage starting could be employed profitably to a much wider extent than it has been in the past; and each exception will warrant an investigation to see if full-voltage starting can be employed. Methods other than full-voltage starting are used broadly to improve an actual or imagined bad effect on system voltage from motor starting. In general use are the autotransformer type of starting, the reactor type, resistance type, and part-winding. The objective of reduced-voltage starting is to minimize the disturbance on the electric system by limiting the starting kilovolt-amperes. Most electric utilities have had definite restrictions on permissible starting kilovolt-amperes; however, it must be recognized that, as their system capacity increases, these permissible starting inrushes can become greater. As a rule, use of reduced-voltage starting equipment will approximately double the cost of the equipment. It also must be noted that any reduction in starting kilovolt-amperes supplied to a given motor, will be accompanied by a considerable reduction in locked rotor and accelerating torque.

Careful consideration should be given to the use of power circuit breakers as motor starting equipment where motors are not started more than two or three times per day. High-voltage, fused, motor controllers are better suited for repetitive operations as they have approximately 100 times longer mechanical life than circuit breakers operating under comparable duty. There are limitations, however, in the use of the fused high-voltage starters, such as the interrupting capacity and maximum horsepower sizes that are available. The horsepower limitations are approximately 700 hp for oil-immersed contactors and 1,500 hp for air-break contactors used in the high-voltage, fused, starting equipment at 2,300 v.

Grouped motor control equipment assembled into integrally built control centers often offers such benefits as lower installation costs, better appearance, and the advantage of factory wiring, testing, and coordination of protective and interlocking devices.

Control centers can be combined and integrally built with transformers to comprise power centers having the advantages similar to those associated with unit substations.

8. Motor Overload Protection

A major cause of motor failure is attributed to the breakdown of the winding insulation. Motor overloads cause excessive heat which is damaging to the insulation and is cumulative, i.e., each occasion of excessive temperature adds to the total deterioration. For this reason, the overload protection device should disconnect the motor before excessive heat can be generated. It has been customary to protect three-phase motors with overload protection in two of the three phases. In recent years, an increased trend toward overload protection in all three phases has been evident, particularly where the service is fed from wye-delta connected transformers. The 1965 National Electrical Code, Section 430–37, now requires overload protection in each phase of a three-phase unattended motor. The Code definition of an unattended motor is one lacking the presence of a person capable of exercising responsible control over the motor. In view of the fact that a large percentage of motors would fall under this classification and considering further the desirability of this protection, overload protection in all three phases should be thought of as a necessity.

Initially, the protection device should be sized or set for the stamped nameplate data of the motor it protects. The setting should follow a careful study of the ambient temperatures of the starter enclosure and that of the motor, as well as motor amperage, temperature rise rating, and service factor. Once properly selected, these settings rarely should be increased, though the operator may wish to reduce them when actual operating amperages are found to be substantially lower than motor nameplate ratings.

There is increased interest in the testing of overload protection devices in the field to ensure that the desired protection will be provided when needed. A phantom overload current is imposed across the device and tripping time recorded (12). A defective or improperly set device can be detected before it permits motor damage.

For motors of 300 hp and larger, consideration should be given to additional and more elaborate overload protection devices. Among these are time delay overcurrent relays, phase protection relays, differential relaying, and temperature detectors imbedded in the motor windings.

9. Controls

Controls should be simple, direct, and reliable. Large stations increasingly employ centralized control systems that automatically start and stop the pump units and associated valves and auxiliaries after initiation of push-button stations or automatic sensory devices. Centralized control panels or consoles usually comprise indicating lights, control switches or pushbutton stations, and a line of instrumentation for operation and record purposes. Instrumentation may consist of pressure gages, pressure or level controls and recorders, flow indicators and recorders, and electrical devices such as ammeters, wattmeters, and voltmeters. The outstanding success of local centralized control has resulted in the use of remote control, which is accomplished by adding interposing starting and stopping relays operated by supervisory equipment. Supervisory control equipment can operate over two-wire telephone lines, radio, or carrier. Telemetering equipment may be employed, in addition, to transmit instrument readings over the same or additional communication channels. The channel requirements for supervisory equipment have been reduced, operating procedures have been simplified, flexibility increased; and, with the proper protective devices applied to the lines, reliability has been increased. Although practically all the supervisory control systems commercially available can perform the same functions, they vary widely in overall performance because of the different operating principles involved and the number of refinements incorporated into their design.

Control equipment also includes apparatus and accessory devices for starting, stopping, regulating, and protecting motors and other equipment. Applications employing automatic and sequence controls based on liquid levels should be designed with sufficient well or reservoir capacity to prevent the starting and stopping of equipment too often. If more than four to six starts per hour are necessary, special care should be taken in selecting both the motor and the controls. Liquid-level controls generally employ probes, floats, or pneumatic devices for sensing levels. Electrode type detectors, employing the liquid as an electrolyte and equipped with a magnetic relay, are used widely. However, there usually is need for special application when there are many control points or when the control points are very close together. The float type controller may be a more suitable choice in such cases. Metal floats do not normally last long in sewage, whereas ceramic ones resist chemical action and give reasonable life. All floats are subject to fouling from accumulations of grease and scum. Indeed, floats may require continuous flushing systems for satisfactory service.

The air purge or bubbler system is suitable for use where there are a large number of switching points, the switching points are close together, or an indication or recording of liquid level is required. The system should include a spare bubbler tube at wall passages for use during cleaning.

Troubles may be experienced with any liquid-level control and consideration should be given to frequent cleansing for reliable operation. The pneumatic type is probably the most trouble-free and electrodes, the most troublesome.

10. Cable

Cable selection becomes quite an involved procedure if all the possible combinations of conductors, insulation, jackets, shielding, armor, etc., are considered. All cables are made up of some of the following components: conductor, tape, insulation, jacket, armor.

Conductors are selected for their conductivity, cost, and weight, and are stranded to provide the desired flexibility. Tapes are applied in numerous places in the construction of cables. Some are merely binders to hold the components together during the application of jackets, whereas others, particularly 4-kv and higher voltage cables, use conducting tapes to distribute voltage stresses equally. Although unshielded cable is available in the 5-kv class, it is recommended that for all installations of 4,160 v and above shielded cable be used.

Insulations used in various cables include paper, varnished cambric, natural and synthetic rubber, thermoplastics, and asbestos compounds. Impregnated paper and varnished cambric insulations normally are applied on systems of above 15 kv. Natural and synthetic rubber insulations are acceptable for all applications of 15 kv and lower. In locations subject to contamination from oil or gasoline, natural rubber insulation should be used only if protected by an oil resistant jacket. Thermoplastic insulations such as polyvinyl chloride (PVC) are used on lighting and small power circuits up to 600 v, and in many areas have advantage over rubber.

Jackets serve to protect the cable from mechanical damage during installation and from moisture, oil, and chemicals after installation. Common jacket materials are neoprene, polyvinyl chloride, polyethylene, nylon, asbestos, and lead.

Metallic armor is used as a jacketing material where the cables are required to have a high tensile strength or resistance to mechanical damage.

11. Conduit and Fittings

Galvanized or sherardized rigid metal conduit, ordinarily used in wastewater treatment works, may be used under nearly all atmospheric conditions. Aluminum conduit may be used in lieu of steel for exposed runs but should not be cast in concrete, because it causes spalling, or installed in soil (13). Epoxy resin or PVC-coated steel conduit or conduit made of special alloys is used where severely corrosive conditions are expected. Rigid metal conduit must be used in all hazardous locations.

PVC conduit is suitable in extremely corrosive, but not hazardous, atmospheres. PVC, asbestos-cement, and fibrous conduit are suited to underground service either direct buried or encased in concrete. These conduits are corrosion resistant, immune to electrolysis, non-inductive, and will not support combustion. Thin-wall fibrous conduits are designed for use where encased in concrete. Rigid or heavy wall conduits are designed to be buried or exposed.

Conduit fittings should conform to the requirements and be of the same material as the conduit. The use of two different metals in a conduit system should be avoided, especially in the presence of moisture, to prevent galvanic action.

H. PREFABRICATED PUMPING STATIONS

Prefabricated or factory-built pumping stations are available in several basic types. A common one includes conventional non-clog pumps, motors, controls, and piping mounted in a vertical cylindrical steel structure. Another comes fitted with pneumatic ejectors, compressors, and controls. Pumping capacities are normally rather limited, but units of the first type have been furnished with as many as three pumps, each with a rated capacity of 4,200 gpm (265 l/sec) (14).

Normally such stations include dehumidification and ventilation equipment, sump pumps, and sometimes elevators where the depths are great. Air conditioning equipment may be supplied with the larger stations for the dissipation of the heat from motor operation. Cathodic protection in the form of a sacrificial anode normally is provided for the outer steel shell. The operating agency should keep a record of the anode's location so that it may be checked periodically and replaced as necessary.

I. CHLORINATION

1. General

Chlorination sometimes is used at pumping stations. The equipment is generally automated and is substantially the same as that utilized for odor control and effluent disinfection at wastewater treatment plants (4).

2. Application Points

When local odor control or structural protection is the objective, the application point of the chlorine solution should be sufficiently far upstream, on or near the site, as to permit effective mixing with the sewage prior to release of the sulfide gases. With gravity inflow, this becomes difficult since the escape of chlorine to the atmosphere may be experienced in small pipelines due to insufficient submergence. Frequently, therefore, the application point arbitrarily becomes the wet well. If, however, the influent sewer is large, and if a submergence of at least 24 in. (61 cm) can be maintained, a multiport diffuser may be used in the pipe invert without serious fuming. If flows are delivered to the facility essentially by force mains, it becomes entirely feasible to introduce the solution into such mains through a single insert tube.

For odor control or structural protection downstream, the application point is normally the wet well or, for low discharge heads, the manifold of the discharge force main.

Diffusers installed in the invert of a gravity sewer are usually multiport to achieve rapid mixing. They are oriented with center line parallel to pipe center line as in Figure 94a. They must be anchored securely to the pipe and be streamlined sufficiently to retard accumulation of debris. A wet-well diffuser of the multiport type also is preferred. It should be fastened several diameters above the wet-well floor at a point which will be scoured by the moving sewage flow when pumps are running. Rubber lined and covered metallic pipe, having ports drilled on alternate sides, have been used successfully for this purpose. Single and multiple rubber hoses occasionally have been used in place of diffusers, but they achieve less dispersion of the solution and may encourage fuming.

Force main diffusers are of two general types. For small applications, a commercial diffuser unit is available, consisting of a silver tube inserted through a stuffing box mounted on a corporation cock which is installed in the pipe wall. The probe can be withdrawn behind the cock for cleaning and inspection, leaving the main line in service. For larger installations, a flanged wye having a flanged reducing outlet is installed in the run of pipe. A flanged, probe-type diffuser is inserted through the outlet. Its end is

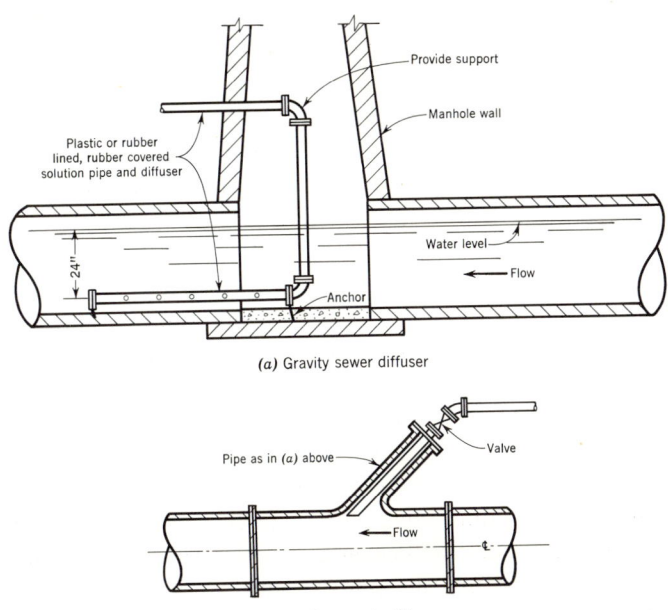

FIGURE 94.—Chlorine solution diffusers for pipelines.

beveled so as to be parallel to the center line of pipe (Figure 94b). In both cases, the main should remain full of liquid at all times.

3. Equipment and Controls

Chlorinator control may be by one of the following methods:

(a) Manually preset,
(b) Multi-step, and
(c) Proportioned to flow.

Under the manually preset procedure, the application rate is set on the machine, after which the chlorinator may operate continuously or on an "on-off" basis when at least one pump is running. Continuous operation is likely to produce fuming during low flows and insufficient dosage at high flows. The "on-off" operation will reduce the tendency of fuming. This procedure is accomplished at lowest equipment cost but provides relatively poor chlorination control. The chlorinator itself is standard with the usual manual rate valve. There has been minor use of this procedure using time clocks only, to provide intermittent dosages irrespective of pump operation.

The "multi-step" procedure offers increased control capability in that several different application rates can be preset and the machine rate valves electrically wired so each becomes operative when a given pump or pump combination is operating. The chlorine application rate then approximates a constant dosage with various pumping rates. The chlorinator is standard with the inclusion, as auxiliaries, of the required additional rate valves.

The most refined, and correspondingly the most costly, procedure is chlorination proportioned to sewage flow. It requires a flow signal produced by a flow meter (Venturi tube, Parshall flume, magnetic meter, etc.). The signal is received by appropriate equipment on the chlorinator and in turn paces the machine application rate in proportion to flow. The signal may be by air, vacuum, or electric transmission, depending on equipment chosen.

Chlorination controls should include an electrically-controlled, pneumatically-operated, chlorine pressure reducing and shutoff valve. This valve should open only after proof of adequate air pressure and injector water supply pressure.

The chlorinator should be housed in a separate room provided with an outside entrance and having a minimum of two vents, one at floor level. Mechanical ventilation should be provided. The room should be separated from the motor and pump rooms of the facility by a gas-tight partition to prevent chlorine leaks from causing extensive equipment damage. It is desirable to locate within the motor room the remote control for the chlorinator. It also is desirable to have a large, gas-tight glass panel between the chlorinator room and the motor control room.

4. Safety

A self-contained air supply mask should be readily available. Personnel chosen for the supervision and servicing of unattended chlorination equipment should be experienced, well trained, and have demonstrated good judgment in the handling of chlorine. Equipment should be inspected daily without exception while operating.

Chlorine leads and cylinders should be protected from vandals and accidental impact. All chlorine gas leads used at unattended facilities should be replaced annually. Fixed chlorine gas piping should be pressure tested annually. Gas piping should be sealed immediately from the atmosphere when temporarily removed from service.

An emergency phone number of the operating agency should be displayed on the exterior of the facility. Service personnel as well as others likely to respond to emergency calls should be equipped properly with chlorine gas masks, air masks, tools, and lighting.

J. APPURTENANCES

1. Wash Rooms

Wash room and toilet facilities should be provided in all stations which are to be attended regularly. Wash rooms for the larger stations should include showers. Foot operated valves should be used on lavatories. At least a wash basin and soap dispenser should be provided at every pumping station.

2. Meters and Gages

For medium-sized and large sewage pumping stations, it is desirable to have a flow meter which indicates and records the discharge from the station and a recording and indicating gage which shows the station header discharge pressure. This information is not only important from an operational standpoint but is of inestimable value if it should be necessary to enlarge the pumping capacity. It is helpful to have an elapsed-time clock mounted on each motor starter. If the facility is unattended, consideration should be given to the installation of monitoring equipment for the remote reporting of malfunctions to an operations center. As a minimum, an external alarm should be provided.

3. Water Supply and Water-Seal Equipment

A potable water supply is desirable in all pumping stations. A positive separation should be maintained between the potable system and any lines subject to contamination. Warning signs should be posted at all water taps not directly connected to the potable supply. Hose bibbs are desirable in pump pits and screen rooms.

Clear water should be provided for sealing stuffing boxes of volute-type pumps wherever possible. As a connection between the pump stuffing boxes and a potable water supply is a possible source of contamination, some means of preventing backflow of sewage into the potable water supply is required. This can be accomplished by providing a positive break with an air gap between the source of supply and the water-seal system. Backflow devices consisting of a system of check valves and relief valves which provide protection against backflow also are used sometimes. The latter method of preventing backflow, however, is not approved by some regulatory agencies. Therefore, before this method is adopted, approval should be obtained for its use. A positive break between the potable water supply and the water-seal system assures that no backflow can take place.

In water-seal systems where an air gap is provided, the potable water normally is discharged into a constant-level tank that is located above any possible flood level. Some public agencies require that the potable water line terminate not less than 6 in. (15 cm) above the top of the tank. Discharge of the potable water to the constant-level tank normally is controlled by a float valve. The required gland seal pressure may be provided by locating the tank high enough in the structure to create the required head by gravity or by re-pumping from the tank.

If a pumped system is used, the water-seal pumps may discharge either directly into the water-seal piping or into a pneumatic pressure tank. For direct pumping systems it is customary to provide a pressure relief valve to control maximum system pressure and return excess water to the sump tank. Direct pumpage systems are designed either to run continuously or are electrically interlocked to run when any sewage pump operates. For systems discharging to a pneumatic pressure tank, the seal-pump operation normally is controlled from the tank pressure. Pump operation will be intermittent and recirculation losses will be eliminated with this type of system.

Where it is desired to conserve water, or where graphite impregnated pump packing is to be used, solenoid valves interlocked with sewage pump starters can be installed in the individual water-seal lines so that water is supplied to a unit only when that pump is running.

A supply of approximately 3 gpm (0.2 l/sec) for each pump gland has been found to be adequate for estimating water-seal supply requirements. Generally accepted practice has been to provide water-seal pressure at the packing gland at not less than 5 psi (0.35 kg/sq cm) greater than the pump shutoff head. Some pump manufacturers recommend that gland water-seal supply pressure be determined on this basis while others recommend considerably lower pressures. If the gland seal system is designed to deliver the water at a pressure of 5 psi (0.35 kg/sq cm) above pump shutoff and a pressure reducing valve is provided in the water-seal supply line, the water-seal pressure can be adjusted in each instance as recommended by the pump manufacturer.

4. Hoisting Equipment

Hoists should be provided in all stations for the handling of equipment and materials which cannot be lifted readily or removed from the station by manual labor. It is advisable to provide overhead bridge cranes in larger stations. Moreover, it has been found in the larger stations that they usually pay for themselves in the handling of equipment and materials in the original installation. Hoists should be provided over basket screens and in other locations where it is necessary to lift pieces of equipment or containers which cannot be handled by two men. The provision of heavy lifting eyes in the ceiling over pumps and heavy valves and equipment will ease materially the close handling of such equipment.

5. Fencing

Stations and station facilities should be fenced where necessary to protect against vandalism or hazard to persons who otherwise may come in contact with electrical transformers and switching equipment.

6. Landscaping

The station site should be landscaped as necessary to make the facilities fit in with the surrounding area. Appropriate landscaping is essential for stations located in residential areas. When a house is located extremely close, a buffer of closely planted shrubbery 6- to 8-ft (1.8- to 2.4-m) high will reduce noise complaints materially and provide increased privacy to the residence.

7. Decoration and Finish

The interior of the station should have surfaces which can be cleaned and maintained easily.

8. Heating and Ventilating

Adequate automatic heating facilities should be provided in all stations to prevent freezing in cold weather and maintain a comfortable temperature in attended stations.

Proper ventilation is essential. Most health authorities consider six air changes per hour for continuous ventilation and a two-minute turnover for intermittent ventilation as necessary for all rooms located below ground level (15). Ventilation should be positive and take suction near the lowest floor level. Electrical interlocking of the ventilation controls should be provided to the extent that the ventilating equipment is operated whenever the lights are turned on in rooms located below the ground level. There should be no interconnection between the wet-well and dry-well ventilation systems.

Ventilation also must be provided to dissipate the heat from electric motors, especially during hot weather. The air required for this purpose will be approximately 30 cfm/hp (0.84 cu m/min/hp) for each motor operated simultaneously, assuming a 10°F (5.6°C) rise in air temperature.

9. Building Drainage

Adequate floor drains are a necessity and floors should be sloped to the drains a minimum of ⅛ in./ft (1 cm/m). Floor drains should be placed in sumps approximately 3 ft (0.9 m) in diam and depressed below the general floor elevation approximately ¾ in. (1.9 cm) to provide positive drainage. For small- and medium-sized structures, good drainage may be obtained by providing 2-in. by 4-in. (5- by 10-cm) gutter along all four walls, draining to the sump. Floors should be sloped from the center of the room toward the gutters in this case.

Sump pumps almost always are required to remove drainage from the pit floor, and they should discharge above any possible high water in the station wet well to preclude backflow of sewage into the pump pit.

10. Lighting

A minimum light intensity of 10 ft-c (110 lumens/sq m) is required for all corridors, stairways, platforms, and landings. Approximately 20 ft-c (215 lumens/sq m) should be provided in areas where machinery or electrical equipment is located. Glares and shadows should be avoided in the vicinity of machinery and floor openings. In deep compartments, two or more levels of fixtures may be required to achieve the recommended intensity for walkways, stairways, etc.

11. Safety Features

WPCF Manual No. 1, "Safety in Wastewater Works," (16) covers the subject in considerable detail. Features mentioned below should be included in all stations when constructed.

Railings should be provided around all manholes and openings where the covers may be left off during operation and at other locations where there are differences in levels or where there is danger of the operator falling.

Guards should be placed on and around all mechanical equipment where the operator might come in contact with belt drives, gears, chain drives, rotating shafting, or other moving parts of equipment.

Rubber mats of a suitable type should be provided in front of all electrical equipment where there is any hazard from electric shock. Electrical equipment and wiring should be insulated and grounded properly. Switches and controls should be of the non-sparking type. Adequate lighting should be provided in all locations, especially where there is moving equipment or openings in the floors. Wiring and devices, except polyphase motors, in hazardous areas should be explosion-proof. All receptacles should be of the three-wire grounding type outlet and a ground test of each conducted after installation. Portable power tools, extension cords, and trouble lights including their outlets should be of the three-wire grounding type. Lighting toggle switches should be "T" rated. When sump pumps are utilized with a flexible plug-in power cord, the cord and attachment cap should be of the three-wire grounding type; the attachment cap should be locked to the outlet with a locking device; and the outlet should be protected by a combination toggle switch having built-in overload protection (fractional hp manual starter). It is good practice to provide a prominent sign on the major gear stating the service voltage.

Stairways should be used in preference to ladders or manhole steps, with straight-run stairways preferable to the circular ones. Stair treads should be of the non-slip type. Vertical ladders should be provided with safety cages. Service elevators are desirable in larger stations with relatively deep pump pits. The use of manhole steps should be restricted to areas of infrequent use, and, if at all possible, to level changes not exceeding 8 ft (2.4 m). Grab bars should be provided approximately 4 ft (1.2 m) above the level of the landing adjacent to the top step.

The possibility of slipping on oil or biological growths which may develop near leaks presents a considerable safety hazard. Thus, proper drainage is important.

A telephone is a highly desirable feature since it will permit the operator, especially where only one is employed, to maintain regular contact with the main office. Also, in case of injury, fire, or equipment difficulty, it will enable him to obtain assistance as rapidly as possible.

Equipment should be available for testing the atmosphere, particularly in the wet well. Explosive mixtures also may be encountered because it is almost impossible to eliminate completely the discharge of flammable substances into sewers in larger systems.

Fire extinguishers of suitable type should be located strategically throughout the station. The use of carbon tetrachloride types should be avoided due to the toxicity of this liquid.

12. Paints and Protective Coatings

A detailed discussion of paints and protective coatings and their applications for wastewater treatment plants may be found in WPCF Manual of Practice No. 17, "Paints and Protective Coatings for Wastewater Treatment Facilities" (17). This publication includes a recommended uniform color code for piping. A uniform system of colors for pipe will minimize the possibility of cross connections, designate the use of each pipe, establish safe and danger areas, etc.

References

1. Weller, L. W., "That Attractive House May Be a Pumping Station." *Pub. Works,* **95,** 4, 86 (1964).
2. "Standards of Hydraulic Institute." 10th Ed. Hydraulic Institute, New York (1959).
3. Karassik, I. J., "Vertical Pumps." *Water and Sew. Works,* **107,** 56 (1960).
4. "Sewage Treatment Plant Design." Manual of Practice No. 8, Water Poll. Control Fed., Washington, D.C.; also known as Manual of Engineering Practice No. 36, Amer. Soc. Civil Engr., New York (1959).
5. Benjes, H. H., "Sewage Pumping." *Jour. San. Eng. Div., Proc. Amer. Soc. Civil Engr.,* **84,** SA3, 1665-1 (1958).

6. Williams, G. S., and Hazen, A., "Hydraulic Tables." 3rd Ed., John Wiley & Sons, Inc., New York (1963).
7. King, H. W., and Brater, E. F., "Handbook of Hydraulics." 5th Ed. McGraw-Hill Book Co., Inc., New York (1963).
8. Boggis, M. C., and Potthoff, E. O., "Motor Drives for Water Pumps." *Jour. Amer. Water Works Assn.,* **45,** 1023 (1953).
9. Potthoff, E. O., "Motor Drives for Sewage Pumping." *Wastes Eng.,* **23,** 6, 296 (1952).
10. Potthoff, E. O., "Adjustable-Speed Pump Drives." *Water Works and Wastes Eng.,* **2**; 2, 41; 4, 47; 7, 47 (1965).
11. Rich, G. R., "Water Hammer." In "Handbook of Applied Hydraulics." 2nd Ed., McGraw-Hill Book Co., Inc., New York (1952).
12. Thayer, R. C., and Smith, J. E., "Relay Test for Motor Protection." *Water and Sew. Works,* **103,** 265 (1956).
13. "Spalled Concrete Traced to Conduit." *Eng.-News Rec.,* **172,** 11, 28 (1964).
14. Weller, L. W., "Record Size Factory-Built Sewage Pumping Station." *Pub. Works,* **96,** 2, 104 (1965).
15. "Recommended Standards for Sewage Works." Great Lakes-Upper Mississippi River Board of State Sanitary Engineers (1960).
16. "Safety in Wastewater Works." Manual of Practice No. 1, Water. Poll. Control Fed., Washington, D.C. (1967).
17. "Paints and Protective Coatings for Wastewater Treatment Facilities." Manual of Practice No. 17, Water Poll. Control Fed., Washington, D.C. (1969).

INDEX

Air entrapment, effect..........111
Air jumpers 157
Asbestos-cement pipe 172

Backfill 210, 267
Backwater curve 73, 92
Backwater gate 161
Bar racks 293
Bedding factor 215
Bedding for sewers 210, 211
Bends 108
Bid proposal 240
Boring machines 271, 272
Brick masonry sewers............173
Building sewer 150

Cast iron pipe 176
Catch basins 156
Channel
    construction 284
    erosion 134
    freeboard 134
    improved 133
    natural 133
    velocities, limiting 135
Check valves 152
Chezy formula 76
Chlorination 324
Chlorinators 326
Chlorine application points.......324
Chlorine safety 327
Clay pipe 173
Cleanouts 149
Cohesion coefficient 203
Comminutors 295
Computation sheets, sewers.......138
Concrete pipe 174
Construction
    channels 284
    records 285
    sewer 247
    site preparation 251
    surveys 247
    tunnel 250
Continuity equation 70
Contract forms 241
Contracts 229
Corrosion protection, sewers 180
Corrugated metal pipe 177, 219
Crack strength..........209, 221, 222
Critical depth 73, 91
Critical flow 73, 91
Culverts, type of control.........100
Culverts, velocity in 100

Darcy-Weisbach formula 76
Definitions 1
Deflection formula, flexible pipe
    218, 223
Design flows, sewers 16
Design frequency, storms 44
Design period, sewers............15
Dewatering, excavations 258
Diversion, stormwater 160
Drawdown curve 73, 92

Drop connections 111
Drop manholes 149
Ductile iron pipe 177

Ejectors 300
Electrical
    cable 323
    controls 322
    equipment 315
    grounding 319
    switchgear 319
    transformers 319
Electric motors 310
Embankments, sewers in 186, 193
Energy dissipators 110
Energy equations 71
Energy grade line 71, 73
Energy losses 72
Encasement of sewers 214
Excavation, trench 251
Exfiltration 130

Fixture units 34
Flap gate 161
Flexible pipe 217, 223
Force mains 124, 132
Freeboard, channels 134
Friction
    conduit shape, effect 85
    construction material, effects....83
    factors 82
    pipe size, effects 85
    Reynolds number, effects 85
Froude number 73

Hazen-Williams coefficient 80
Hazen-Williams nomograph 83
House sewer 150
Hydraulic-elements graph 86
Hydraulic jump 73, 98
Hydraulic symbols 68
Hydrogen sulfide 124, 159
Hydrograph methodology 55, 61

Impact factor 205, 207
Imperfect trench 187, 197, 217
Industrial wastewater, flow allowance
    25
Infiltration 29, 130
Influence coefficients, trench..205, 206
Inlet time 47
Inlets
    combination 100
    depressed, undepressed 100
    grate, gutter 100
    spacing 122
    stormwater 100, 153
Intensity-duration
    annual duration 45
    curves 46
    data collection 45
    extended duration 45
    partial duration 45
    variation with area 49
Interceptor sewers 136

Jacked sewers 200
Joints, sewer 181
Junctions, head losses 109
Junctions, sewer 148

Kutter's formula 78

Leaping weirs 112, 160
Load factor 210, 215
Loads on sewers 185, 205, 223
Lump sum bid 229

Manholes
  channels 144
  construction 281
  drop 149
  frame and cover 142
  joints 146
  large 145
  leakage 29
  shallow 146
  shapes 142
  spacing 122
  steps 144
Manning's coefficient 78, 86
Manning's formula 78
Manning's formula, nomograph ... 81
Marston's formula
  186, 187, 188, 194, 197, 201, 207
Masonry sewers 173
Measuring devices 112, 165
Metering 165
Momentum equation 71
Motors 310
Motor overload protection 321
Motor starting 320

Negative projecting conduit
                                  186, 197, 216
Non-uniform flow 91

Ocean outfalls 163
Outfall sewers 281
Open-channel flow 73
Organization, sewer projects 1
Overflows, stormwater 136, 160
Overflow weir 160
Overland flow 55

Palmer-Bowlus flume 165
Parshall flume 167
Pipe beams 162
Pipe flow analysis 72
Plan preparation 230
Plans, construction 230
Plans, scale 231
Plastic pipe 179, 220, 223
Positive projecting conduit
                                    186, 193, 215
Projection ratio 198
Pumping stations
  buildings 287, 329
  capacity 288
  floor drainage 330
  heating 329
  hoists 328
  landscaping 329
  paints, coatings 331
  piping, valves 315
  prefabricated 324
  safety features 330
  suction piping 290
  ventilation 288, 329

wash rooms 327
water seals 327
water supply 327
wet wells 289
Pumps
  casings 297
  centrifugal 296
  construction 299
  curves 305, 306, 307
  drives 310
  impellers 297
  non-clog 299
  selection 300
  types 296

Railroad crossings 208, 280
Rainfall frequency 44
Rainfall intensity-duration 45
Rational method, formula 42
Rational method, runoff coefficient
                                                        49
Regulating devices 112
Relative roughness 77
Relief overflow 152
Relief sewer 137
Retention basins 61
Reynolds number 77, 85
Runoff, hydrographs, methodology
                                               55, 61
Runoff, inlet method for estimating
                                                       57
Runoff, rational method, for determining 42
Runoff coefficient
  soil type, effect 50
  surface depressions, effect 50
  time, effect 51
  vegetation, effect 50

Safe supporting strength, sewers
                                     221, 222, 223, 225
Safety factor ... 210, 221, 222, 223, 225
Screening devices 293
Service connections 150
Settlement ratio 195, 198
Sewage flow estimates 15
Sewers
  above ground 162
  alignment 248
  asbestos-cement 172
  bedding 210, 211
  bends 147
  brick 173
  cast-in-place 261, 265
  cast iron 176
  clay 173
  combined 120
  combined, capacity 136
  combined, pollution 120
  concrete 174
  construction 247, 262
  corrosion protection 180
  corrugated metal 177
  curved 122
  definitions 1
  depth 127
  depth of flow 130
  design computations 138
  design flows 16
    commercial 24
    industrial 25
    institutional 27

INDEX

Sewers—*continued*
    minimum 32
    peak 32
    residential 17
  design period 15
  field strength 210, 221, 222, 223
  force mains 124, 132
  foundations 132, 260
  function 67
  hydraulic grade line 67, 73
  hydraulic types 67
  intercepting capacity 136
  jacking 270
  joints 181
  junctions 148
  layout 121
  line and grade 248
  materials of construction 171
  outfall 281
  plastic 179
  prohibited use 8
  proximity to water supply 122
  railroad crossings 208, 280
  relief 137
  safe supporting strength
                    221, 222, 225
  self-cleaning 119
  separation from water mains ...122
  service, house connections......127
  siphon 156
  standard details 237
  steel 177, 178
  steep 132
  structural requirements 185
  sulfide generation 124, 159
  supports 162
  supporting strength
              209, 215, 217, 221, 222
  three-edge bearing strength
                    209, 221, 222
  trench bracing 254
  trench excavation 251
  tunnel 270
  types 123
  ultimate strength 209, 221, 222
  underwater 163
  use 8
  velocities, limiting 128
  ventilation 123
Shape factor 215
Sheeting 190, 226, 254
Siphons 156
Soil, pipe load from backfill
                    187, 190, 223
Specifications 237
Steel pipe 177, 178
Storm frequency, design 44

Stormwater
  inlets ....15$
  overflow ....112, 16(
  quality ....4]
  retention basins 61
Stream crossings 281
Subcritical flow 74
Suction piping 290
Sulfide control 124, 159
Supercritical flow 74
Superimposed loads 205
Supporting strength, sewers
              209, 215, 217, 221, 222
Surveys 9, 247
Switchgear 319
System head 305

Three-edge bearing strength
                    209, 221, 222
Tide gates 161
Time of concentration 46
Transformers 319
Transitions
  contractions 107
  enlargements 107
  head loss 107
  shape effect 106
Trench
  backfilling 267
  bracing 254
  conditions 186
  dewatering 258
  width effects 190
Tunnel shafts 279
Tunnel ventilation 280
Tunnels 131, 200, 204, 250, 270

Ultimate strength, sewers. 209, 221, 222
Underwater sewer 163
Uniform flow 70, 73
Unit hydrograph 61
Unit price bids 229

Velocity
  equal cleansing 89
  limits in channels.............135
  limits in sewers 128
  scouring 88
  self-cleansing 88
Ventilation, pumping stations. 288, 329
Ventilation, sewers 123
Ventilation, tunnels 280
Venturi, water 169
Vitrified clay sewers 173
Voltage selection 316

Weirs, leaping 112, 160
Weirs, measuring 165
Weirs, side-overflow 113, 160

DESIGN AND CONSTRUCTION OF SANITARY AND STORM SEWERS

Joint Committee of American Society of Civil Engineers
and Water Pollution Control Federation

1969

This Manual of Practice is designed as an aid to the practicing engineer in the design and construction of sanitary and storm sewers as represented by acceptable current procedure. It is not intended as a substitute for engineering experience and judgment or as a replacement for standard texts and reference material. It covers all phases of investigation, plan and specification preparation, contract documents, and construction procedures.

Keywords: catch basins, construction, contracts, design, drawings, force mains, hydraulics, manholes, (Manual of Practice), pipes (tubes), planning, (plans), pumping stations, sanitary sewers, sewers, siphon (sewage), specifications, storm sewers, (stormwater), (studies), surveys, tunneling (excavation), tunnels, (WPCF publication), wastewater.